Genome Mapping

The Practical Approach Series

SERIES EDITOR

B. D. HAMES
Department of Biochemistry and Molecular Biology
University of Leeds, Leeds LS2 9JT, UK

★ indicates new and forthcoming titles

Affinity Chromatography
★ Affinity Separations
Anaerobic Microbiology
Animal Cell Culture
(2nd edition)
Animal Virus Pathogenesis
Antibodies I and II
★ Antibody Engineering
★ Antisense Technologies
★ Applied Microbial Physiology
Basic Cell Culture
Behavioural Neuroscience
Biochemical Toxicology
Bioenergetics
Biological Data Analysis
Biological Membranes
Biomechanics—Materials
Biomechanics—Structures and
Systems
Biosensors
★ Calcium-PI signalling
Carbohydrate Analysis
(2nd edition)
Cell–Cell Interactions
The Cell Cycle

Cell Growth and Apoptosis
Cellular Calcium
Cellular Interactions in
Development
Cellular Neurobiology
Clinical Immunology
★ Complement
Crystallization of Nucleic
Acids and Proteins
Cytokines (2nd edition)
The Cytoskeleton
Diagnostic Molecular Pathology
I and II
Directed Mutagenesis
★ DNA and Protein Sequence
Analysis
DNA Cloning 1: Core
Techniques (2nd edition)
DNA Cloning 2: Expression
Systems (2nd edition)
★ DNA Cloning 3: Complex
Genomes (2nd edition)
★ DNA Cloning 4: Mammalian
Systems (2nd edition)
Electron Microscopy in Biology

Electron Microscopy in
  Molecular Biology
Electrophysiology
Enzyme Assays
★ Epithelial Cell Culture
Essential Developmental
  Biology
Essential Molecular Biology I
  and II
Experimental Neuroanatomy
★ Extracellular Matrix
Flow Cytometry (2nd edition)
★ Free Radicals
Gas Chromatography
Gel Electrophoresis of Nucleic
  Acids (2nd edition)
Gel Electrophoresis of Proteins
  (2nd edition)
Gene Probes 1 and 2
Gene Targeting
Gene Transcription
★ Genome Mapping
Glycobiology
Growth Factors
Haemopoiesis
Histocompatibility Testing
HIV Volumes 1 and 2
Human Cytogenetics I and II
  (2nd edition)
Human Genetic Disease
  Analysis
★ Immunochemistry 1
★ Immunochemistry 2
Immunocytochemistry
In Situ Hybridization
Iodinated Density Gradient
  Media

Ion Channels
Lipid Analysis
Lipid Modification of Proteins
Lipoprotein Analysis
Liposomes
Mammalian Cell Biotechnology
Medical Bacteriology
Medical Mycology
Medical Parasitology
Medical Virology
★ MHC 1
★ MHC 2
Microcomputers in Biology
Molecular Genetic Analysis of
  Populations
Molecular Genetics of Yeast
Molecular Imaging in
  Neuroscience
Molecular Neurobiology
Molecular Plant Pathology
  I and II
Molecular Virology
Monitoring Neuronal Activity
Mutagenicity Testing
★ Neural Cell Culture
Neural Transplantation
★ Neurochemistry
  (2nd edition)
Neuronal Cell Lines
NMR of Biological
  Macromolecules
Non-isotopic Methods in
  Molecular Biology
Nucleic Acid Hybridization
Nucleic Acid and Protein
  Sequence Analysis

Oligonucleotides and
  Analogues
Oligonucleotide Synthesis
PCR 1
PCR 2
Peptide Antigens
Photosynthesis: Energy
  Transduction
Plant Cell Biology
Plant Cell Culture (2nd edition)
Plant Molecular Biology
Plasmids (2nd edition)
★ Platelets
Pollination Ecology
Postimplantation Mammalian
  Embryos
Preparative Centrifugation
Prostaglandins and Related
  Substances
Protein Blotting
Protein Engineering
★ Protein Function (2nd edition)
Protein Phosphorylation
Protein Purification
  Applications

Protein Purification Methods
Protein Sequencing
★ Protein Structure
  (2nd edition)
★ Protein Structure Prediction
Protein Targeting
Proteolytic Enzymes
Pulsed Field Gel
  Electrophoresis
Radioisotopes in Biology
Receptor Biochemistry
Receptor–Ligand
  Interactions
RNA Processing I and II
★ Subcellular Fractionation
Signal Transduction
Solid Phase Peptide
  Synthesis
Transcription Factors
Transcription and
  Translation
Tumour Immunobiology
Virology
Yeast

Genome Mapping

A Practical Approach

Edited by

PAUL H. DEAR

*Medical Research Council Laboratory of Molecular Biology,
Hills Road, Cambridge CB2 2QH, UK*

OXFORD UNIVERSITY PRESS
Oxford New York Tokyo

Oxford University Press, Great Clarendon Street, Oxford OX2 6DP

Oxford   New York

Athens   Auckland   Bangkok   Bogota   Bombay   Buenos Aires
Calcutta   Cape Town   Dar es Salaam   Delhi   Florence   Hong Kong
Istanbul   Karachi   Kuala Lumpur   Madras   Madrid   Melbourne
Mexico City   Nairobi   Paris   Singapore   Taipei   Tokyo   Toronto   Warsaw

and associated companies in
Berlin   Ibadan

Oxford is a trade mark of Oxford University Press

Published in the United States
by Oxford University Press Inc., New York

A catalogue record for this book is available from the British Library

Library of Congress Cataloging in Publication Data
Genome mapping : a practical approach / edited by Paul H. Dear.
(The practical approach series)
Includes bibliographical references and index.
1. Gene mapping—Methodology.   I. Dear. Paul H.   II. Series.
QH445.2.G467   1997 572.8'633—dc21   97-12590

ISBN 0 19 963631 1 (Hbk)
ISBN 0 19 963630 3 (Pbk)

Typeset by Footnote Graphics, Warminster, Wilts
Printed in Great Britain by Information Press, Ltd, Eynsham, Oxon.

Preface

No area of biology has undergone as explosive a growth as that witnessed in genome analysis over the last few years. The 1987 human linkage map, an impressive achievement of its time, gave the locations of 403 markers throughout the genome. More than that number of markers have now been placed on the smallest human autosome.

Several factors have contributed to this rate of advance. Automation has been an increasing factor in making large-scale projects feasible, as has the trend toward larger research groups and well co-ordinated collaborative projects: hundred-author papers are no longer the sole province of particle physics. A second factor has been the development and implementation of new techniques and strategies, each yielding a new generation of human genome maps. A further catalyst has been the sharing of materials and data within the genome community, and the use of shared resources such as clone libraries and mapping panels. The value of these resources increases as their use becomes more widespread, multiplying the benefits to the researchers who use them.

The mapping of non-human genomes (with the notable exceptions of a few model organisms) has so far lagged behind that of the human, but is surely set to undergo a similar revolution over the next few years. Not only are genome maps of agriculturally important species invaluable in their own right, but the need for maps of many plant and animal genomes will become increasingly apparent as the human genome project moves towards its ultimate goal of a complete sequence. Much of the information content of the human genome will be irrelevant or uninterpretable except in the context of comparative molecular genetics.

The methods of genome analysis do not fall into any natural sequence. I have tried as far as possible to arrange them in order of increasing resolution, from linkage analysis to restriction mapping. Two comprehensive appendices give details of resources—both biological and computational—available to the genome community.

Finally, I would like to thank the contributors for producing uniformly excellent chapters, my colleagues for their help and advice, and my wife Denise for her patience as much as for her proof-reading.

Cambridge
June 1997

P. H. D.

Contents

List of Contributors   xix

Abbreviations   xxiii

1. Human linkage mapping   1

Julie L. Curran

1. An introduction to linkage mapping   1

2. Planning linkage studies   2
   Preliminary considerations   2
   Drawing family trees   3

3. Hypervariable microsatellites—highly informative markers
for gene mapping   4
   Dinucleotide repeats   4
   Tri- and tetranucleotide repeats   5

4. Isolation of polymorphic markers from specific regions   6
   Sources of regional DNA for marker isolation   7
   Screening libraries for microsatellites   8
   Isolation and sequencing of inserts   10
   PCR primer design   13

5. Evaluation of the candidate marker   14
   Preliminary testing and optimization of PCR conditions 14
   Determination of heterozygosity and PIC values   15

6. Genotyping   16
   Genotyping using radiolabelling and autoradiography   17
   Genotyping using fluorescently labelled primers   21

7. Genotype scoring and linkage analysis   23

References   25

2. Linkage mapping of plant and animal genomes   27

Ross Miller

1. Introduction   27

2. Genome data for non-human species   28

3. Marker isolation and genotyping   29
   Randomly amplified polymorphic DNA (RAPD) markers   30

5′ and 3′ SINE PCR markers 33
Amplified fragment length polymorphisms—AFLPs 36

4. Problems and opportunities in plant and animal genome mapping 39
Resources for plant and animal genome mapping 39
Opportunities afforded by directed breeding programmes 40
Transfer of markers between species 40
Comparative mapping 42

5. Conclusions and questions 43

References 45

3. Linkage mapping of quantitative trait loci in plants and animals

49

Chris S. Haley and Leif Andersson

1. Introduction 49

2. Principles of QTL mapping 50
One marker and a cross between inbred lines 50
Interval mapping 53
Outbred populations 55

3. Populations for QTL mapping 56

4. Maps and markers for QTL analysis 57

5. Setting significance thresholds 58

6. Predicting the power of an experiment to detect QTLs 59

7. Reducing background 'noise' 61
Careful design 61
Identifying and allowing for 'noise' in analysis 62
Progeny testing 63
Use of recombinant inbreds, dihaploids, and clones 64

8. Selective genotyping and DNA pooling 64

9. Pitfalls in interpretation 65
Overestimation of QTL effects 65
Linked QTLs 66
Interaction between genotype and environment 66

10. Towards positional cloning of QTLs 67

11. Software 68

12. Conclusions 70

References 70

4. Radiation hybrid mapping   73

Elizabeth A. Stewart and David R. Cox

1. Introduction   73

2. Preliminary considerations   76

3. Creation of radiation hybrid panels   77

4. Preliminary evaluation of panels and determination of optimal radiation dose   81

5. Existing human radiation hybrid panels   83
     G3 panel   83
     TNG panel   83
     Genebridge4 panel   83

6. Selection of STSs for radiation hybrid mapping   83

7. PCR typing of radiation hybrid panels   84

8. Analysis of radiation hybrid data   86
     Calculation of two-point lod scores   87
     Map building   90
     Resources for the analysis of radiation hybrid data   91

References   92

5. HAPPY mapping   95

Paul H. Dear

1. Introduction   95

2. Preparing the mapping panel   97
     'Clean' conditions for avoiding PCR contamination   97
     Preparation of genomic DNA   97
     Preparation of aliquots   99

3. Panel pre-amplification and screening   108
     Pre-amplification and screening using nested PCR   109
     Pre-amplification and screening using IRS-PCR followed by sequence-specific PCR   111
     Pre-amplification using 'whole-genome' PCR   118

4. Data entry   119

5. Data analysis and mapmaking   119
     Pairwise distance estimates   119
     Map construction   120

6. Conclusions and future developments   122

Acknowledgements   123

References   123

6. Construction and use of somatic cell hybrids 125

Susan L. Naylor

1. Somatic cell hybrids and genome research 125

2. Choice of cell lines and selection systems 127
- Donor cells 128
- Recipient cells 128
- Selection systems 129

3. Production of somatic cell hybrids 132
- Production of whole-cell hybrids 132
- Microcell-mediated chromosome transfer 136
- Hybrid clone isolation 138

4. Verification of chromosome content 139
- Karyotyping of human–rodent somatic cell hybrids 139
- FISH with total human DNA 142
- Chromosome painting 142
- PCR karyotyping 143
- PCR analysis of known markers 145
- Isozyme analysis 147

5. Maintenance and stability of clones 148

6. Mapping with somatic cell hybrids 148
- Southern analysis of somatic cell hybrids 148
- PCR typing of somatic cell hybrids 151

7. Isolation of donor DNA from hybrid cells 153
- Screening of hybrid-derived clones for donor sequence motifs 153
- Repeat element-mediated PCR of hybrid DNA 153
- Isolation of donor hnRNA sequences 154

8. Existing hybrid mapping panels 154
- Whole-chromosome hybrid panels 154
- Regional hybrid mapping panels 156

9. Summary 159

Acknowledgements 159

References 159

7. The use of flow-sorted chromosomes in genome mapping 165

Mark T. Ross and Cordelia F. Langford

1. General introduction 165

2. Flow-sorting of chromosomes 165
- Establishing the flow karyotype 165

Contents

Choice of cell line and chromosome preparation 166
Chromosome sorting 169

3. The uses of flow-sorted material in genome mapping 169
Direct cloning of flow-sorted material using cosmid vectors
with two *cos* sites 169
Direct cloning using other systems 177
PCR amplification of flow-sorted DNA 179

Acknowledgements 182

References 182

8. The use of microdissected chromosomes in genome mapping 185

Uwe Claussen, Gabriele Senger, and Ilse Chudoba

1. Introduction 185

2. Instruments and procedures for microdissection 186
Instrumentation 186
Preparation and banding of metaphase spreads on coverslips 188
Microdissection 190

3. DNA amplification of microdissected fragments using DOP-PCR 191

4. Fluorescence *in situ* hybridization 193

5. Cloning of microdissected fragments 196

References 197

9. Fluorescence in situ hybridization 199

Margaret A. Leversha

1. Introduction 199

2. Basic FISH techniques 200
Preparation of target cells 200
Probe preparation 203
Optional slide pre-treatment 206
Probe hybridization 207
Detection of hybridization signals 209
Chromosome banding 212

3. Interpretation of metaphase results 213
Chromosome band assignments 213
Clone ordering on metaphase chromosomes 213

4. Interphase FISH mapping 215

Isolation of nuclei for interphase FISH 215
Preparations for interphase FISH 216
Assembling interphase mapping data 217
Interpretation of interphase results 221

5. High-resolution mapping on DNA fibres 222
DNA haloes 222
Nuclear lysis 222

6. Microscopy 223

Acknowledgements 223

References 224

10. Contig assembly by fingerprinting 227

Simon G. Gregory, Carol A. Soderlund, and Alan Coulson

1. Introduction 227

2. Principle of restriction fingerprinting 228

3. Expected coverage and rate of completion of contigs 230

4. Fingerprinting strategies 231
Fingerprinting of whole-chromosome cosmid libraries 232
Assembly of large-insert bacterial clone contigs on marker 'scaffolds' 238
Regional fingerprinting 240

5. Gap closure 241

6. *Alu* PCR fingerprinting of YACs 246

7. Data entry and analysis 247
Autoradiograph scanning and data entry 247
Overlapping and analysis 248
Incorporation of marker data by *FPC* 250
Regional assembly 252
Complications 252

8. Conclusions 253

Acknowledgements 253

References 253

11. Chromosome walking 255

Jiannis Ragoussis and Mark G. Olavesen

1. Introduction 255

2. Vectorette PCR 256

Contents

Isolation of end-probes from PACs, BACs, and cosmids by
vectorette PCR   258
Isolation of end-probes from YACs by vectorette PCR   261
3. Bluescript-mediated isolation of end-fragments   265
4. Identification of end-fragments by hybridization with
oligonucleotides   267
5. End-fragment rescue by vector religation   270
Choice of restriction enzymes   271
6. Isolation of internal sequences from YAC clones   273
7. Hybridization of end-probes to clone filters or Southern blots 277
Acknowledgements   279
References   279

12. Long-range restriction mapping of genomic DNA

281

Wilfried Bautsch, Ute Römling, Karen D. Schmidt, Akhtar Samad, David C. Schwartz, and Burkhard Tümmler

1. Principle of long-range restriction mapping   281
2. Preparation of agarose-embedded genomic DNA   282
3. Choice of restriction endonuclease   288
General guidelines   288
Bacteria   289
Yeast   290
Mammals   290
4. Restriction digestion of agarose-embedded DNA   291
5. One-dimensional pulsed-field gel electrophoresis   293
PFGE techniques   293
Size markers for pulsed-field gel electrophoresis   294
Electrophoresis   295
6. Blotting and hybridization of pulsed-field gels   296
7. Two-dimensional pulsed-field gel electrophoresis   300
Partial–complete mapping   302
Reciprocal digest mapping   303
8. Alteration of recognition specificity for long-range
restriction mapping   308
Use of DNA methyltransferase to modify restriction digestion 308
CpG methylation: use of 5-acacytidine to reveal cryptic
restriction sites   310
9. Perspective: Optical Mapping   311
References   312

Appendix 1 Non-commercial resources for genome mapping 315

Ramnath Elaswarapu

1. Introduction 315

2. Vertebrates 316
Human 316
Non-human primates 320
Pig 320
Sheep 321
Mouse 321
Zebrafish 323
Pufferfish (*Fugu*) 323

3. Invertebrates 323
Drosophila 323
Caenorhabditis elegans 325

4. Plants 325
Rice 325
Maize 326
Wheat 326
Oats 326
Arabidopsis thaliana 326

5. Fungi 327
Saccharomyces cerevisiae 327
Schizosaccharomyces pombe 328

Acknowledgements 328

Appendix 2 Bioinformatics for genome mapping 329

Peter M. Woollard and Gary Williams

1. Introduction 329

2. The Internet and World Wide Web 329

3. Hardware and software requirements 330

4. Starting points for navigating the WWW 331
General search facilities 331
Bioinformatics search facilities 332
Online bioinformatics centres 332

5. Data resources 335
Human genome-wide mapping databases 335
Human chromosome-specific and mitochondrial databases 338
Rodent genome databases 338
Genome databases for other species 342

Contents

Comparative genome databases 343
Mutation databases 344
PCR primer and probe databases 345
Clone catalogues 346
Sequence databases 346

6. Software resources 347
Linkage analysis software 350
Contig and fragment assembly software 350
Software for designing primers for PCR and sequencing 351

References 351

Appendix 3 List of suppliers 353

Index 359

Contributors

LEIF ANDERSSON
Swedish University of Agricultural Sciences, Biomedical Center Box 597, S-7512 Uppsala, Sweden.

WILFRIED BAUTSCH
Insitut für Medizinische Mikrobiologie, OE 5210, Medizinische Hochschule, D-30623 Hannover, Germany.

ILSE CHUDOBA
Institute of Human Genetics and Anthropology, University of Jena, Kolligiengasse 10, D-07740, Jena, Germany.

UWE CLAUSSEN
Institute of Human Genetics and Anthropology, University of Jena, Kolligiengasse 10, D-07740, Jena, Germany.

ALAN COULSON
The Sanger Centre, Wellcome Trust Genome Campus, Hinxton, Cambridge CB10 1SA, UK.

DAVID R. COX
Stanford Human Genome Center, 855 California Avenue, Palo Alto, California 94305, USA.

JULIE L. CURRAN
Academic Unit of Anaesthesia and Molecular Medicine Unit, Department of Clinical Medicine, St James's and Seacroft University Hospitals, Leeds LS9 7TF, UK.

PAUL H. DEAR
Medical Research Council Laboratory of Molecular Biology, Hills Road, Cambridge CB2 2QH, UK.

RAMNATH ELASWARAPU
UK MRC Human Genome Mapping Project Resource Centre, Hinxton, Cambridge CB10 1SB, UK.

SIMON G. GREGORY
The Sanger Centre, Wellcome Trust Genome Campus, Hinxton, Cambridge CB10 1SA, UK.

CHRIS S. HALEY
Roslin Institute, Roslin, Midlothian EH25 9PS, UK.

Contributors

CORDELIA F. LANGFORD
The Sanger Centre, Wellcome Trust Genome Campus, Hinxton, Cambridge CB10 1SA, UK.

MARGARET A. LEVERSHA
The Sanger Centre, Wellcome Trust Genome Campus, Hinxton, Cambridge CB10 1SA, UK.

ROSS MILLER
The Babraham Institute, Babraham Hall, Cambridge CB2 4AT, UK.

SUSAN L. NAYLOR
Department of Cellular and Structural Biology, University of Texas Health Center at San Antonio, 7703 Floyd Curl Drive, San Antonio, TX 78284-7762, USA.

MARK G. OLAVESEN
Division of Medical Molecular Genetics, UMDS, Guy's Hospital, London SE1 9RT, UK.

JIANNIS RAGOUSSIS
Division of Medical Molecular Genetics, UMDS, Guy's Hospital, London SE1 9RT, UK.

UTE RÖMLING
Klinische Forschergruppe, OE 4350, Medizinische Hochschule, D-30623 Hannover, Germany.

MARK T. ROSS
The Sanger Centre, Wellcome Trust Genome Campus, Hinxton, Cambridge CB10 1SA, UK.

AKHTAR SAMAD
Department of Pathology, Cornell Medical College, The New York Hospital, New York, NY 10021, USA.

KAREN D. SCHMIDT
Klinische Forschergruppe, OE 4350, Medizinische Hochschule, D-30623 Hannover, Germany.

DAVID C. SCHWARTZ
Department of Chemistry, W. M. Keck Laboratory for Biomolecular Imaging, New York University, New York, NY 10003, USA.

GABRIELE SENGER
Institute of Human Genetics and Anthropology, University of Jena, Kolligiengasse 10, D-07740, Jena, Germany.

CAROL A. SODERLUND
The Sanger Centre, Wellcome Trust Genome Campus, Hinxton, Cambridge CB10 1SA, UK.

Contributors

ELIZABETH A. STEWART
Stanford Human Genome Center, 855 California Avenue, Palo Alto, California 94305, USA.

BURKHARD TÜMMLER
Klinische Forschergruppe, OE 4350, Medizinische Hochschule, D-30623 Hannover, Germany.

GARY WILLIAMS
UK MRC Human Genome Mapping Project Resource Centre, Hinxton, Cambridge CB10 1SB, UK

PETER M. WOOLLARD
UK MRC Human Genome Mapping Project Resource Centre, Hinxton, Cambridge CB10 1SB, UK.

Abbreviations

^{5m}C	5-methylcytidine
^{6m}A	6-methyladenosine
AFLP	amplified fragment length polymorphism
ATCC	American Type Culture Collection
azaC	5-azacytidine
BAC	bacterial artificial chromosome
BLAST	Basic Local Alignment Search Tool
BrdU	5-bromodeoxyuridine
BSA	bovine serum albumin
CAD	carbamoyl phosphate synthase/aspartate transcarbamoylase/dihydroorotase
CCD	charge-coupled device
cDNA	complementary deoxyribonucleic acid
CDP-Star	disodium 2-chloro-5-(4-methoxyspiro[1,2-dioxetane-3,2'-(5'-chloro)-tricyclo[3.3.1.1^{3,7}]decan]-4-yl)phenyl phosphate
CEPH	Centre d'Etudes du Polymorphisme Humain
CHEF	contour-clamped homogeneous electric field
CHLC	Co-operative Human Linkage Centre
CHO	Chinese hamster ovary
CIAP	calf intestinal alkaline phosphatase
CISS	chromosomal *in situ* suppression (hybridization)
cM	centiMorgan
cR	centiRay
Cot	concentration × time (DNA reassociation kinetics)
DAPI	4,6-diamidino-2-phenylindole
ddGTP	dideoxy guanosine triphosphate
ddTTP	dideoxy thymidine triphosphate
DIG	digoxigenin
DNA	deoxyribonucleic acid
dNTP	any of dATP, dCTP, dGTP, or dTTP; or a mixture of all four
DOP	degenerate oligonucleotide primer (or primed)
DTE	dithioerythritol
DTT	dithiothreitol
EBV	Epstein–Barr virus
ECACC	European Collection of Animal Cell Cultures
EDTA	ethylenediamine tetraacetic acid (disodium salt unless specified)
EGTA	ethylene glycol *bis*(β-aminoethyl ether) $N,N,N,'N'$-tetraacetic acid
EMBL	European Molecular Biology Laboratory

EST	expressed sequence tag
F_1, F_2	first/ second filial generation
FAM	5-carboxyfluorescein
FITC	fluorescein isothiocyanate
F(B)CS	foetal (bovine) calf serum
FISH	fluorescence *in situ* hybridization
FSC	flow-sorted chromosome
FTP	file transfer protocol
GART	glycinamide ribonucleotide formyltransferase
Gb	billion (10^9) bytes (computing)
GCG	Genetics Computing Group
GDB	Genome Data base
HAT	hypoxanthine/aminopterin/thymidine
Hepes	*N*-2-hydroxyethylpiperazine-*N*'-2-ethanesulfonic acid
HGMP (RC)	Human Genome Mapping Project (Resource Centre)
hnRNA	heterogeneous nuclear RNA
HPRT	hypoxanthine phosphoribosyl transferase
HTML	hypertext markup language
IMAGE	Integrated Molecular Analysis of Genomes and their Expression
IRS	interspersed repeated sequence
JOE	6-carboxy-2',7'-dimethoxy-4',5'-dichlorofluorescein
LINE	long interspersed repeat element
LMP	low-melting point (agarose)
Lod	logarithm (base 10) of odds
Mb	million base pairs (DNA) *or* million bytes (computing)
MAS	marker-assisted selection
MES	2-(*N*-morpholino)-ethanesulfonic acid
MMCT	microcell-mediated chromosome transfer
OMIM	Online Mendelian Inheritance in Man
PAC	P1-derived artificial chromosome
PACE	programmable autonomously-controlled electrophoresis
PBS	phosphate-buffered saline
PCR	polymerase chain reaction
PEG	polyethylene glycol
PFG(E)	pulsed-field gel (electrophoresis)
PHA	phytohaemagglutinin
PI	propidium iodide
PIC	polymorphism information content
PMSF	phenylmethylsulfonyl fluoride
QFH	quinacrine/ fluorescence/ Hoechst 33258
QTL	quantitative trait locus
RAPD	randomly amplified polymorphic DNA
rDNA	ribosomal DNA

REM PCR	repeat element-mediated PCR
RFLP	restriction fragment length polymorphism
RH	radiation hybrid
ROX	6-carboxy-X-rhodamine
SDS	sodium dodecyl sulfate
SHMT	serine hydroxymethyl transferase
SINE	short interspersed repeat element
SSC	standard saline citrate
SSCP	single-strand conformational polymorphism
STR(P)	short tandem repeat (polymorphism)
STS	sequence-tagged site
TAFE	transverse alternating field electrophoresis
TAMRA	$N,N,N'N'$—tetramethyl-6-carboxyrhodamine
T_{ann}	annealing temperature (of DNA probe or oligonucleotide)
TEMED	N,N,N',N'-tetramethylethylenediamine
TIFF	tagged information file format
TK	thymidine kinase
T_m	melting temperature (of DNA probe or oligonucleotide)
tRNA	transfer ribonucleic acid
UMP	universal mapping probe
UMPS	uridine monophosphate synthase
UV	ultraviolet
WWW	World Wide Web
YAC	yeast artificial chromosome

1

Human Linkage Mapping

JULIE L. CURRAN

1. An introduction to linkage mapping

The basic principles of linkage mapping have been used by scientists for many years, to establish whether a particular disease is associated with a specific chromosomal region. Linkage amongst simple Mendelian traits is often observable at the phenotypic level, and the chromosomal rearrangements associated with some disorders could be located cytogenetically. But even in combination, these two approaches gave an extremely sparse map by today's standards. One of the first loci to be mapped was that for zonal, pulverulent cataracts (CAE1) on 1q21-q25, linked to the Duffy blood group (FY) at the same chromosomal location in 1969 (1).

For many years, genetic mapping was hampered by the lack of available markers. With the advent of recombinant DNA technology, Botstein *et al.* (2) proposed in 1980 a way of defining a large number of marker loci, albeit in an arbitrary manner. They suggested the use of restriction-fragment length polymorphisms (RFLPs) as simple Mendelian co-dominant genetic markers. Although such markers normally have no phenotypic correlates (and do not generally lie within a gene of interest), they have the major conceptual advantage of requiring no prior knowledge about the DNA sequence in which they lie, or its function. RFLPs were first used in genetic analysis in 1974 (3) to locate mutations on a physical map of restriction fragments in adenovirus. Botstein's paper suggested that a map with 150 markers spaced at approximately 20 cM (approximately 20 Mb) would be sufficient to allow any given locus to be located by linkage, provided that the markers were sufficiently polymorphic. (Linkage maps are generally measured in centi-Morgans [cM]; one centiMorgan is the distance over which recombination occurs once in every hundred meioses and, in humans, corresponds to around 1 Mb on average). In practice this spacing is rather too wide, and a 10 cM map is more effective for the initial placement of loci throughout the genome.

It is obviously desirable that the markers used to build genetic linkage markers should be sufficiently polymorphic, and two terms have been devised to describe the degree of polymorphism. The first of these, 'hetero-zygosity', is simply the proportion of individuals in a sample which are heterozygous for the marker (4). Heterozygosity is obviously greatest when many alleles exist at comparable frequencies, and least when there are fewer alleles or when a small number of alleles predominate. The second term, 'polymorphism information content' or PIC, is more complex. It takes account of the fact that heterozygous offspring of identically-heterozygous parents are not informative for linkage, and better reflects a marker's value in pedigree analysis (2, 5; see also Section 5.2). A heterozygosity value of greater than 60%, or a PIC value exceeding 0.60, indicates a potentially usefully informative marker.

Since Botstein's seminal paper (2), the field of molecular genetics has moved extremely rapidly, and there are now a number of dense genetic maps available (see Section 2). In recent years, RFLPs have been largely superseded by microsatellites (tandem repeats of short sequences, in which the number of repeat elements is polymorphic) as genetic markers. Microsatellites, like RFLPs, usually have no phenotypic manifestations and serve as arbitrary markers. They have, however, several advantages over RFLPs, as will be discussed in later sections of this chapter.

One aim of this chapter is to provide an introduction to the principles and execution of linkage analysis. Initial consideration is therefore given to the planning of linkage studies aimed at locating a particular (typically disease-related) gene. In many cases, the existing genome-wide linkage maps are too sparsely populated for the closing stages of such gene-hunting, and additional markers will need to be generated and added to the map in the region of interest. The bulk of this chapter, therefore, concerns the isolation and evaluation of new polymorphic markers from chosen regions of the genome.

2. Planning linkage studies

2.1 Preliminary considerations

Before embarking on any linkage study there are certain essential points which, if adhered to, will make the path much smoother :

(a) Ensure the disease under study does have a genetic component! Kuru, for example, was believed to be inherited in an autosomal dominant manner, but was later discovered to be a transmissible disease propagated by intrafamilial cannibalism. Such circumstances are hardly common, but many other non-genetic diseases can manifest apparently heritable components in a less dramatic manner.

(b) It must be possible to ascertain disease status of individuals from whom DNA is available. It is not possible to do a linkage study on a disease which can only be diagnosed *post-mortem*, when there are no samples available from the deceased individuals.

(c) In order to detect linkage reliably, the family size must be large enough

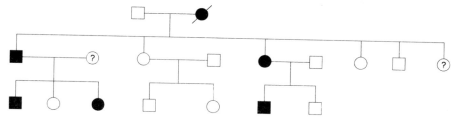

Figure 1. A three-generation pedigree of the minimum size necessary to obtain a significant lod score—ten potentially informative meioses. Other family patterns with a similar number of individuals of known disease status would of course be equally suitable. Females are indicated by circles, males by squares. Filled symbols indicate affected individuals; a question mark within a symbol indicates disease status unknown; a line through the symbol means that the individual is dead (not necessarily from the disease in question).

to give a statistically significant answer. In practice, this means there must be a minimum of ten potentially informative meioses (i.e. those in which potential cross-overs could be recognized) within a family in order to give a lod score of +3, which is the minimum necessary to declare linkage (see *Figure 1*; further discussion of lod scores, and how to calculate them, can be found in Section 7). It is advisable to draw a family tree and calculate the theoretical maximum possible lod score for each family prior to starting any experimental work. This helps to give a feel for the data as it comes through, suggesting where it might be advisable to pursue missing or ambiguous data on one or two individuals, and where such data are unlikely to be of worthwhile value.

(d) If there is any evidence of locus heterogeneity or possible multi-gene effects, it is not a valid procedure to simply sum the results (lod scores) from different families. Thus if only small families are available (perhaps four or five individuals) and there is evidence for several distinct loci, it will not be possible to prove by linkage mapping which loci are involved. The reasons for this are further discussed in Section 7.

2.2 Drawing family trees

It is usually easiest to start at the top of the family tree and work downwards, rather than the reverse. Standardization of the symbols makes for easier checking, and many diseases will already have standard notations. There are a number of computer programs available for pedigree drawing, and these have many advantages over the old methods of pen and ink or general-purpose computer drawing programs. Pedigree-drawing packages include *Cyrillic* (Cherwell Scientific) and Pedraw (David Curtis). The author uses *Cyrillic* v2.01, which has a number of useful features including the ability to draw coloured haplotype bars, to scale pedigree drawings to fit on one page, and to export data directly to analysis programs such as *LINKAGE*. The programs come complete with comprehensive instructions and tend to be user-friendly. An example of a suitable three-generation pedigree that could potentially give a lod score of more than +3 is shown in *Figure 1*.

3. Hypervariable microsatellites—highly informative markers for gene mapping.

Microsatellites, first described in 1989 (6, 7) are small blocks of tandemly repeated DNA, in which the repeated element is usually a di-, tri-, or tetra-nucleotide sequence (e.g. [CA]n, [CAG]n, [AGAT]n). The number of repeat elements in these blocks is often highly polymorphic, and shows simple Mendelian inheritance. Genotyping is straightforward: the number of repeats is commonly measured as the length of sequence amplified (using the polymerase chain reaction) between primers flanking the repeat region.

3.1 Dinucleotide repeats

The most common microsatellite is the dinucleotide repeat (CA)n, being present in all eukaryotic genomes tested, and comprising around 0.5% of the human genome (8). In humans, the number of repeats in each block ranges from 10 to 60, but is most commonly 15–30. It is estimated that there may be as many as 50 000 (CA)n microsatellites in the genome, giving an average spacing of approximately 30 kb. Microsatellites are more uniformly distributed than other marker types, and are the only type of polymorphic marker which is abundant around centromeres.

The 1996 Genethon human genetic linkage map (9) gives the location of more than 5000 (CA)n repeat markers with an average heterozygosity of 70%. It spans the entire genome, with only 22 gaps of more than 10 cM, and more than half of all points in the genome lie at a distance of 1 cM or less from one of its markers.

There are three classes of dinucleotide repeats :

- perfect repeats with no interruptions in the runs of dinucleotides (e.g. [CA]n)
- imperfect repeats with one or more interruptions (e.g. [CA]n–CCA–[CA]m)
- compound repeats with adjacent tandem repeats of different sequences (e.g. [CA]n[TG]m)

The class of repeat has no effect on the level of heterozygosity. However, the number of repeat elements is important, and the PIC of dinucleotide microsatellites (perfect, imperfect or compound) tends to vary from 0 where n (or n + m) $\leq$ 10, to $\geq$ 0.8 where n (or n + m) $\geq$ 24 (8). For microsatellites to be useful as genetic markers, only those with more than 16 dinucleotide repeat units will have a reliably high PIC, but some microsatellites with

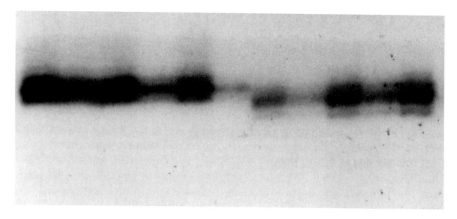

Figure 2. Insufficient heterozygosity of a microsatellite. Members of two families (first six and last five lanes, respectively) were genotyped for this marker. As can be seen, within each family every individual is homozygous for the same allele.

between 10 and 16 repeat units may be useful. *Figure 2* shows a microsatellite with 14 (CA) repeat units, and as can be seen in the figure the level of heterozygosity was too low for this to be considered worth using as a mapping marker.

The function of these repeats is unknown. It has been suggested that they may be 'hot-spots' for recombination, though the linkage disequilibrium seen between them argues against this. Their length polymorphism may be the result of slippage mispairing during replication, as suggested by experiments which mimic some aspects of chromosome replication *in vitro* (10). Several different mutation mechanisms have been suggested, but all tend to invoke the idea of strand slippage occurring during replication, repair, or recombination. A similar process appears to occur during PCR amplification of microsatellites (particularly dinucleotide repeats) resulting in a number of 'shadow bands' both smaller and larger than the true alleles by integral numbers of repeat units. *Figure 3* shows this effect, which can be a major drawback in the use of these markers: in this example it has not been possible to type the individuals with this marker due to the strong shadow bands.

3.2 Tri- and tetranucleotide repeats

Trinucleotide repeats are around ten times less common than the dinucleotide variety, occurring on average once every 300-500 kb. The most common form is [AAB]n (where B = C, G, or T). Tetranucleotide repeats are rarer still, and the most common form of these is [AAAB]n (B as above). One group has mapped more than 850 tetranucleotide markers across the genome with an average distance between loci on the autosomes of 7.4 cM, and an

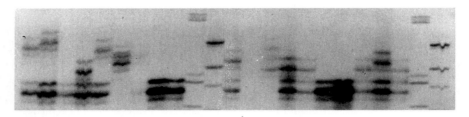

Figure 3. The effect of 'shadow bands' on genotyping for a dinucleotide repeat. Optimization of magnesium concentration and annealing temperature in the PCR failed to reduce the intensity of the shadow bands, which make it impossible to reliably assign alleles.

average heterozygosity of over 70% (11). Around 80% of [AAB]n and [AAAB]n microsatellites are associated with interspersed repetitive *Alu* elements, and these can occasionally lead to PCR problems if the primers used for amplification lie in the conserved *Alu* sequence.

Many of the comments in Section 3.1 apply to these classes of repeat also, although not those regarding the number of repeat units required for reasonable levels of polymorphism. The repeat sequence appears to be important in determining the PIC of such microsatellites, with AT-rich repeats (such as [AAT]n, [ACT]n, [AATG]n, and [AGAT]n, tending to be more polymorphic for a given number of repeat units. In general, though, repeats of less than seven tri- or tetranucleotide sequences are rarely useful.

Tri- and tetranucleotide repeats are more stable during PCR, so the resulting genotypes are much easier to read, lacking the many shadow bands characteristic of dinucleotide repeat markers. They are also easier to read when using fluorescently labelled primers and automated detectors such as the Applied Biosystems 373 system. The software used with such detectors can assign alleles incorrectly when analysing dinucleotide microsatellites, but is much more accurate with the tri- and tetranucleotide markers due both to the near absence of shadow bands and to the greater size differences between alleles. In favourable cases, these longer repeats can even be typed on high percentage agarose gels, obviating the need for radioactive or fluorescent labelling. These features make tri- and tetranucleotide repeats far easier to use than their di-nucleotide counterparts, though their lower frequency in the genome limits their widespread use.

4. Isolation of polymorphic markers from specific regions

If the existing maps do not have a sufficiently dense coverage of markers in a region of interest, it is necessary to isolate further regional markers oneself. (It is always wise to consult the on-line versions of chromosomal maps since the printed versions have frequently been improved since publication—see Appendix 2.) As existing maps give sufficient coverage for initial mapping studies, I will assume that these markers will be for a specific chromosome or region thereof. Most of the following procedures, however, are also applicable to genome-wide searches in human or, with minor modifications, to other species.

There are essentially two levels of genetic maps: framework and comprehensive. Framework maps tend to have low resolution, but there is high confidence in the location of each marker, at odds of greater than 1000:1. It is desirable that the physical locations of these markers are also known. Comprehensive maps are of high resolution, but the order of the most closely spaced markers is less certain, so there is often a need to fine-tune the local order of the markers. It is the addition of markers to, and annotation of these maps that is described here.

4.1 Sources of regional DNA for marker isolation

The American Type Culture Collection (ATCC) is a commercial source of chromosome-specific libraries, and they can be purchased already digested and subcloned into suitable vectors. Such libraries provide the ideal 'hunting ground' for chromosome-specific markers. Use of these libraries saves a considerable amount of work but, if cost is an important consideration, there are other sources.

Monochromosomal somatic cell hybrids are another source of chromosome-specific material, and for UK users can be obtained free of charge from the UK HGMP resource centre, provided that data generated is returned to the centre to benefit the mapping community (this does not of course preclude publication of results). Application forms are available from the Hinxton centre (see Appendix 1), and DNA is usually sent for PCR use with full documentation. Other sources of somatic cell hybrids will be found in Chapter 6 and in Appendix 1.

The most convenient way to isolate the human component from such hybrids is to amplify their genomic DNA using the polymerase chain reaction with primers specific for the human-specific *Alu* repeat elements. This amplifies only those sequences which lie between closely-spaced (typically < 1–2 kb) *Alu* elements, which can then be cloned and screened for suitable microsatellite sequences. Many protocols exist for *Alu* PCR, and an example is given in Chapter 6. One drawback of this approach is that the non-random distribution of *Alu* elements leads to an underrepresentation of some chromosomal regions.

Flow-sorted chromosomes and microdissected chromosome fragments (Chapters 7 and 8) provide other sources of localized DNA for microsatellite identification, and circumvent the need to use *Alu* PCR to isolate the human component.

Whatever the source of human DNA, it will normally need to be cloned as a prelude to screening for suitable microsatellite motifs and sequencing of likely candidates. The method used for cloning will be dictated by personal preference and by the source of the DNA, but one common requirement is that the insert size should not be more than about one kilobase. If the cloned inserts are larger than can be conveniently sequenced (even by double-ended sequencing from both ends of the insert), then there is the risk that the microsatellite sequence will lie in an inaccessible central part of the insert. Cloned PCR products, or DNA digested with frequently-cutting restriction enzymes such as *Sau*IIIa normally present no problems in this respect. We favour cloning into the Lambda FIX-II system (Stratagene), but many alternatives exist. Chapters 7 and 8 give details for the construction of libraries from flow-sorted and microdissected chromosomes. More general information on DNA cloning will be found in ref. 12.

4.2 Screening libraries for microsatellites

Once a suitable chromosome- or region-specific library has been obtained, one can move on to the isolation of microsatellite markers. This is normally done by screening the library with a hybridization probe which will detect suitable microsatellite sequences. Detailed instructions are given in *Protocols 1* and *2* below.

Protocol 1. Transfer of clones to hybridization membranes

Equipment and reagents

- Agar plates carrying the clones at a suitable density (i.e., as high as possible whilst allowing individual clones to be identified).
- Hybond-N+ membranes, of the same size as the library plates (Amersham)
- Sterile syringe needles
- Denaturing solution: 1.5 M NaCl, 0.5 M NaOH
- Neutralizing solution: 1.5 M NaCl, 0.5 M Tris-HCl pH8.0
- Rinsing solution: 0.2 M Tris-HCl pH 7.5, 2 × SSC (prepare SSC as 20x stock: 3 M NaCl, 0.3 M Sodium Citrate pH 7.0)
- UV transilluminator
- Filter paper (Whatman 3MM)

Method

1. Chill the agar plates carrying the clone library for at least 2 h at 4°C.[a]

2. Carefully lay a membrane on top of the plate without trapping air bubbles. Using a syringe needle, prick an asymmetric pattern of holes around the periphery of the membrane (penetrating through into the agar) so that the membranes may be orientated with the plates later. Leave the membrane in contact with the plate for 2 min.[b] When removing the membrane, ensure that it does not 'drag' across the surface of the plate. Return the plates to the fridge.

3. Submerge the membrane in denaturing solution for 2 min.
4. Submerge the membrane in neutralizing solution for 5 min.
5. Rinse the membrane for 20 sec in rinsing solution.
6. Place the membrane clone-side uppermost on dry 3MM briefly to remove surplus liquid.
7. Cross-link the DNA by placing the membrane on a UV transilluminator for 2 min.[c]

[a] This prevents the agar sticking to the membrane.
[b] It is essential to wear gloves during this procedure, and the membranes should be handled with forceps.
[c] It may be necessary to adjust this timing to suit one's own transilluminator, as UV bulbs tend to get less efficient as they get older.

The membranes are now ready to probe for the presence of microsatellite repeats. The probe must be a sequence corresponding to the type of microsatellite sought. For example, poly (CA)/(GT) will detect (CA)n repeats. Probes for many dinucleotides are available commercially (e.g. Pharmacia); for others, custom synthesized oligonucleotides may be used. Probes are best labelled by using a random priming kit (available from many molecular biology suppliers) following the manufacturer's instructions. However, the random hexamers normally used for priming synthesis need not be included in the reaction mixture, provided that both complementary strands are labelled together, as they can be effectively labelled by self-priming (13). Protocol 2 provides details of hybridization conditions.

Protocol 2. Hybridization of microsatellite repeat probes

Equipment and reagents

- 20 × SSC: 3 M NaCl, 0.3 M sodium citrate pH 7.0
- Prehybridization buffer: 6 × SSC, 20 mM NaH$_2$PO$_4$, 0.4% (w/v) SDS, 5X Denhardt's reagent (Denhardt's is prepared as 50 × stock: 1% w/v Ficoll 400, 1% (w/v) polyvinylpyrrolidone, 1% w/v bovine serum albumin—sterilize by filtration and store at −20°C in 25 ml aliquots)
- Hybridization oven and bottles[a]
- Denatured, sonicated salmon sperm DNA (10 mg/ml; Sigma)
- Hybridization buffer (as pre-hybridization buffer but without Denhardt's reagent)
- SaranWrap food film
- X-ray film and cassettes
- Filter paper (Whatman 3MM)
- Labelled probe DNA (see text)

Method

1. Heat the pre-hybridization solution to 50°C. Each membrane will require approximately 3 ml of solution.
2. Denature the salmon sperm DNA in a boiling water-bath for 10 min and add it to the warm pre-hybridization buffer in the hybridization bottle to a final concentration of 500 µg/ml.

9

Protocol 2. *Continued*

3. Place the membranes in this solution, ensuring each one is completely wetted before adding the next.

4. Pre-hybridize at the appropriate temperature for at least 2 h.[b]

5. Pour off this solution.

6. Prepare the hybridization solution as in steps 1 and 2.

7. Add labelled probe (approx. 1–5 × 10⁶ c.p.m./ml of hybridization solution).

8. Hybridize at the appropriate temperature[b] overnight.

9. Wash with 6 × SSC for 5 min at room temperature. Repeat.

10. Wash with 1 × SSC for 5 min at 37 °C.

11. Wash with 0.1 × SSC for 5 min at 65 °C. Repeat.[c]

12. Blot on 3MM paper to remove excess liquid.

13. Wrap the membrane in SaranWrap.[d]

14. Expose to X-ray film overnight.

15. The developed autoradiograph should reveal a small proportion of positive colonies. If the background of non-specific hybridization (present on all or most colonies) is too high, the membranes should be re-washed and re-exposed as in steps 11–14. If necessary, higher washing temperatures may be used.

16. If it is desired to reprobe the membranes (for example with a different probe), incubate them at 100 °C in 0.1% SDS for 5 min and allow them to cool to room temperature. Rinse thoroughly with distilled water, blot on Whatman 3MM paper, and autoradiograph overnight to ensure that no residual probe remains.

[a] As an alternative to the use of a hybridization oven, incubations may be performed in heat-sealed polythene bags in a shaking water-bath at the appropriate temperature. In either case, the membranes should be gently agitated during all incubations and washes, whether by means of the roller bottles or otherwise. The volumes of solutions may need to be increased when using bags rather than bottles.

[b] For dinucleotides, this temperature should be 10 °C below the T_m calculated using formula: $T_m = 69.3 + 0.41 \times (G + C)\% - 650/L$, where $(G + C)\%$ is the percentage of G + C bases in the probe and L is the length of the probe before labelling. Those for other repeat motifs can be found in ref 14.

[c] This is a high stringency wash, and should detect only repeats greater than 30 base pairs in length, most of which should be polymorphic (8).

[d] Membranes should never be allowed to dry completely. If this happens, probe will bind irreversibly to the membranes, preventing re-washing or stripping and re-probing.

4.3 Isolation and sequencing of inserts

If the hybridization described in *Protocol 2* has been successful, there should be discrete black spots on the X-ray film corresponding to plaques containing microsatellite repeats. It is possible to identify the corresponding colonies by aligning the needle holes in the membrane with those on the plate. DNA may be prepared from these colonies using a standard miniprep method (for example, see ref. 15) but it is more convenient to prepare DNA for sequencing by direct PCR amplification of the insert. Four PCR primers are required, two of which flank the insertion site for the vector used. These may be designed following the guidelines in Section 4.4, and with reference to the sequence of the vector. The other two primers required are internal primers which amplify from the repeat sequence. Each internal primer consists of 15 bases corresponding to the repeat sequence (or its complement), with the addition at the 5′ end of either five Gs (on one primer) or five Cs (on the other). If the repeat motif includes a G or C residue, then the primer should be designed to end with a C or G, rather than an A or T. For example, suitable primers for [CA]n microsatellites would be:

$$5'GGGGGCACACACACACACAC3'$$
and $\qquad$ $5'CCCCCGTGTGTGTGTGTGTG3'$

whilst for [CAG]n repeats, appropriate primers would be:

$$5'GGGGGCAGCAGCAGCAGCAG^{3'}$$
and $\qquad$ $5'CCCCCGTCGTCGTCGTCGTC^{3'}$

Three PCR reactions are performed for each positive clone: one amplifies the entire insert using the two vector-specific primers, and the other two each amplify half of the insert, between the microsatellite motif and either the left or right vector ends. These latter two products can be useful if the full-length product is too long to enable complete sequencing. *Protocol 3* describes the amplification of inserts from positive clones.

Protocol 3.   Amplification of inserts

Equipment and reagents

- PCR primers (left vector primer; right vector primer; microsatellite forward primer; microsatellite reverse primer; see discussion above) each at 50 μM
- *Taq* DNA polymerase, 5 U/μl (Promega)
- 10 × reaction buffer: 500 mM KCl, 100 mM Tris–HCl pH 9.0, 1% Triton X-100
- MgCl$_2$ solution (25 mM)[a]
- 100 × dNTP solution: 20 mM each dATP, dCTP, dGTP, dTTP (Pharmacia)
- Sterile, double distilled water[b]

- Light mineral oil (Sigma)
- Agarose (molecular biology grade)
- 5 × TBE buffer: 0.45 M Tris, 0.45 M boric acid, 0.125 M EDTA
- Sample dyes: 10% glycerol, 0.05%(w/v) bromophenol blue, 2 × TBE
- Horizontal mini-electrophoresis apparatus and power pack
- Thermocycler, and compatible reaction tubes or microtitre plates
- Sterile glass Pasteur pipettes

Method

1. Stab each positive clone with a sterile Pasteur pipette, to excise a small plug of agar. Transfer each to a 1.5 ml microcentrifuge tube containing 100 μl of water and leave to soak overnight.

11

Protocol 3. *Continued*

2. Prepare three reaction tubes (or microtitre wells[c]) for each plaque to be analysed, containing:
   (a) 1 μl each of left vector primer and microsatellite reverse primer.
   (b) 1 μl each of right vector primer and microsatellite forward primer.
   (c) 1 μl each of left vector primer and right vector primer.

3. Add 5 μl of the liquid containing the plaque DNA (step 1) to each tube.

4. Prepare a master mix consisting of the following components (quantities are per PCR reaction):
   - 2 μl 10 × reaction buffer
   - 0.2 μl 100 × dNTP solution
   - 1.2 μl $MgCl_2$
   - 4.6 μl water

   Add 8 μl of master mix to each tube.

5. Overlay each tube (or microtitre well) with a drop of light mineral oil, unless using an 'oil-free' thermocycler.

6. Incubate at 95°C for 5 min.

7. To each reaction, add 0.2 μl (1 U) of *Taq* polymerase diluted in 5 μl of water.

8. For the vector–insert reaction (a and b), perform three cycles[d] of:
   - 95°C × 30 sec
   - 37°C × 45 sec
   - 72°C × 5 min

followed by nine cycles of:
   - 95°C × 30 sec
   - 60°C × 45 sec
   - 72°C × 5 min

For the vector-vector reaction (c), perform 12–18 cycles[e] of:
   - 95°C × 30 sec
   - 60°C × 45 sec
   - 72°C × 5 min

9. Add 3 μl of sample dyes to a 3 μl sample of each PCR product, and analyse by electrophoresis through 0.8% agarose gels in 0.5 × TBE buffer.

[a] Magnesium chloride is extremely hygroscopic, so it is best to buy this solution already prepared in order to be sure of the molarity. It is supplied free by many suppliers of *Taq* polymerase, or may be purchased from Sigma (M1028).

[b]Water quality can be an important factor in PCR success. If the laboratory water is of doubtful quality then it may be better to buy water (BDH 44384 7D).
[c]Note that the vector–vector amplification (c) requires different cycling conditions from the other two reactions, and cannot therefore be performed in the same microtitre plate.
[d]The low annealing temperature used in the first three cycles is necessary because only 15 of the primer bases are homologous to the microsatellite repeat. Products of these early cycles will contain the additional five G or C residues from the 5' end of the primer, and may therefore be further amplified at the higher annealing temperature.
[e]The number of cycles needs to be determined empirically for each clone, although 12 cycles is often sufficient.

The positive PCR products may then be directly sequenced, for example using the Sequenase v2.0 kit (USB). It is normally desirable to sequence all three PCR products for any clone (i.e. the full-length vector–vector product and the two products amplified between the microsatellite motif and either end of the vector), to ensure that adequate sequence is obtained on either side of the microsatellite motif. A crucial factor in the success of sequencing directly from PCR products is the cleanliness of the template. If this proves to be a problem, the amplification reactions should be scaled up to 100 μl volume (increasing the volume of all reagents fivefold), substituting a biotinylated primer for either the left or right vector primer in each reaction. Streptavidin-conjugated magnetic beads (Dynabeads, Dynal Ltd.) are then used according to the manufacturer's instructions to purify the product. Either the biotinylated or non-biotinylated strand may be sequenced, but it is the author's experience that the non-biotinylated strand tends to give the better results.

4.4 PCR primer design

After sequencing the clones containing microsatellites, the next stage is to design PCR primers which flank the repeat region, and which will be used for genotyping. There are many computer programs written for both PCs and Macs, such as *Primer Express* (Applied Biosystems) to aid this process. Other primer design programs, available via the Internet, are detailed in Appendix 2. However, primer design is not difficult, and if the following points are borne in mind can easily be done without the aid of these programs.

(a) Primers should be between 16 and 24 bases long.

(b) As far as possible, the G + C content of each primer should be approximately 50%.

(c) Avoid long runs of any single base, in particular the tetranucleotide GGGG due to its frequent occurrence throughout the human genome.

(d) To avoid excessive primer dimer formation, ensure that the ends of the primers are not complementary (either each primer to itself, or one primer to the other).

(e) If analysis is, or might be, carried out on a ABI automated DNA analyser, then the 5′ end of one of the primers should be tagged with a fluorescent dye such as JOE, ROX, or TAMRA (see Section 6.2).

(f) The T_m of the two primers should preferably be within 2 °C of each other. The T_m is best calculated using the formula $T_m = 69.3 + 0.41 \times (G + C)\% - 650/L$, where L = length of the primer in bases and T_m is in degrees centigrade (16). (The more frequently used $T_m = 4 [C + G] + 2 [A + T]$ is rather inaccurate.)

The length of the amplimer should in general be between about 80 and 200 bp. If several sets of primers are being designed, it is often advantageous to choose amplimers with a range of sizes, so that multiplex analysis (the loading of two or more PCR products on the same gel track) can be performed. The size differences between the amplimers must, of course, be sufficient to avoid any risk of overlap between any of the alleles of one microsatellite and those of another.

Occasionally it will be the case that primers fail to work, even when the guidelines above have been followed, and despite numerous attempts at modifying the conditions for PCR (see Section 5.1). In these cases a simple and usually effective solution is to redesign primers a few base pairs up- or down-stream of the original sites.

5. Evaluation of the candidate marker

5.1 Preliminary testing and optimization of PCR conditions

Before proceeding to more detailed evaluation of the marker it should be verified that the primers amplify the desired region successfully from genomic DNA, and optimal conditions for amplification should be determined. *Protocol 4* describes this procedure.

Protocol 4. PCR primer testing and optimization

Equipment and reagents

- Thermocycler and compatible reaction tubes or microtitre plates
- PCR primers (Section 4.4) at 50 μM each
- MgCl₂ solution (25 mM)
- Genomic DNA (50 ng/μl in water or in 1 mM Tris pH 7.5, 0.1 mM EDTA)ᵃ
- 100 × dNTP solution: 20 mM each dATP, dCTP, dGTP, dTTP (Pharmacia)
- 10 × reaction buffer: 500 mM KCl, 100 mM Tris–HCl (pH 9.0 at 25°C), 1.0% Triton X-100
- *Taq* DNA polymerase (Promega, 5 U/μl)
- Sterile double distilled H₂O
- Light mineral oil (Sigma)
- Sample dyes and electrophoresis apparatus as for *Protocol 3*

Method

1. For each primer pair to be tested, prepare a reaction mixture as follows:

- 1 μl 10 × reaction buffer
- 1 μl genomic DNA
- 0.5 μl each PCR primer
- 0.1 μl *Taq* polymerase
- 0.5 μl 100 × dNTP solution
- 4.8 μl water

2. Aliquot 8.4 μl of this mixture into each of four reaction tubes (or wells of a microtitre plate), and add:

- 0.4 μl of MgCl$_2$ and 1.2 μl of water
- 0.8 μl of MgCl$_2$ and 0.8 μl of water
- 1.2 μl of MgCl$_2$ and 0.4 μl of water
- 1.6 μl of MgCl$_2$

(These quantities will give final MgCl$_2$ concentrations of 1, 2, 3, and 4 mM respectively.)

3. Overlay each reaction with one drop of mineral oil (unless using an oil-free thermocycler).

4. Cycle under the following conditions for 28–32 cycles:[b]

- 90° C × 60 sec

- Annealing × 45 sec (at the T_m determined in Section 4.4)[c]

5. Add 10 μl of loading dyes to each reaction and analyse the products by electrophoresis in a 1.5% agarose gel in 0.5 × TBE, with suitable size standards.

6. One or more of the four magnesium concentrations should yield a PCR product of the expected size.[d] If this is not the case, repeat the experiment substituting in step 4 an annealing temperature 4°C higher or lower than that previously used.

[a]This can be from any individual or cell line known to contain the microsatellite in question. Human and some other genomic DNAs are available from Sigma.
[b]28 cycles will be adequate in many cases. The number of cycles should be minimized to reduce non-specific PCR products.
[c]For amplimers of less than 400 bp, an extension step at 72°C is not normally required. Omission of the extension step may also help to reduce the occurrence of non-specific amplification products.
[d]Agarose gels will not reveal the separate alleles of most microsatellites.

5.2 Determination of heterozygosity and PIC values

To be of any use in linkage mapping, a marker must be sufficiently polymorphic. Initial evaluation requires the screening of 50 unrelated individuals to determine not only PIC and level of heterozygosity in the candidate marker, but also to evaluate allele frequencies. As allele frequencies can vary substantially between populations it is important that this 'random' population is of the same ethnicity as the population one intends to assess for linkage. If this is not done, there is a very real risk of obtaining biased results during linkage analysis. This same caveat applies when using published markers with poorly defined allele frequencies, or markers for which allele frequency data has been obtained on a substantially different population from the one under study.

When choosing the 50 unrelated individuals, remember that all persons marrying into the study families qualify for this role. If members of the same ethnicity as the study population are available locally, random individuals are easily obtained by collecting the spare portion of blood remaining after haematological analysis in antenatal clinics. It is usually possible to ascertain ethnic origin from the forms and, provided that the samples are anonymized (and only that portion of blood that would have been disposed of is used) local ethics committees seldom object. If one has access to the CEPH panel of DNAs, it is useful for the wider scientific community to include several reference individuals in this preliminary assessment. Their genotypes can then be submitted to the relevant databases (see Appendix 2). It is possible to register with CEPH and obtain aliquots of DNA, but only if you clearly have the intention (and skills required) to make a considerable number of new markers. Certain of the CEPH DNAs are also included in some gene mapping kits e.g. Perkin-Elmer's fluorescently labelled marker set.

The individuals should be genotyped following the protocols described in Section 6, and the alleles scored for each individual. Heterozygosity is determined simply as the proportion of individuals which are heterozygous for the microsatellite in question. The PIC value is calculated from the following equation:

$$\text{PIC} = 1 - \sum_{i=1}^{n} P_i^2 - \sum_{i=1}^{n-1}(\sum_{j=i+1}^{n} 2P_i^2 P_j^2) \quad\quad [1]$$

where each allele present in the sample population is assigned an (arbitrary) number from 1 to n, and P_i or P_j is the relative frequency of the specified allele in the sample (i.e., the number of occurrences of that allele divided by twice the number of individuals tested; homozygous individuals, of course, count as two occurrences of the relevant allele).

Current published markers vary substantially in their PIC and heterozygosity values, but Genethon's dinucleotide repeats have an average heterozygosity of 70% (9), as do the Utah tetranucleotide markers (11). However even if the new marker isolated has rather a lower value (e.g. 50–60%) it may still be of considerable value if it lies in one's region of interest.

6. Genotyping

For genotyping, whether for preliminary marker assessment or as part of a linkage study, the microsatellite is labelled either radioactively, or fluorescently by using one PCR primer which carries a dye. Analysis is then performed on denaturing polyacrylamide gels, the PCR products being detected either by autoradiography or by a fluorescence-based laser scanning detection system, as appropriate.

Alleles are distinguished by the lengths of the amplified microsatellite, either in absolute terms (which is preferable) or relative to one another, the length differences being multiples of the repeat unit. It is vitally important when genotyping families to ensure continuity of allele assignment, that is, that any given allele is typed consistently for all members. This may be achieved in a variety of ways:

(a) By running the whole family on one gel.

(b) By including common 'reference' individuals on every gel. These may either be members of the CEPH panel, or individuals that were used in the initial determination of allele frequencies, making sure that several different alleles are represented.

(c) By using radiolabelled size markers. This has the advantage of determining allele lengths in absolute terms, rather than relative to one another. This is particularly important in cases where different alleles have different population frequencies, as the 'absolute' identity of the allele then becomes significant.

In addition, negative controls (in which water is substituted for template DNA) should be included in every experiment.

6.1 Genotyping using radiolabelling and autoradiography

When genotyping using radiolabelling followed by autoradiography, it is best to analyse the results for the entire family on one gel if possible to ensure internal consistency. *Protocol 5* describes the incorporation of radiolabelled nucleotides into amplimers during the PCR.

Protocol 5.   Incorporation of radiolabelled dNTPs during PCR

Equipment and reagents

- Thermocycler and compatible tubes or microtitre plates
- Forward and reverse PCR primers (50 μM)
- MgCl$_2$ solution (25 mM)
- Template DNAs at 50 μg/μl in 10 mM Tris pH 7.5, 1 mM EDTA
- 'Low A' dNTP mix[a]: 5 μl each of dCTP, dGTP, and dTTP (each at 25 mM), 1.25 μl dATP (2.5 mM), 10 μl sterile double distilled H$_2$O (dNTPs from Promega)
- [α^{33}P]dATP (1 μCi/μl; Amersham)[b]
- 10 × reaction buffer: 500 mM KCl, 100 mM Tris–HCl (pH 9.0 at 25°C), 1.0% Triton X-100
- *Taq* DNA polymerase (5 U/μl; Promega)
- Sterile double distilled H$_2$O[c]
- Light mineral oil (Sigma)
- Formamide loading buffer: 950 μl formamide, 200 μl 5x TBE, 0.05% (w/v) bromophenol blue, 0.05% (w/v) xylene cyanol

Method

1. Aliquot 0.5 μl of DNA into a sterile reaction tube or microtitre well.

17

Protocol 5. *Continued*

2. Prepare a 2 × reaction mixture. This is best scaled up, and aliquoted out to each tube; the recipe below is for one reaction:

   - 1 µl 10x reaction buffer
   - 0.1 µl 'low A' dNTP mix
   - 0.4–1.6 µl MgCl$_2$ (as determined in *Protocol 4*)
   - 0.5 µl forward primer
   - 0.5 µl reverse primer

   Adjust to 5 µl volume with sterile double distilled water, and add 5 µl to each reaction tube or well.

3. Add a drop of mineral oil to the reaction tube to prevent evaporation of the reaction mixture, unless using an 'oil-free' thermocycler.

4. Heat samples at 95 °C for 5 minutes.

5. During this time prepare the other part of the reaction mixture. Again, the recipe is for one reaction, but a multiple of this should be prepared to avoid errors in pipetting small volumes:

   - 0.5 U *Taq* DNA polymerase
   - 0.02 µl [α^{33}P]dATP
   - 4.4 µl sterile double distilled water

6. After the initial denaturation step has finished, add this mixture to the reaction. There is no need to place the pipette tip under the oil layer, as the mixture will sink under the hot oil easily of it's own accord (this avoids the need to use a fresh tip for each sample).

7. Place in thermocycler and perform 28 cycles of:

   - 90 °C × 60 sec
   - annealing temperature × 45 sec (where the annealing temperature is the same as in *Protocol 4*)

8. Add 10µl of formamide loading buffer to each sample.

9. Denature samples at 90 °C for 5 min prior to analysis on denaturing polyacrylamide gels (see below).

---

[a] If a different radiolabel is used, modify the mixture accordingly so that the concentration of the unlabelled counterpart of the radiolabelled dNTP is reduced. It is recommended to use either radiolabelled dATP or dCTP.
[b] Other radiolabels may be substituted, such as [α^{32}P]dATP or [α^{35}S]dATP.
[c] Water quality can be an important factor in PCR success. If the laboratory water is of doubtful quality then it may be better to buy water (BDH 44384 7D).

Provided that suitable PCR conditions were established during the preliminary assessment of the marker, there is no advantage in checking the labelled PCR products on agarose gels, and one should proceed directly to analytical acrylamide electrophoresis.

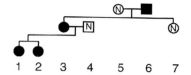

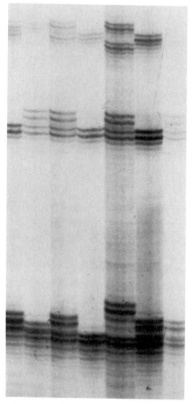

Figure 4. Multiplex genotyping of a three-generation family with an autosomal dominant disorder. Three pairs of primers were combined in the PCR, amplifying the chromosome 15 loci GABRA5, D15S11, and GABRB3 (respectively, from top to bottom). A common pitfall in multiplex PCR is that the more efficient primer pairs out-compete the less efficient; this is seen particularly in the case of the GABRA5 locus for individuals 2 and 7. Note the presence of 'shadow bands', which can make interpretation difficult. The alleles (numbered with respect to reference families included on the same gel but not shown here) are as follows (a dash indicates no result):

Individual	GABRA4	D15S11	GABRB3
1	2, 3	4, 4	4, 5
2	–, –	3, 4	5, 6
3	1, 3	3, 4	4, 6
4	2, 3	4, 4	6, 6
5	1, 4	3, 4	3, 6
6	3, 4	4, 4	4, 6
7	–, –	4, 4	4, 6

Samples are analysed on conventional sequencing gels, the pouring and running of which is dealt with in a previous volume in this series (12) and will not be described in detail here. Before autoradiography, it is helpful to scan the gel with a Geiger counter in order to estimate how long to expose the film for. This is very much a trial and error process, but the following may serve as an initial guide:

(a) > 500 c.p.m.: develop after 6 h or overnight if more convenient (overnight exposure may be too dark and require a 6 h exposure the following day).

(b) 300–500 c.p.m.: 24 h exposure.

(c) 100–300 c.p.m.: 2–3 days exposure.

(d) < 100 c.p.m.: a trial exposure of a 7–10 days may result in some readable data, but it may be necessary to repeat the experiment.

Protocol 5 may be modified to allow multiplex genotyping of two or more loci simultaneously. The amplimers of the different loci must not overlap in their size ranges, and the PCR primer pairs for all loci must operate under identical conditions (*Figure 4*). All primers are then combined in the PCR (*Protocol 5*, step 2), the volume of water being reduced to allow for the extra volume of the primers.

If, after autioradiography, the alleles are not clearly distinguishable from one another due to the presence of shadow bands (particularly likely to be a problem when using dinucleotide repeats; see *Figure 3*) then it may be necessary to further refine the PCR by further adjusting either magnesium concentration or annealing temperature. If these fail to resolve the problem, an alternative is to prepare and use an end-labelled PCR primer for amplification (rather than incorporating label during amplification), to ensure that only one of the two DNA strands is labelled. Unfortunately one cannot normally judge the efficacy of these solutions until the products are run on denaturing acrylamide gels: the resolving power of agarose gels is insufficient to detect this sort of base stuttering.

Protocols 6 and 7 describe the end-labelling of PCR primers, and their use in a modified PCR.

Protocol 6. End-labelling of PCR primers

Equipment and reagents

- Primer to be labelled (one of the two microsatellite primers)
- T4 polynucleotide kinase (Promega M4101)
- 10 × kinase buffer: 700 mM Tris–HCl pH 7.6, 100 mM MgCl$_2$, 50 mM DTT
- [γ-^{35}S]dATP (> 1000 Ci/mmol, Amersham SJ1318)
- Heat blocks or water-baths (37°C and 65°C)
- Vacuum centrifuge
- Sterile double distilled water

Method

1. Place 200–500 pmol of one of the PCR primers in a 0.5 ml micro-centrifuge tube and evaporate to dryness in a vacuum centrifuge.

2. Add the following :
   - 5 μl 10 × kinase buffer
   - 8 μl [γ-^{35}S]dATP
   - 12 U T4 polynucleotide kinase

   Adjust to 50 μl with water and pipette up and down to mix and to dissolve the primer.

3. Incubate at 37 °C for 35 min.

4. Incubate at 65 °C for 5 min to stop the reaction.

5. Spin for 5 seconds at full speed (approx. 13 000 r.p.m.) in a micro-centrifuge to remove residual liquid from the walls and cap of the tube.

6. Freeze at −20 °C or proceed directly to *Protocol 7*. (It is not necessary to remove unincorporated [γ-^{35}S]dATP from the labelled oligo-nucleotide.)

Protocol 7.   Use of end-labelled primers in microsatellite PCR

Equipment and reagents

- As *Protocol 5*
- Substitute the dNTP mix from *Protocol 3* for that described in *Protocol 5*
- [α-^{33}P]dATP is not required
- Replace one of the unlabelled primers with an equal amount of the corresponding end-labelled primer (*Protocol 6*)

Method

1. As for *Protocol 6*, making the reagent substitutions noted above and omitting the labelled nucleotide in step 5.

6.2 Genotyping using fluorescently labelled primers

An alternative to the use of radioactive detection for genotyping is to use a fluorescence-based laser scanning detection system, such as the Applied Biosystems 373 or 377 machines. These are essentially automated gel readers, in which fluorescently labelled DNA molecules are detected during electrophoresis as they travel past a fixed point on an otherwise conventional acrylamide gel. Software then generates a simulated image of the gel, with the lengths of the DNA molecules being determined from their time of arrival at the scanning point (and further refined by comparison with fluorescently labelled size standards). Four different dye labels may be used and detected independently, enabling four populations of DNA molecules to be super-imposed in each gel lane.

The amplification of microsatellites for fluorescent genotyping differs little from that for radioactive detection. The radioactive nucleotide is omitted (and equimolar amounts of all four unlabelled bases are used), and one of the two PCR primers is labelled with one of the four fluorescent dyes known as JOE, FAM, TAMRA, and ROX. These dyes must be incorporated into the primer at the time of synthesis, and this represents a major expense of this approach. Post-synthetic incorporation of dyes into the PCR primers or the amplimer is not feasible in most cases; ABI have recently introduced fluorescently labelled dUTP (to replace dTTP in the dNTP mixture used for PCR), but this approach works well only for tri- and tetranucleotide repeats. (Note that fluorescently labelled primers can, if desired, also be used in conventional PCRs in which the products are labelled radioactively.) The PCR reactions are performed as for radioactive labelling, though 22–23 cycles of amplification normally give optimal results (*Protocol 8*).

Protocol 8. Use of fluorescently labelled primers in microsatellite amplification

Equipment and reagents

- As for *Protocol 5*, except there is no requirement for a radiolabel, and the following dNTP mix should be substituted in place of the 'low A' mix: 4 μl each of dATP, dCTP, dGTP, dTTP (each at 25 mM concentration; Pharmacia), 4 μl sterile double distilled water.

- In addition, either the forward or reverse primer (not both) must have a 5' fluorescent tag

Method

1. As for *Protocol 5*, but with the following modifications:
   (a) Omit the radiolabelled nucleotide, and substitute the dNTP mixture given above.
   (b) The number of PCR cycles should be reduced to 22 or 23.

By using a combination of amplimers which have different lengths (see Section 4.3) and which are labelled with different fluorescent dyes, multiplex genotyping of nine or more different microsatellites can be performed in a single gel lane. (One of the four dye colours is normally reserved for use by internal DNA size standards.) Software for fluorescent genotyping using ABI machines is available from ABI for use on Macintosh computers. It facilitates comparison between gel tracks on the same or on different gels, and can automatically identify the alleles present in each gel lane, greatly reducing the risk of errors in data entry. Largely because of this, it is not as important to analyse the entire family on a single gel as it is when using radioactive detection, provided that one or more positive controls consisting of amplimers from the same individuals are included on each gel.

7. Genotype scoring and linkage analysis

It is best to score the alleles without reference to the pedigrees, then to fit the results onto the family trees. This method avoids subconscious bias of results, particularly when scoring results manually. Genotyping errors may manifest themselves as apparent non-Mendelian inheritance, though not all errors will be apparent in this way. If there is doubt over particular individuals it can often be helpful to repeat the gel, running the 'problem individuals' together. It is then much easier to determine whether or not they all share the same allele. Bear in mind that a single incorrect typing can significantly affect the subsequent results of linkage analysis.

In rare cases, apparent non-Mendelian inheritance of alleles can arise through mutation of the microsatellite from one generation to the next. (This, of course, is the process by which microsatellite polymorphism arises.) In such cases, the initial assumptions of incorrect genotyping or other errors must first be eliminated by repeating the analysis; the possibility of non-paternity may be excluded by genotyping several other polymorphic markers on the same individuals. The expansion of particular trinucleotide motifs is common in so-called 'trinucleotide repeat diseases' such as myotonic dystrophy, Huntingdon's disease, and fragile X syndrome. In other (non-disease-related) cases it seems to occur more frequently with compound repeats than with other types.

Once one has generated genotype data, the next step is to analyse it and, in the case of a new marker that has been shown to be of sufficient heterozygosity, integrate it into the currently available chromosome maps. The common difficulty in collecting data on human pedigrees is that of small family size, and that few generations are generally available for analysis. This means that linkage cannot be determined by simple Mendelian analysis, but that informative meioses—those in which it is possible to ascertain which allele has been donated by which parent—must be looked for instead. The standard technique for doing this is by the likelihood method of linkage analysis (17). For any two markers, the probability of their pattern of segregation arising by chance alone (i.e. assuming them to be unlinked) is compared to the probability of its arising if they were linked. The ratio of the latter probability to the former, expressed as a logarithm (base 10) is then the 'lod score' between those two markers. For example, a lod score of 2 means that the observed segregation pattern is 100 times more likely to have arisen as a result of linkage between the two markers than by the chance segregation of unlinked markers.

The conventional threshold for claiming linkage is a lod score of 3.0, corresponding to 1000:1 odds that the loci are linked. However, any two randomly chosen loci in the human genome have only a 1 in 50 *a priori* chance of being truly linked—these are the 'prior odds' against linkage, and they weigh against the 1000:1 odds in favour of linkage. In practise, then, a lod score of 3 corresponds to 20:1 odds (1000:1 'for' divided by 50:1 'against') that the two loci are truly linked. For any given data-set, higher lod scores reflect 'tighter' linkage, the score being highest between the most closely linked loci. It should be noted that the summation of lod scores from independent experiments to reach the 'significance threshold' of 3 is rarely a valid procedure, and can produce misleading results: such summation simply increases the background 'noise' of small, random statistical fluctuations.

Lod scores can be easily calculated on any desktop PC. Small analyses can be performed on relatively low power machines, but larger analyses are time-consuming. A number of computing centres allow users to submit linkage data for analysis; details of these will be found in Appendix 2.

The original *LINKAGE* program, written in Turbo Pascal, was compiled by J. Ott (18) and accepts genotype data either as a direct input or from the Cyrillic pedigree drawing software. *LINKAGE* calculates lod scores between a chosen marker (or a disease locus) and one or more other markers, and gives an output which indicates where the chosen marker lies relative to the others. The reliability with which the marker is placed relative to the others (i.e. the odds against inversion between it and flanking markers) will depend upon the disposition of the surrounding markers. The HGMP resource centre at Cambridge, UK, runs courses in *LINKAGE* (contact the Course Manager, UK HGMP Resource Centre, Hinxton, Cambridge CB10 1SB), and these are essential if there is no one locally to ask for help. Further details of linkage analysis software and courses will be found in Appendix 2.

The main problem is that most of the linkage programs available such as *LINKAGE* are not intended for map generation. As such, a large amount of human intervention and repetitive calculation is required, increasing the potential for operator errors. There are some newer computer programs available which claim to overcome the excessively large amounts of computer time required to construct maps, and are more suited to the incorporation of new markers into existing linkage maps. These are based on novel algorithms and include *VITESSE* (19) and MultiMap (20), and differ in a fundamental way in their approach to map construction. As with *LINKAGE*, a certain amount of expertise is required to use these programs effectively, and potential users should contact their local computing or bioinformatics centre (Appendix 2) for detailed advice.

Genetic maps may also be finally linked with the many physical maps available, commonly based on yeast or bacterial clones and having better resolution than genetic maps. There has been a recent report of an integrated map linking the two approaches by using high-resolution FISH to give a kilobase scale along the whole of chromosome 19 (21). This is a major advance, as it can be very difficult to align genetic and physical maps of the same region, there often being very few common reference points. It is likely that, due to the speed of the entire human mapping program, this will be the first of many such integrated maps of increasing resolution.

References

1. Renwick, J. H. (1969). *Br. Med. Bull.*, **25**, 65.
2. Botstein, P., White, R. L., Skolnick, M., and Davis, R. W. (1980). *Am. J. Hum. Genet.*, **32**, 314.
3. Grodzicher, T., Williams, J., Sharp, P., and Sambrook, J. (1974). *Cold Spring Harbor Symp. on Quan. Biol.*, **39**, 439.
4. Wier, B. S. (1990). *Genetic data analysis*. Sinauer Associates Ltd., Sunderland, Mass.
5. Ott, J. (1991). *Analysis of human genetic linkage*. John hopkins University Press, Baltimore.
6. Weber, J. L. and May, P. E. (1989). *Am. J. Hum. Genet.*, **44**, 388.
7. Litt, M. and Luty, J. A. (1989). *Am. J. Hum. Genet.*, **44**, 397.
8. Tautz, D. and Renz, M. (1984). *Nucleic Acids Res.*, **12**, 4127.
9. Dib, C., Fauré, S., Fizames, C., Samson, D., Dronot, N., Vignal, A., *et al.* (1996). *Nature*, **380**, 152.
10. Schlotterer, C. and Tautz, D. (1992). *Nucleic Acids Res.*, **20**, 211.
11. The Utah Marker Development Group (1995). *Am. J. Hum. Genet.*, **57**, 619.
12. Alphey, L. (1995). In *DNA cloning 1: a practical approach* (ed. D. M. Glover and B. M. Hames), p. 241. IRL Press, Oxford.
13. Taylor, G. R., Haward, S., Noble, J. S. and Murday, V. (1992). *Anal. Biochem.*, **200**, 125.
14. Gaustier, J. M., Pulido, J. C., Sunden, S., Barcroft, C. L., Kiousis, S., and Chamberlain, J. S. (1995). *Hum. Mol. Genet.*, **4**, 1829.
15. Maniatis, T., Fritsch, E. F., and Sambrook, J. (ed.) (1982). *Molecular cloning: a laboratory manual*, p. 368. Cold Spring Harbor Laboratory Press, N. Y.
16. Marmur, J. and Doty, P. (1962). *J. Mol. Biol.*, **5**, 109.
17. Ott, J. (1974). *Am. J. Hum. Genet.*, **26**, 773.
18. Lathrop, G. M., Lalouel, J. M., Julier, C., and Ott, J. (1984). *Proc. Natl. Acad. Sci. USA*, **81**, 3443.
19. O'Connell, J. R. and Weeks, D. E. (1995). *Nature Genetics*, **11**, 402.
20. Matise, T. C., Perlin, M., and Chakravarti, A. (1994). *Nature Genetics* **6**, 384.
21. Ashworth, L. K., Batzer M. A., and Brandriff, B. (1995). *Nature Genetics*, **11**, 422.

2

Linkage mapping of plant and animal genomes

ROSS MILLER

1. Introduction

The human genome project has provided increasingly detailed physical (1–3) and genetic maps (4). A large number of other genomes are also being mapped: animal genomes under study include those of mouse (5), rat (6), sheep (7), pig (8–10), cow (11, 12), dog (13), chicken (14), the pufferfish *Fugu* (15), zebrafish (16), honey bee (17), *Drosophila* (18), and the nematode *C. elegans* (19), whilst plant genomes being studied include rice (20), wheat (21), maize (22), *Arabidopsis* (23), spruce (24), tomato (25) and potato (25). These lists are intended merely to illustrate the breadth of the field rather than to be exhaustive.

The main goals in the analysis of the human genome (and that of some model organisms) are an understanding of genomic organization and it's relation to function; assembly of gene catalogues; identification of genes responsible for disease phenotypes; provision of DNA segments as overlapping contigs; and, eventually, determination of the complete sequence of the genome—the ultimate physical map. In contrast, the analysis of most plant and animal genomes is oriented much more towards providing a means for locating genes of economic importance, a theme which is pursued in the following chapter.

The human genome is frequently and sometimes successfully used as a paradigm for other animal genomes. Consequently, many techniques and resources developed for the human genome have been transposed to the mapping of other genomes. Mapping data, too, can flow between species: there is a growing realization within the animal genome community that comparative mapping will enable information to flow from the gene-dense human map to the relatively gene-poor maps of domestic animals (26). By this means the domestic animal genome projects can 'piggy back' on the human project. For example, no one has yet suggested that large scale EST identification and chromosomal assignment should be carried out on domestic animal genomes. Likewise it is difficult to envisage any requirement for widespread sequencing of domestic animal genomes. However, there almost certainly will be requirements for intensive sequencing of localized areas in the search for genes of importance.

The situation regarding plants is different, as the human genome ceases to be a paradigm (though there does seem to be some conservation of synteny between man and rice) (20). Two plant genomes appear to have acquired 'model organism' status. *Arabidopsis* is a flowering plant with a genome of only 100 Mb, and an intensive project is underway to construct a comprehensive physical map including provision of YAC contigs. Rice is a member of the grass family which includes sorghum, wheat, maize, barley, and oats. At only 430 Mb, the rice genome is far smaller than those of cereals: wheat, for example, is 17000 Mb. As there is significant colinearity of gene location between rice and wheat (27), rice would seem to be an ideal model for the study of the wheat genome and perhaps those of other cereals (28).

Further down the scale of complexity, there is a strong mapping and sequencing programme for the nematode *C. elegans* (19), an animal with differentiated tissues whose cell lineage has been precisely described (29). Amongst unicellular organisms there are sequencing programmes for *B. subtilis* (30) and *E. coli* genomes (31), whilst the sequences of *Saccharomyces cerevisiae* (32), *Haemophilus inlfuenzae* (33), and *Mycoplasma genitalium* (34) have recently been completed, thus allowing gene prediction by computer analysis.

This chapter will attempt to highlight the problems, opportunities, and techniques relevant to non-human genomes, such as the lack of extensive resources, the possibilities afforded by directed breeding programmes, and novel PCR techniques such as RAPD (randomly amplified polymorphic DNA) and AFLP (amplified fragment length polymorphism) analysis.

2. Genome data for non-human species

Table 1 presents information on genome characteristics for a number of different organisms, both animal and plant. It should be borne in mind that many of the genetic sizes given in the table have been deduced from partial, rather than complete, genetic maps and hence should be regarded as minimum estimates. Likewise, the physical sizes of many genomes have not been accurately determined. The number of genes has been estimated for only a few higher organisms (man, *C. elegans*, *Arabidopsis*); sequencing studies have provided rather more precise estimates for the three bacterial genomes listed and for yeast. What should be apparent from the table are the occasional occurrence of organisms such as *Arabidopsis* (23) and *Fugu* (15) which have significantly smaller genomes than other members of their phylum. These small genomes, which appear to contain few repeat elements, offer clear advantages in mapping and sequencing studies. It is also noteworthy that there are enormous variations in gene density, ranging from around one gene

Table 1. Genome data for plant and animal species[a]

Species	Genome size (Mb)	Genetic size (cM)	Number of genes[b]	Number of chromosomes
Human	3000	3700	75 000	$2n = 46$
Mouse	2700	1400		$2n = 40$
Rat		1700		$2n = 42$
Sheep		>2700		$2n = 54$
Pig	2700	2000		$2n = 38$
Cow	3000	3400		$2n = 60$
Dog				$2n = 78$
Fugu	400			
Zebrafish	1700	2500		$2n = 50$
Chicken	1200	2600-3000		$2n = 78$
Honey bee	180	3450		$n = 16$
Drosophila	165			$2n = 8$
C. elegans	100		13,100	$2n = 12$
Rice	430	>1575		$2n = 24$
Wheat	17 000			$2n = 6 \times = 42^c$
Arabidopsis	100		20-25 000	$2n = 10$
Tomato	950			$2n = 24$
Potato				$2n = 24$
Maize	2500	1700		$2n = 20$
White Spruce	10 000	2500		$2n = 24$
S. cerevisiae	15		7000	$n = 16$
E. coli	4.7		4000	$n = 1$
H. influenze	1.8		1743	$n = 1$
Mycoplasma genitalium	0.58		470	$n = 1$

[a] Physical and genetic sizes, gene number and chromosome number are indicated where these are known, or where available data allows reasonable estimates to be made.
[b] Gene number is commonly assumed to be similar between related species, though this assumption has not been entered in the table.
[c] Note that wheat ($2n = 6 \times = 42$) is diploid for each of three ancestral genomes from which it is derived, and is hence pseudo-hexaploid.

per kilobase in bacteria to about one per 30 kilobases in man; that of wheat is presumably far lower still.

3. Marker isolation and genotyping

Microsatellites (also known as short tandem repeats or STRs) and to a lesser extent restriction fragment length polymorphisms (RFLPs) have been the two classes of marker predominantly used to map mammalian genomes. In addition, single-strand conformation polymorphisms (SSCPs) (35), capable of detecting single base substitutions, have proven valuable in placing expressed sequences on genetic maps (36). As the use of these markers has been extensively described in Chapter 1 and elsewhere in the context of human genome mapping, no further information on them will be presented here.

Animal genome mappers may be surprised to learn that, in the various plant species mapped, microsatellites have not been widely used. Microsatellite motifs exist in plants, but at significantly lower densities than in mammalian genomes. Two separate studies have shown that (AT)n repeats comprise the most abundant class of plant microsatellites (37, 38). The most common mammalian dinucleotide repeat motif, (CA)n, occurs significantly more rarely in plants: its frequency in the *Arabidopsis* genome was estimated to be one per 430 kb—some 14-fold lower than in humans. Due in part to the above considerations and also for historical reasons, the types of marker which have predominated in the construction of plant genetic maps are RAPDs (randomly amplified polymorphic DNA) (39, 40) and RFLPs. This is exemplified by the map of the rice genome containing 1383 RFLP and RAPD markers (20).

The following sections describe the use of RAPD and other marker types which are commonly used in the mapping of plant and animal genomes.

3.1 Randomly amplified polymorphic DNA (RAPD) markers

RAPDs are genomic sequences amplified by PCR using a single short primer of arbitrary sequence at low stringency. Under these conditions, a number of PCR products are generated from random locations within the genome. Of these, a proportion will be polymorphic, reflecting sequence variations which create or destroy binding sites for the arbitrary primer (*Figure 1*).

RAPDs have proved popular markers in a variety of plants, but not as yet in mammalian genomes. They have the enormous advantages of being rapid to generate and, unlike markers such as microsatellites, of not requiring any prior DNA sequence information. Design of the oligonucleotide primers follows simple criteria: ten bases in length; G + C content of 50–80%; no palindromic motifs of six or more nucleotides. Many such primers are

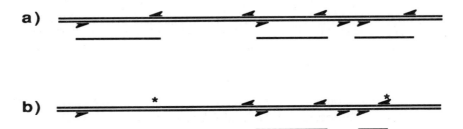

Figure 1. RAPD mapping. Two haplotypes (a and b, double heavy lines) are shown, each of which carries a number of binding sites for a short, arbitrary PCR primer (arrowheads). The expected amplification products are shown as single lines beneath the haplotypes. Sequence polymorphisms (asterisks) will alter the presence or position of primer binding sites, thereby changing the electrophoretic 'fingerprint' seen when the PCR products are analysed.

synthesized and tested on the mapping population to ascertain whether or not their PCR products are polymorphic. This means that a panel of markers suitable for mapping can rapidly be assembled, without the cloning, screening and sequencing normally required to design primers for microsatellite markers.

A surprisingly high proportion of RAPD products are polymorphic. A study on haploid DNA from spruce tree seeds demonstrated that 43 out of 100 RAPD primers detected at least one polymorphic band segregating in a panel of DNAs from five seeds (24). RAPD mapping on honey bees found that within a collection of 132 primers used in mapping, 40% of the PCR bands generated were polymorphic (17).

RAPDs do, however, suffer from certain limitations. Because the PCR reactions are carried out at low stringency, their products can vary with only minor changes in reaction conditions; this can lead to inconsistencies between laboratories. A more serious problem is that RAPD markers are typically dominant rather than co-dominant: many sequence polymorphisms are reflected simply as the presence or absence of a given RAPD marker, rather than as a length variation as in the case of microsatellites. This, in turn, means that it is difficult to distinguish a homozygote from a heterozygote with one 'null' allele. The use of haploid populations for mapping (as in the production of linkage maps of the honey bee (17) and zebrafish (16)) circumvents this problem. In another study, haploid DNA was extracted from the megagametophyte tissue of individual seeds, allowing the production of a genetic map of a single spruce tree (24). This approach represented an elegant solution to the problem of deriving a genetic map from an organism which requires 15–20 years to attain sexual maturity.

One further drawback to the RAPDs lies in the fact that these markers do not specify sequence-tagged sites (STSs). When a microsatellite marker detects an interesting linkage, the marker can immediately be used to screen a resource such as a YAC library or subchromosomal hybrid cell panel. When an RAPD detects such a linkage, cloning and sequencing of the RAPD band will be required in order to convert it into a conventional STS.

Protocol 1 describes the RAPD procedure. Initially, RAPD analysis should be performed on a panel of 10–20 unrelated individuals in order to ascertain which primers yield polymorphic markers. Genotyping may then be performed on family members using the same protocol.

Protocol 1. Genotyping by use of RAPD markers

Equipment and reagents

- Thermocycler and compatible reaction tubes or microtitre plates—a high quality machine which gives uniform and reproducible heating across the entire block is essential for this and other analytical PCR-based procedures

- RAPD primer (see text) at 5 μM
- 10 × PCR buffer: 100 mM Tris–HCl pH 8.3, 500 mM KCl, 20 mM MgCl$_2$
- 10 × dNTP mix: 1 mM each dATP, dCTP, dGTP, dTTP (Pharmacia)

Protocol 1. *Continued*

- *Taq* DNA polymerase, 5 U/ μl (e.g. Amplitaq, Perkin-Elmer)
- Light mineral oil (Sigma)
- Genomic DNAs at 5 ng/ μl
- Sterile, double distilled water
- TBE electrophoresis buffer, 10 × stock: 108 g Tris base, 55 g boric acid, 9.3 g Na$_2$EDTA, made to 1 litre with deionized water
- Horizontal gel electrophoresis apparatus
- Sample buffer: 10% (w/v) glycerol, 0.1 mg/ml bromophenol blue
- High-resolution agarose (e.g. Nusieve 3:1, FMC Bioproducts)
- Ethidium bromide stock solution (0.5 mg/ml)[a]
- UV transilluminator and gel photography apparatus

Method

1. Prepare a master mix containing all PCR components excluding the DNA template. Sufficient mix should be prepared for each genomic DNA to be analysed, plus one negative control. Ingredients per reaction are:
   - 1.25 μl 10 × PCR buffer
   - 1.25 μl 10 × dNTP mix
   - 0.05 μl (0.25 U) *Taq* polymerase
   - 1.25 μl RAPD primer
   - 7.7 μl water

2. Aliquot 11.5 μl of the mixture into each PCR reaction tube (or microtitre plate well) and add 1 μl of the appropriate genomic DNA[b] or, for the negative control, 1 μl of water.

3. Overlay each reaction with one or two drops mineral oil.

4. Perform 45 thermal cycles as follows:
   - 94°C × 1 min
   - 35°C × 1 min
   - Ramp over 2 minutes to 72°C
   - 72°C × 1min

5. Add 8 μl of sample buffer to each reaction.

6. Analyse products by electrophoresis (typically in a 3% agarose gel) in 1 × TBE, following the recommendations of the supplier of the gel tank and of the agarose to achieve the highest resolution. To facilitate interpretation of marker segregation in families, samples should be loaded on the gel in an order corresponding to their position in the pedigree (e.g. offspring in the centre flanked on either side by parents).

7. Stain the gel in ethidium bromide solution (0.5 μg/ml)[a] and photograph with UV transillumination.

[a] Ethidium bromide is a mutagen, and should be handled and disposed of with care.
[b] The exact quantity of template is not critical; consistent patterns of amplification have been observed with template quantities of between 30 pg and 7.5 ng (39, 40).

3.2 5′ and 3′ SINE PCR markers

5′ and 3′ SINE PCR (41, 42), used in mammalian genome analysis, is similar in some ways to RAPD genotyping in that it detects arbitrary polymorphisms scattered throughout the genome, rather than screening for specific poly-morphisms previously characterized by other means. It relies on the fact that short interspersed repeat elements (SINEs, such as the ubiquitous *Alu* ele-ments in man) are frequently associated with a microsatellite—or other length polymorphic motif (43, 44). Hence, PCR amplification between a primer at the 3′ (or 5′) end of the SINE and a unique but arbitrarily chosen primer lying 3′ (or 5′) to the SINE and to the associated length polymorphism, will gener-ate a product whose length is polymorphic (*Figure 2*). Nucleotide substitutions which create or destroy targets for either primer will be reflected as dominant polymorphisms, in which the 'null' allele produces no band in the fingerprint.

The frequency with which polymorphisms may be detected using 5′ or 3′ SINE PCR is extremely high. In one study, 12 out of 15 'unique' primers gave at least one polymorphic product when used in combination with a 5′ SINE primer; when used with a 3′ SINE primer, 30 out of 31 'unique' primers gave at least one polymorphic product (42).

The method is distinct from simple SINE PCR in that the concentration of the SINE primer is roughly tenfold lower than that of the unique primer, so that the yield of unwanted SINE–SINE products (i.e. those resulting from amplification between adjacent SINEs) is greatly reduced. The unique primer is any conventional PCR primer (often previously synthesized for another purpose) whose length is between 20 and 30 nucleotides.

The PCR is performed under reasonably high-stringency annealing con-ditions, under which the 'unique' primer still primes at a sufficient number of sites (some of which will be adjacent to targets for the SINE primer) to give several amplification products. (In a conventional PCR, the effects of this non-specific priming are not usually seen, as *both* unique primers must prime in close proximity for amplification to occur.) This use of high stringency

Figure 2. 3′ SINE PCR. SINE elements (hatched rectangles) in the genome (double lines) act as targets for suitable 3′ SINE primers (hatched arrowheads), whilst a 'unique' primer (solid arrowheads) primes fortuitously at various points in the genome. Length polymor-phisms between the SINE and unique primer (dotted line), or sequence polymorphisms which create or destroy targets for either primer, will change the pattern of PCR amplifi-cation products (single lines). A similar approach may be used with a primer lying at the 5′ end of the SINE. (Originally published in: Miller, J. R. and Archibald, A. L. (1993), 5′ and 3′ SINE-PCR allows genotyping of pig families without cloning and sequencing steps. *Mammalian Genome*, **4**, 243–6. Reproduced by kind permission of Springer-Verlag, who hold copyright.)

conditions contrasts with RAPD where oligo lengths are shorter and the stringency of PCR is reduced.

SINE primers should be designed with reference to the appropriate sequences in EMBL and Genbank databases (see Appendix 2), selecting well-conserved regions close to the 5' or 3' end of the SINE element, and following the conventional guidelines for the design of PCR primers (length 18–28 bases; approximately 50% G/C content; no runs of more than four consecutive bases; no regions of complementarity of > 4 bases either within one molecule or between two molecules). Note that the primers must, of course, be oriented such that extension from them will occur away from, rather than into, the SINE element.

This technique has been used in both the human and pig genomes and has the benefit that the only prior sequence information required is that of the SINE element. One advantage of this method is that it is very different from standard microsatellite-based genotyping, and is thus liable to locate genetic markers in regions devoid of conventional microsatellites. It also has several advantages over RAPD analysis. First, it is far more reproducible because amplification is carried out under stringent conditions. Secondly, 5' and 3' SINE PCR polymorphisms are often co-dominant (i.e., they appear as length polymorphisms in the PCR product) rather than dominant. And thirdly, such polymorphisms often have far more alleles (and are hence more informative) than RAPD markers, which typically appear only as 'positive' and 'null' alleles. Recently, a 19-allele polymorphism was detected at the centromere of human chromosome 5 (45) using a 3' *Alu* primer in combination with an alpha satellite primer. This was the first highly polymorphic marker to be localized to a human centromere.

Just as with RAPD analysis, a large number of primer combinations (a single SINE oligo with many different unique oligos) can be tested for their ability to detect polymorphisms against reference families or population DNAs. The method is described in *Protocol 2*, and examples of 5' and 3' SINE PCR polymorphisms are shown in *Figures 3* and *4*. We routinely use a labelled SINE primer in combination with many different, unlabelled 'unique' primers in turn for detection of the amplification products. The low concentration of the SINE oligo ensures that amplification between SINEs (i.e. SINE–SINE PCR) does not occur to any significant degree.

Protocol 2. Genotyping by use of 5' or 3' SINE-PCR

Equipment and reagents

- Thermocycler (as in *Protocol 1*) and compatible reaction tubes or microtitre plates
- 10 × kinase buffer: 500 mM Tris–HCl pH 7.5, 70 mM MgCl$_2$, 100 mM β-mercaptoethanol
- 10 × PCR buffer (protocol 1)
- 10 × dNTP mix: 2 mM each dATP, dCTP, dGTP, dTTP (Pharmacia)
- *Taq* DNA polymerase, 5 U/ μl (e.g., Amplitaq, Perkin-Elmer)
- 5' or 3' SINE primer (see text), 10 μM
- Unique primer (see text), 2 μM

- T4 polynucleotide kinase (PNK), 10 U/ μl
- [γ-^{32}P]ATP, > 5000 Ci/mmol
- Sterile, double-distilled water
- Genomic DNAs at 10 ng/ μl
- Light mineral oil (Sigma)
- Sequencing gel apparatus capable of giving an even heat distribution across the gel (e.g. BRL Model S2)
- TBE electrophoresis buffer (*Protocol 1*)
- Sequencing gel mix: 6% acrylamide, 0.3% bisacrylamide, 7 M urea in 1 × TBE

- *N,N,N′,N′*-tetramethylethylenediamine (TEMED)
- Ammonium persulfate (APS), 10% (w/v)
- Fixative: 10% (v/v) methanol, 10% (v/v) acetic acid
- 3MM paper (Whatman)
- Vacuum gel dryer
- X-ray film and autoradiography cassette
- Formamide dye mix: 99% formamide, 1 mg/ml xylene cyanol, 1 mg/ml bromo-phenol blue

Method

1. Prepare the end-labelling reaction for the SINE primer:
   - 10 μl SINE primer
   - 2 μl 10 × kinase buffer
   - 5 μl [γ-^{32}P]ATP
   - 1 μl PNK
   - 2 μl water

   Incubate at 37°C for 45 min, then denature the enzyme at 68°C for 10min. Note that the SINE primer concentration is now 5 μM. No further purification is required.

2. Prepare a 'master mix' for the genotyping reactions, consisting of the following ingredients per reaction:
   - 5 μl 10 × PCR buffer
   - 5 μl 10 × dNTP mix
   - 0.4 μl labelled SINE primer (step 1)
   - 10 μl unique oligo
   - 0.125 μl *Taq* polymerase
   - 24.5 μl water

3. Prepare the genotyping reactions by adding 5 μl (50 ng) of genomic DNA to 45 μl of master mix in a suitable reaction tube (or microtitre well), and overlaying with one to two drops of mineral oil.

4. Transfer to the thermocycler and perform PCR as follows:
   - 94°C × 2 min
   - 55°C × 1 min
   - 72°C × 2 min

   followed by 31 cycles of:
   - 94°C × 30 sec
   - 55°C × 1 min
   - 72°C × 2 min

   followed by:
   - 72°C × 5 min

5. To a 6 μl aliquot of each reaction product, add 4 μl of formamide dye mix. Denature the samples at 70°C for 2min, then chill rapidly on wet ice.

Protocol 2. *Continued*

6. Prepare a sequencing gel following standard procedures, adding 80 μl of TEMED and 800 μl of APS for every 50 ml of gel mix to initiate polymerization. Flush the sample wells with the running buffer (1 × TBE), and immediately load the samples. Run the gel under denaturing conditions (e.g. 60 W for the BRL model S2) until the bromophenol blue dye reaches the bottom of the gel.

7. Remove one glass plate and carefully place a sheet of dry 3MM paper over the gel. Carefully remove the paper, to which the gel should adhere.

8. Immerse the paper carrying the gel in fixative for 20 min.

9. Transfer the gel, still on the 3MM paper, to a gel dryer and dry under vacuum.

10. Expose to X-ray film, overnight initially, or longer if required.

3.3 Amplified fragment length polymorphisms—AFLPs

Recently a major new class of markers has been described (46). AFLPs are similar in some respects to conventional RFLPs but, like RAPDs, have the advantage of not requiring extensive prior sequence information or unique, marker-specific primers.

The method involves three distinct steps: digestion and ligation; PCR amplification; and gel analysis. First, genomic DNA is digested to completion with a frequently cutting restriction enzyme. An oligonucleotide 'adapter' is ligated to these fragments, such that the adapter and the few terminal bases of the fragments (which correspond to the restriction site) serve as a 'universal' target for amplification of the fragments. PCR amplification based upon this 'universal' target alone would amplify all fragments—too many to be analysed. Instead, therefore, the primer consists of the 'universal' sequence *plus* a few arbitrarily chosen extra 'selective' bases at its 3′ end. Then, only that subset of fragments whose terminal sequences happen to match these selective bases will be amplified (*Figure 5*). By choosing a suitable number of selective bases, the amplified subset can contain only a few fragments which can then be resolved as a 'fingerprint' on an acrylamide gel. Polymorphisms in either the restriction sites or in the adjacent bases corresponding to the selective bases of the primer will then be reflected as polymorphisms in the fingerprint.

The frequency of polymorphism amongst AFLP products may be expected to be similar to that of RFLPs; however, because each AFLP experiment yields a fingerprint of many different bands, the probability of finding polymorphisms is correspondingly increased. Moreover, many different primers (each with a different 'selective' 3′ region, and each amplifying a different subset of the fragments) can be tested in turn.

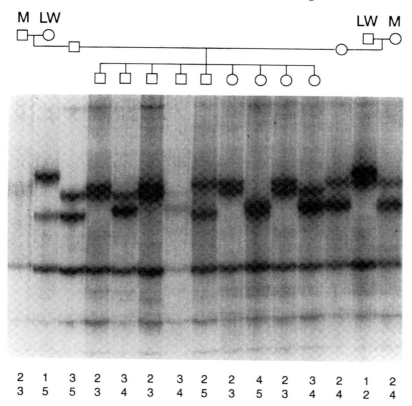

Figure 3. A five-allele length polymorphism detected in a three-generation pig pedigree by amplification between a 'unique' oligo and the 3′ end of the porcine SINE sequence. The pedigree is indicated above the figure (M = Meishan, LW = Large White). The polymorphic bands lie just above the centre of the image; the alleles present in each individual are indicated along the bottom. (Originally published in: Miller, J. R. and Archibald, A. L. (1993), 5′ and 3′ SINE-PCR allows genotyping of pig families without cloning and sequencing steps. *Mammalian Genome*, **4**, 243–6. Reproduced by kind permission of Springer-Verlag, who hold copyright.)

Although the AFLP method undoubtedly involves a large amount of preliminary effort, it has the immense advantage of using stringent reaction conditions for PCR, which enhances reproducibility. This is borne out by the observation, made in the original study, that addition of an extra selective nucleotide always yielded a fingerprint which was a subset of the original fingerprint. This ability to vary the complexity of the products by addition or removal of selective nucleotides is of great significance. AFLP fragments originate from unique locations in the genome and can thus be used as landmarks in genetic and physical maps. The method appears to be applicable to a wide range of genomes. Would-be users should note that the technique has

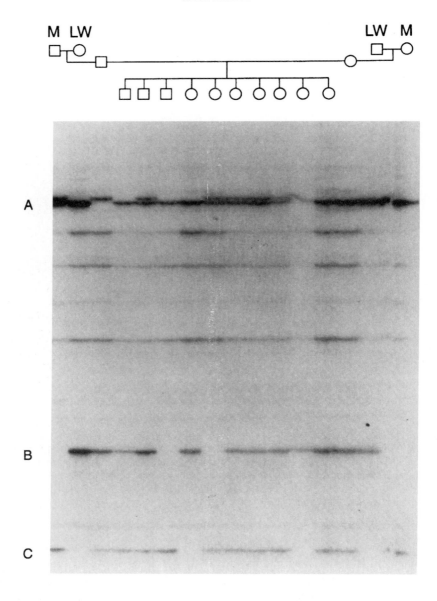

Figure 4. Detection of multiple polymorphisms in a three-generation pig pedigree by amplification between a 'unique' oligo and the 5' end of the porcine SINE sequence. The pedigree is indicated along the top; Meishan and Large White grandparents are indicated by M and LW, respectively. Amplimer 'A' is a two-allele, co-dominant polymorphism, reflecting a length polymorphism in one of the 5'-SINE PCR products. Amplimers 'B' and 'C' show dominant polymorphism, in which target sites (for either the SINE or 'unique' primer) are absent in the 'null' allele.

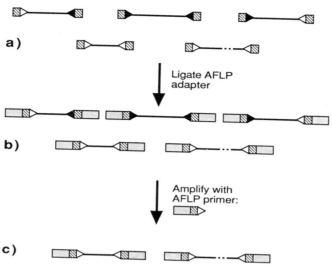

a)

Ligate AFLP
adapter

b)

Amplify with
AFLP primer:

c)

Figure 5. AFLP detection. (a) Genomic DNA is first digested to completion with a restriction enzyme, to give fragments which all carry the restriction half-site (hatched squares), but with various different sequences (filled and empty triangles) internal to these. (b) An AFLP adapter (shaded rectangle), is then ligated to all fragments. Amplification is then performed with an AFLP primer consisting of sequences complementary to the adapter and to the restriction half-site, plus one or more 'selective bases' (white triangle). (c) Only those fragments whose internal sequences at each end happen to match the selective bases of the AFLP primer will be amplified. These amplified fragments are then resolved as an electrophoretic 'fingerprint'. Length polymorphisms (dotted line), or mutations which affect the distribution of restriction sites or the adjacent 'selective' bases, will be reflected as changes in the fingerprint.

been protected by patents owned by Keygene N.V. Fingerprinting kits for research are available from Life Technologies and from Perkin-Elmer.

4. Problems and opportunities in plant and animal genome mapping

4.1 Resources for plant and animal genome mapping

Although large-insert yeast artificial chromosome libraries have recently been constructed for the cattle, sheep, and pig genomes (47–49), no intensive contig construction programme has commenced, due to the labour-intensive nature of the process. Animal genome projects can of course hope to import data via the comparative map from the human project, but this is much less of a possibility for plant genome projects. It is noteworthy that an *Arabidopsis* YAC library exists and that large scale contig construction is underway (50) in this species and in rice. RFLP markers were derived from rice YAC clones; subsequent linkage analysis in both rice and barley revealed colinearity of marker order (51).

Radiation hybrid panels now exist for the human genome (2; see also Chapter 4), enabling STSs to be mapped not only to a chromosome but in most instances to an unambiguous location on a robust framework map. Similar radiation hybrid panels are under construction for the cattle and pig genomes. There also exist cattle, sheep, and pig somatic cell hybrid panels (Chapter 6) which can be used to give a chromosomal localization; however subchromosomal localization is not yet possible. Many other resources, from clone libraries to cell lines, are described in Appendix 1.

4.2 Opportunities afforded by directed breeding programmes

Mapping of non-human genomes allows the use of selective breeding strategies which, although often costly, can be specifically designed for the individual project. Divergent lines of animals can be crossed to increase the polymorphism of the markers used. For example the EU PiGMaP project used a number of Chinese Meishan × European Large White pedigrees as part of its mapping population (8). Similarly *Bos taurus* × *Bos indicus* crosses have been used to derive genetic maps of cattle (11).

Mouse genome mapping has benefited enormously from the availability of both a large number of inbred mouse strains and interspecific backcrosses between inbred laboratory strains and distantly related *Mus* species. In an interspecific cross, most genes or DNA fragments examined are polymorphic, as the cross exploits the genetic diversity between the inbred laboratory strain and its distant relative (52). Consequently, large numbers of genes or anonymous DNA fragments can be readily placed on the genetic map. One such backcross, the European Collaborative Interspecific Backcross (EUCIB), comprises 1000 progeny from a *Mus domesticus* C57BL/6 × *Mus spretus* intercross. Such a cross should provide genetic resolution of 0.3 cM with 95% confidence. A cross between two inbred laboratory strains has been exploited by the MIT Genome Centre/Whitehead Institute to generate a genetic map comprising more than 4000 simple sequence length polymorphisms (53). A similar strategy using an intercross between two inbred strains has resulted in a genetic linkage map of the rat genome (49, 54). The power of the mouse interspecific backcross can be illustrated by the observation that use of interspersed repetitive element PCR (IRS-PCR) on backcross DNAs yielded large numbers of polymorphic products (55). A similar IRS-PCR study using a porcine reference family derived from a Chinese × European cross yielded far fewer polymorphisms (56).

Directed breeding programmes also offer improved strategies for QTL detection, which is discussed in Chapter 3.

4.3 Transfer of markers between species

It would be extremely desirable to be able to transfer markers and map information between marker-rich (human, mouse) and marker-poor (agricultural) species. Although the microsatellites which are so popular in mammalian linkage mapping are widely regarded as anonymous sequences, it has been demonstrated that the location of CA repeats tends to be conserved between mammalian species (57). However, the sequences immediately flanking the microsatellite motif, which serve as primer binding sites, are much less often conserved. In practice, roughly 50% of the bovine microsatellite markers tested in the sheep genome (and vice versa) have proved to be polymorphic and map to the equivalent place on the genetic map (7). Nevertheless, interspecific portability of microsatellites has proved to be the exception rather than the rule. In another study, 48 primer pairs which gave specific polymorphic products with bovine DNA were tested against DNAs from sheep, horse, and human (58). With sheep DNA, 27 of the pairs gave specific products of which only 20 were polymorphic; with horse DNA only three primer pairs gave specific products, none of which were polymorphic; no primer pairs gave specific PCR products from human DNA. Thus it is evident that microsatellites are portable only between very closely related species. This has major implications, as to date microsatellites have been the marker of choice for construction of mammalian genetic maps.

So-called 'zoo-FISH' experiments (in which non-human chromosomes are hybridized with human chromosome paints) have shown that homologous chromosomal regions could be detected readily in primates (59) and subsequently in more distant species such as pig and cattle (60, 61). The main source of the sequence conservation observed in these interspecific hybridizations appears to be the coding sequences and, to a lesser extent, the control sequences. Thus, genetic markers derived from coding sequences are more likely be transferable between species.

These considerations have led to the classification of mammalian genetic markers into two categories. Type I markers are conserved across different mammalian species; such markers are typically RFLPs or SSCPs, often of limited polymorphism, lying in coding or control sequences. Type II markers, on the other hand, are highly polymorphic sequences (normally microsatellites) which are valuable for producing a genetic map within a species, but which are not useful for comparative studies as they are infrequently conserved between species. Consideration of the merits of Type I markers gave rise to the concept of 'mammalian anchor loci' (26). These loci were derived from existing Type I markers which were then selected on the basis of 5–10 cM spacing in the human and mouse genomes, evolutionary conservation in other genomes and their position relative to the boundaries of synteny groups (see Section 4.4). Anchor loci provide a resource which should ultimately enable information to flow from map-rich to map-poor mammalian species to facilitate the mapping of coding sequences. This flow of information, however, will not take place until framework maps incorporating the anchor loci have been constructed in a range of domestic species.

The same considerations gave rise, independently, to the related concept of universal mapping probes (UMPs) (57). These were defined as short segments of human DNA suitable for physical and genetic mapping in a variety of animal species. Clones containing both a rare-cutter restriction enzyme site (indicative of a CpG island and thus of a gene) and a microsatellite motif were selected from a human genomic library with an insert size of 12–20 kb. The presence of the microsatellite motif ensures that such clones can be placed on the human genetic map, whilst the presence of a coding sequence ensures that the clones can also be localized by FISH on the physical maps of both human and rat and presumably (though this has yet to be demonstrated) those of other species.

4.4 Comparative mapping

Comparative mapping is a valuable tool which will not only enable information to be transferred from map-rich to map-poor species but which will also enable the reconstruction of common ancestral genomes for plant and animal species. The human and mouse genomes have comprehensive and dense gene maps, whereas the domestic animal maps are much more meagrely populated. Information can flow in both directions between these two groups. On the one hand, once a trait has been localized to a particular segment of an animal chromosome then information, such as a catalogue of nearby expressed sequence tags and thus candidate genes, can be imported from the equivalent human chromosomal region. On the other hand, directed breeding programmes or phenotypic measurements such as recording of carcass traits—which could not be pursued in humans—can be used to map loci in domestic animals; this information can then lead to the corresponding human locus. As mentioned above, evolutionary sequence conservation seems largely to be confined to coding sequences (genes) and, to a lesser extent, their control sequences.

A chromosomal region of one species is said to be syntenic with a chromosomal region in another species if the two regions carry two or more homologous genes. If the synteny includes more than two genes and the relative order of the genes is conserved between the species, the situation is referred to as a conserved linkage. Severe caution must be observed when assuming that large syntenic groups represent chromosomal regions which have been tightly conserved in evolution. Synteny mapping, usually performed on somatic cell hybrid panels, takes no account of gene order within a chromosome or subchromosomal region. Several studies have shown that rearrangements in gene order within syntenic groups are common (62, 63). By looking at syntenies between two species, for example human and mouse, or human and cow, it is possible to estimate the minimum number of chromosome rearrangements needed to transform the genome of one species into that of the other. The expectation is that the number of such rearrangements reflects the evolutionary distance between the species. By this logic both the pig and cattle genomes are significantly closer to the human genome in evolutionary terms than is the mouse genome (60, 61). For example, it has been estimated, using the average length of conserved chromosomal segments, that there are 35 chromosomal rearrangements between human and pig, whereas between pig and mouse the corresponding figure is 77 (63). By the same token 'zoo-FISH' can be used to look at the conservation of entire chromosomes between species. A study of pig/human homologies found that the 24 human syntenic groups (chromosomes) are conserved as 47 syntenic groups in the domestic pig (60). A parallel study in cattle found the human karyotype conserved as 50 bovine syntenic groups (61). Both these figures contrast with the observation that more than 101 mouse syntenic groups are required to represent the human karyotype (64).

It is now realized that extensive syntenies also exist between plant species. The genomes of the various grasses (including rice, wheat, and maize) can be considered to have been derived from a common ancestral genome, and the position and order of many genes appears to be conserved. The rice genome (430 Mb) has thus become a popular model for the much larger cereal genomes such as maize (65) and wheat (27). Indeed, the gene maps of sorghum, millet, maize, sugar cane and wheat can be derived from that of rice (66). High density linkage maps have recently been constructed for the tomato and potato genomes, using a common set of markers (RFLPs, isozymes, and morphological markers). The genetic maps of the two species are nearly identical, being distinguished by the occurrence of only five chromosomal inversions (25). Thus, information on the location of coding sequences in one plant species can almost certainly be applied to the other. Some possible syntenies have even been established between rice and human: rice chromosomes 1 and 3 each have three genes whose human equivalents are located on chromosome 10 (20). Whether these are syntenies with real significance will only be learnt in the longer term.

Syntenic relationships between two species are often difficult to visualize. This problem has been addressed by use of the 'Oxford grid' to give a two-dimensional representation of syntenic relationships (28). The Oxford grid is a table divided into cells horizontally to represent the chromosomes of the first species and vertically to represent those of the second. A syntenic relationship is then indicated by an entry in the appropriate cell (*Figure 6*). In more elaborate versions of the Oxford grid, the horizontal and vertical sizes of each cell reflect the sizes of the corresponding chromosomes, and entries are made in the form of symbols whose size and orientation denotes the number of loci shared between the chromosomes, and their orientation with respect to the centromere.

5. Conclusions and questions

Mapping of plant and animal genomes is becoming increasingly geared to the concept of 'model organisms': human (as a model for animal genomes); and rice and *Arabidopsis* (as models for plant genomes). This strategy offers

Human Chromosome

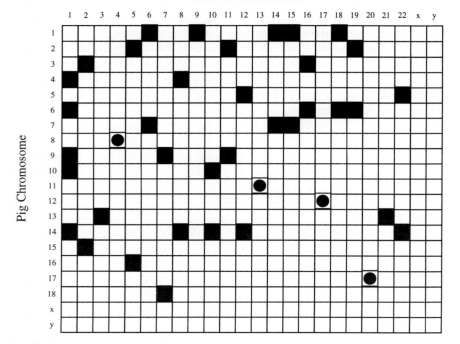

Figure 6. A simple 'Oxford' type grid showing the deduced relationships between human and pig autosomes. Homologies between genes on pig and human chromosomes are denoted by filled boxes. In the four instances where an entire human chromosome is believed to syntenic with an entire pig chromosome, filled circles are used. The data were taken from the study by Rettenberger and colleagues (60).

numerous advantages, including access to well-characterized resources developed for model organisms and a reduction in the effort needed to map massive genomes such as that of wheat (17000 Mb). It remains to be seen, however, how far such an approach can be taken. We know, from chromosome painting experiments, the extensive syntenic relationships between humans, pigs and cattle. But we are still unsure of the degree of small scale shuffling that has taken place or, in other words, just how colinear are two apparently homologous chromosomes.

Going further down the evolutionary scale, we must ask what advantage can be taken from the increasing number of unicellular organisms whose genomes have been completely sequenced. At present this list comprises *Haemophilus influenzae, Mycoplasma genitalium,* and *Saccharomyces cerevisiae*; other organisms will shortly follow. These DNA sequences, complete physical maps, allow for accurate identification of genes. It has been recently suggested that yeast may prove to be a valuable system for elucidation of gene function (67).

One further question would be whether the various plant and animal projects will ever benefit the human project. To date, the flow of information and technology has been almost exclusively from human to non-human genomes, with the development of the RAPD and AFLP techniques being notable exceptions.

The only safe conclusion is that with an ever expanding and hungry human population, the linkage mapping of plant and animal genomes must be viewed as a critical scientific endeavour.

References

1. Hudson, T. J., Stein, L. D., Gerety, S. S., Ma, J., Castle, A. B., Silva, J. *et al.* (1995). *Science*, **270**, 1945.
2. Chumakov, I. M., Rigault, P., Le Gall, I., Bellane-Chantelot, C., Billault, A., Guillou, S. *et al.* (1996). *Nature*, **377**, 175.
3. Gyapay, G., Schmitt, K., Fizames, C., Jones, H., Vega-Czarny, N., Spillet, D. *et al.* (1996). *Hum. Mol. Genet.*, **5**, 339.
4. Dib, C., Faure, S., Fizames, C., Samson, D., Drouot, N., Vignal, A. *et al.* (1996). *Nature*, **380**, 152.
5. Dietrich, W. F., Miller, J., Steen, R., Merchant, M. A., Damron-Boles, D., Husain, Z. *et al.* (1996). *Nature*, **380**, 149.
6. Jacob, H. J., Brown, D. M., Bunker, R. K., Daly, M. J., Dzau, V. J., Goodman, A. *et al.* (1995). *Nature Genet.*, **9**, 63.
7. Crawford, A. M., Dodds, K. G., Ede, A. J., Pierson, C. A., Montgomery, G. W., Garmonsway, H. G. *et al.* (1995). *Genetics*, **140**, 703.
8. Rohrer, G. A., Alexander, L. J., Keele, J. W., Smith, T. P., and Beattie, C. W. (1994). *Genetics*, **136**, 231.
9. Ellegren, H., Chowdhary, B. P., Johansson, M., Marklund, L., Fredholm. M., Gustavsson, I., and Andersson, L. (1994). *Genetics*, **137**, 1089.
10. Archibald, A. L., Haley, C. S., Brown, J. F., Couperwhite, S., McQueen, H. A., Nicholson, D. *et al.* (1995). *Mamm. Genome*, **6**, 157.
11. Barendse, W., Armitage, S. M., Kossarek, L. M., Shalom, A., Kirkpatrick, B. W., Ryan, A. M. *et al.* (1994). *Nature Genet.*, **6**, 227.
12. Bishop, M. D., Kappes, S. M., Keele, J. W., Stone, R. T., Sunden, S. L. F., Hawkins, G. A. *et al.* (1994). *Genetics*, **136**, 619.
13. Ostrander, E. A., Mapa, F. A., Yee, M., and Rine, J. (1995). *Mamm. Genome*, **6**, 192.
14. Burt, D. W., Bumstead, N., Bitgood, J. J., Ponce De Leon, F. A., and Crittenden, L. B. (1995). *Trends Genet.*, **11**, 190.
15. Brenner, S., Elgar, G., Sandford, R., Macrae, A., Venkatesh, B., and Aparicio, S. (1993). *Nature*, **366**, 265.
16. Postlethwait, J. H., Johnson, S. L., Midson, C. N., Talbot, W. S., Gates, M., Ballinger, E. W. *et al.* (1994). *Science*, **264**, 699.
17. Hunt, G. J. and Page, Jr., R. E. (1995). *Genetics*, **139**, 1371.
18. Hartl, D. L. and Lozovskaya, E. R. (1992). *Comp. Biochem. Physiol.*, **103B**, 1.
19. Coulson, A., Huynh, C., Kozono, Y., and Shownkeen, R. (1995). In *Methods in cell biology* (ed H. F. Epstein and D. C. Shakes), vol. 48, p. 533. Academic Press, London.

20. Kurata, N., Nagamura, Y., Yamamoto, K., Harushima, Y., Sue, N., Wu, J. *et al.* (1994). *Nature Genet.,* **8**, 365.
21. Devos, K. M. and Gale, M. D. (1992). *Outlook on Agriculture,* **22**, 93.
22. Ahn, S. and Tanksley, S. D. (1993). *Proc. Natl. Acad. Sci. USA,* **90**, 7980.
23. Lister, C. and Dean, C. (1993). *Plant J.,* **4**, 745.
24. Tulsieram, L. K., Glaubitz, J. C., Kiss, G., and Carlson, J. E. (1992). *Bio/Technology,* **10**, 686.
25. Tanksley, S. D., Ganal, M. W., Prince, J. P., de Vicente, M. C., Bonierbale, M. W., Broun, P. *et al.* (1992). *Genetics,* **132**, 1141.
26. O'Brien, S., Womack, J. E., Lyons, L. A., Moore, K. J., Jenkins, N. A., and Copeland, N. G. (1993). *Nature Genet.,* **3**, 103.
27. Kurata, N., Moore, G., Nagamura, Y., Foote, T., Yano, M., Minobe, Y. *et al.* (1994). *Bio/Technology* **12**, 276.
28. Edwards, J. H. (1994). *Curr. Opin. Genet. Dev.,* **4**, 861.
29. Sulston, J. E., Schierenberg, E., and Thomson, J. N. (1983). *Dev. Biol.,* **100**, 64.
30. Devine, K. (1995). *Trends Biotechnol.,* **13**, 210.
31. Daniels, D. L., Plunkett III, G., Burland, V., and Blattner, F. R. (1992). *Science,* **257**, 771.
32. Dujon, B. (1996). *Trends Genet.,* **12**, 263.
33. Fleischmann R. D., Adams. M. D., White, O., Clayton, R. A., Kirkness, E. F., Kerlavage, A. R. *et al.* (1995). *Science,* **269**, 496.
34. Fraser, C. M., Gocayne, J. D., White, O., Adams, M. D., Clayton, R. A., Fleischmann, R. D. *et al.* (1995). *Science,* **270**, 397.
35. Orita, M., Iwahana, H., Kanazawa, H., Hayashi, K., and Sekiya, T. (1989). *Proc. Natl. Acad. Sci. USA,* **86**, 2766.
36. Avramopoulos, D., Chakravarti, A. and Antonarakis, S. E. (1993). *Genomics,* **15**, 98.
37. Lagercrantz, U., Ellegren, H., and Andersson, L. (1993). *Nucleic. Acids Res.,* **21**, 1111.
38. Bell, C. J. and Ecker, J. R. (1994). *Genomics,* **19**, 137.
39. Welsh, J. and McClelland, M. (1990). *Nucleic. Acids Res.,* **18**, 7213.
40. Williams, J. G. K., Kubelik, A. R., Livak, K. J., Rafalski, J. A., and Tingey, S. V. (1990). *Nucleic. Acids Res.,* **18**, 6531.
41. Charlieu, J.-P., Laurent, A.-M., Carter, D. A., Bellis, M., and Roizès, G. (1992). *Nucleic. Acids Res.,* **20**, 1333.
42. Miller, J. R. and Archibald, A. L. (1993). *Mamm. Genome,* **4**, 243.
43. Epstein, N., Nahor, O., and Silver, J. (1990). *Nucleic. Acids Res.,* **18**, 4634.
44. Economou, E. P., Bergen, A. W., Warren, A. C., and Antonarakis, S. E. (1990). *Proc. Natl. Acad. Sci. USA,* **87**, 2951.
45. Prades, C., Laurent, A.-M., Yurov, Y., Puechberty, J., and Roizès, G. (1996). *Cytogenet. Cell Genet.,* **72**, 69.
46. Vos, P., Hogers, R., Bleeker, M., Reijans, M., van de Lee, T., Hornes, M., *et al.* (1995). *Nucleic. Acids Res.,* **23**, 4407.
47. Libert, F., Lefort, A., Okimoto, R., Womack, J., and Georges, M. (1993). *Genomics,* **18**, 270.
48. Broom, M. F. and Hill, D. F. (1994). *Mamm. Genome,* **5**, 817.
49. Leeb, T., Rettenberger, G., Hameister, H., Brem, G., and Brenig, B. (1995). *Mamm. Genome,* **6**, 37.

50. Schmidt, R., West, J., Love, K., Lenehan, Z., Lister, C., Thompson, H. *et al.* (1995). *Science*, **270**, 480.
51. Dunford, R. P., Kurata, N., Laurie, D. A., Money, T. A., Minobe, Y., and Moore, G. (1995). *Nucleic Acids Res.,* **23**, 2724.
52. Avner, P., Amar, L., Dandolo, L., and Guénet, J. L. (1988). *Trends Genet.,* **4**, 18.
53. Dietrich, W. F., Miller, J. C., Steen, R. G., Merchant, M., Damron, D., Nahf, R. *et al.* (1994). *Nature Genet.,* **7**, 220.
54. Pravenec, M., Gauguier, D., Schott, J.-J., Buard, J., Kren, V., Bíla, V. *et al.* (1996). *Mamm. Genome,* **7**, 117.
55. Cox, R. D., Copeland, N. G., Jenkins, N. A., and Lehrach, H. (1991). *Genomics,* **10**, 375.
56. Miller, J. R. (1994). *Mamm. Genome,* **5**, 629.
57. Hino, O., Testa, J., Buetow, K. H., Taguchi, T., Zhou, J.-Y., Bremer, M. *et al.* (1993). *Proc. Natl. Acad. Sci. USA,* **90**, 730.
58. Moore, S. S., Sargeant, L. L., King, T. J., Mattick, J. S., Georges, M., and Hetzel, D. J. S. (1991). *Genomics,* **10**, 654.
59. Jauch, A., Wienberg, J., Stanyon, R., Arnold, N., Tofanelli, S., Ishida, T. *et al.* (1992). *Proc. Natl. Acad. Sci. USA,* **89**, 8611.
60. Rettenberger, G., Klett, C., Zechner, U., Kunz, J., Vogel, W., and Hameister, H. (1995). *Genomics,* **26**, 372.
61. Solinas-Toldo, S., Lengauer, C., and Fries, R. (1995). *Genomics,* **27**, 489.
62. Nadeau, J. H. (1989). *Trends Genet.,* **5**, 82.
63. Johansson, M., Ellegren, H., and Andersson, L. (1995). *Genomics,* **25**, 682.
64. Copeland, N. G., Jenkins, N. A., Gilbert, D. J., Eppig, J. T., Maltais, L. J., Miller, J.C. *et al.* (1993). *Science,* **262**, 57.
65. Ahn, S. and Tanksley, S. D. (1993). *Proc. Natl. Acad. Sci. USA,* **90**, 7980.
66. Moore, G. (1995). *Curr. Opin. Genet. Dev.,* **5**, 717.
67. Oliver, S. (1996). *Trends Genet.,* **12**, 241.

3

Linkage mapping of quantitative trait loci in plants and animals

CHRIS S. HALEY and LEIF ANDERSSON

1. Introduction

Much of the variation within populations, or between lines or breeds, is quantitative in nature. That is, variation between individuals does not fall into discrete classes in Mendelian proportions but is continuous, showing a gradation from one extreme to the other. Examples include growth rate and leanness in livestock, hypertension and intelligence quotient in man, and yield and flowering time in plants. To the extent that differences between populations are maintained, and that there is a tendency for offspring to resemble parents, there is evidently some genetic component to this variation. The genetic basis of such phenotypic variation lies in the combined effects of variation at several or many loci. Environmental influences are superimposed upon (and may interact with) these genetic contributions to variation.

To distinguish between the genetic and environmental contributions to variation, 'heritability' is defined as the proportion of the observed phenotypic variance which is genetic in origin (1). Trait variation controlled by several or many loci and environmental effects is referred to as 'quantitative', 'polygenic', 'multifactorial', or 'complex', and the individual loci that contribute to such variation have been christened quantitative trait loci or QTLs (2).

Until recently it has been impossible to identify the majority of QTLs. Using phenotypic data alone, it is possible to identify genes of large effect on quantitative traits using various techniques (3). However, even the best of these techniques, segregation analysis, is relatively lacking in power: only those genes of the largest effect can be detected given a reasonable data-set. Segregation analysis is also very vulnerable to incorrect assumptions about the statistical distribution of the data, which can lead to spurious detection of a major gene.

The detection of linkage between a QTL and genetic markers provides a more powerful and robust method of identifying QTLs. Linkage maps, moderately or densely populated with several types of highly informative marker, have now been developed for many species. These maps provide the basic tools with which to study the variation underlying quantitative traits.

The detection and mapping of QTLs is valuable for several reasons. First, it gives insight into the actions and interactions of individual genes. This in turn allows the more realistic modelling of phenotypic variation, responses to selection, and evolutionary processes. Such models augment our understanding of trait variation in man and our ability to predict breeding values and to implement selection in plant and livestock species. Secondly, information on genetic markers associated with a QTL can be used to improve breeding-value estimation (even when the QTL itself is not pinpointed). Selection based on such QTL-linked markers (marker-assisted selection or MAS) may be an effective means of introducing desirable genes from one breed or line to another, or of improving responses to selection within a breed (4, 5). Thirdly, the mapping of a QTL opens the way to positional cloning of the gene, as has been achieved for a number of monogenic traits in the mouse, in humans and in plants. This will allow the study of the molecular causes of existing variation, and may also allow further improved alleles to be produced by direct molecular intervention, for example for use in plant or animal breeding programmes.

The purpose of this chapter is to consider some of the strategies and methods that can be used to map QTLs and to look at the factors which influence the power and accuracy of these approaches. Every QTL mapping experiment is a large and complex study: there is no single protocol that can be applied and each study will be different. Here we try to outline some of the major considerations required in the design and execution of a successful study.

2. Principles of QTL mapping

2.1 One marker and a cross between inbred lines

The basic principle behind mapping QTLs is illustrated in *Figure 1*. In this example two lines of pigs are crossed which differ in performance (size in this example) and also for markers distributed throughout the genome. The diagram focuses on one small part of a chromosome, and shows that at this point the two breeds carry a genetic marker which has different alleles designated '1' and '2'. This marker is *not* the QTL itself: its importance is that it acts as a proxy for the QTL, as a genetic marker which can be easily followed through the pedigree. Each breed is homozygous ('1 1' or '2 2') for this marker.

The F_1 animals inherit one chromosome from each breed and hence one copy of each allele of the marker (i.e. they are '1 2' heterozygotes). In the F_2 generation (produced by crossing two F_1 animals) the marker segregates, giving '1 1' and '2 2' homozygotes and '1 2' heterozygotes. Because of genetic linkage, animals inheriting the marker allele from one breed also tend to inherit the surrounding region of chromosome from the same breed.

3: Linkage mapping of quantitative trait loci in plants and animals

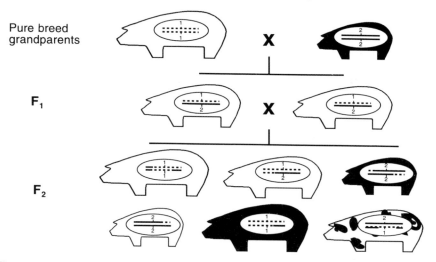

Figure 1. The principle of QTL detection using a genetic marker. In this example we focus on one marker segregating in a cross between two breeds of pig. The marker is used to infer the inheritance of a linked section of chromosome and, in this example, show the association of this section with size in the F_2 population. See text for further details. Reproduced by permission of Roslin Institute.

There is also segregation in the F_2 for the genes that control the size differences between the two breeds, including the QTL we are seeking. In this example we find, in the F_2 animals, an association between the marker alleles they inherit and their size: those with the '1 1' marker genotype tend to be larger than those with the '2 2' genotype, while '1 2' animals are intermediate in size. This association shows that one or more QTLs which contribute size differences between the breeds must linked (and therefore close) to the marker.

The situation is at its simplest when we are dealing with a cross between two inbred lines. In this case, where the marker locus or the QTL differ between the lines only two alleles are possible (as each line is homozygous for one allele) and all F_1 individuals will be heterozygous for that locus. Furthermore, for a marker and linked QTL all F_1 animals will have the same linkage phase. For example, taking the two gametes in an F_1 animal, the gamete inherited from line 1 and carrying marker allele 1 will also carry QTL allele 1; the gamete inherited from line 2 and carrying marker allele 2 will also carry QTL allele 2.

In the F_2, if the QTL is located some distance from the marker, recombination may occur between it and the marker. The result is that inheritance of the marker will correspond less and less to that of the QTL (and hence, in this example, to size) as the distance between them increases. For a cross between inbred lines involving a proxy marker and a QTL of given effect, we

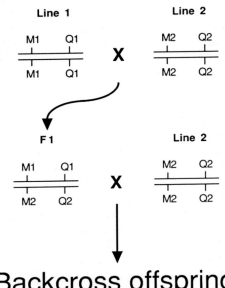

Backcross offspring

Genotype	Frequency	Effect on trait	Mean trait effect for marker group
M1M2 Q1Q2 (Non-recombinant)	(1-r)/2	+d	} (1-r)d-ra
M1M2 Q2Q2 (Recombinant)	r/2	-a	
M2M2 Q1Q2 (Recombinant)	r/2	+d	} rd-(1-r)a
M2M2 Q2Q2 (Non-recombinant)	(1-r)/2	-a	

Difference between means: (1-2r)d + (1-2r)a

Figure 2. The effects of a QTL linked to a marker in a backcross. The QTL (alleles Q1, Q2) has an effect *a* on the trait, such that Q1 homozygotes show effect *a*, Q2 homozygotes effect −*a*, and Q1Q2 heterozygotes effect *d* (the effect of the dominant allele of the QTL), all measured relative to the mean of the trait of the two QTL homozygotes. Line 1 (homozygous for marker allele M1 and for QTL allele Q1) is crossed with line 2 (homozygous for M2 and Q2), and the F1 heterozygous offspring are backcrossed to line 2. In this backcross, recombination between the marker and the QTL occurs at frequency *r*, giving recombinant and non-recombinant offspring at the frequencies shown. In the backcross, offspring are divided into classes based on their marker genotype: the mean trait effect for each class, and the difference between these means, can be calculated as shown. Note that the difference between the means is a function of the effects of the QTL (*d* and *a*) and of the recombination frequency (*r*) between the QTL and the marker. Hence, a small QTL (*a* = 1, *d* = 0) close to the marker (*r* = 0.1) will give the same difference between means (0.8) as a large QTL (*a* = 2, *d* = 0) further from the marker (*r* = 0.3).

can predict the mean phenotypic difference between animals of different *marker* genotypes in segregating generations. *Figure 2* gives an example of this for a backcross. For a QTL of large effect, the *phenotypic* differences between different marker *genotypes* becomes less as the distance between the marker and the QTL increases. When the distance becomes very large (recombination fraction = 0.5), there is no correlation between inheritance of the marker and of the QTL, and hence no phenotypic difference between the different marker genotypes. By the same token, a QTL of small effect lying close to the proxy marker may appear similar to that of a QTL of large effect located further from the marker, as judged by the phenotypic differences between marker genotype classes.

The effect of a QTL and its distance from the proxy marker are thus interrelated. With only a single marker, we cannot use this approach to determine both the size of a QTL and its position: only the combined effect of the two parameters can be estimated. Information to separate the estimates of the size and position of a QTL can, however, be obtained from the phenotypic distribution *within* a marker class. Segregation within a marker class will be greater as the distance between the QTL and the marker increases; thus a gene of large effect will produce a more skewed distribution within the marker class. This information can be retrieved and used by methods such as maximum likelihood, but it is not very good information and there are many other causes of non-normality that can lead to misleading results. Furthermore, even with an estimated distance between a QTL and a marker, the QTL could be either side of the marker.

We should bear in mind also that each QTL we are studying will only explain a proportion of the genetic variance, and usually an even smaller proportion of the total (genetic plus environmental) variance. Except in special cases discussed below, environmental influences often cause more of the total variation than do genetic factors. Thus although a linked QTL will cause a mean trait difference between marker genotype classes, there will be much overlap in the trait scores for individuals of different marker genotypes. This overlap will often occur even if the marker and QTL are completely linked.

2.2 Interval mapping

The difficulty of separately estimating the effect of a QTL and its position is resolved by the availability of complete marker maps and the use of analysis by 'interval mapping' (6). In interval mapping we look for a QTL between *two* markers whose positions are known. Thus we need to estimate the effect and relative position of any QTL between the two flanking markers. There are now four marker classes (assuming two alleles for each of the two markers), and hence four class means of the phenotypic trait. This provides sufficient information to estimate the QTL's effect and its position simultaneously.

This method is usually implemented (in most available software packages, for example) by choosing a given point in the genome and predicting the effect of a QTL at that position. The process is repeated at fixed positions through a chromosome (e.g. at points one centiMorgan apart) and the position at which a QTL would explain most of the variation between marker classes is identified. This point is then the estimated position of the QTL, and the effects of the QTL are those estimated for this position.

The evidence for the presence of a QTL through the chromosome is displayed as a test statistic plotted against the chromosomal position. The test statistic chosen is usually the log likelihood ratio for the test of a QTL at a given position versus no QTL on the chromosome. This provides a curve showing a peak at the most likely position of a QTL (*Figure 3*).

The main advantage of interval mapping when applied to populations derived from crossing inbred lines is its ability to separate the effect of a QTL from its position. There may also be some advantage in terms of power (i.e. the ability to detect QTLs of lesser effect) as the spacing between markers increases but these advantages are only noticeable when markers are more than about 20 cM apart. With dense maps of highly informative markers, the marker-by-marker analysis described in Section 2.1 produces similar results to interval mapping, and the marker which has the most significant effect can be selected as the most likely position of the QTL. Interval mapping has the

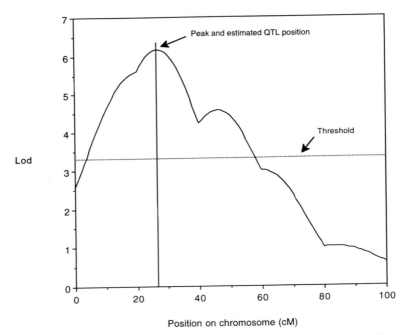

Figure 3. An example interval mapping curve from a simulated backcross. The graph shows the analysis of a single chromosome of 100 cM in length which has six markers spaced at 20 cM intervals. The strength of evidence for the presence of a QTL (in this case as a lod, or $\log_{10}$ of the likelihood ratio) is plotted against chromosomal position. The significance threshold appropriate for a genome-wide search in a mouse backcross (14) is also shown. For a single QTL the peak represents the best estimated position, and the effect of a QTL at this position would be estimated in the analysis.

greatest advantage where markers are not all equally informative and different markers are informative in different individuals. This situation arises in outbred populations, or when some marker genotype information is missing due to technical problems, or even with dominance at marker loci. In such cases, interval mapping has the edge because information from an informative marker in an individual can compensate for that missing from an adjacent marker.

2.3 Outbred populations

The problems of detecting and mapping QTLs are increased if we move away from the inbred line cross and attempt to detect QTLs segregating within populations. In an outbred population markers will be segregating and there may be two or several alleles. However, only a proportion of animals will be heterozygous for a given marker locus. The same applies to any QTL—only a proportion of animals will be heterozygous. If we consider a QTL and linked marker jointly, the probability that an individual is heterozygous for both may not be very high. For example, if the marker and the QTL each have two alleles of equal frequency in the population, the expected frequency of heterozygotes at either locus is 0.5 at Hardy–Weinberg equilibrium. Thus the probability of *both* loci being heterozygous in an individual is $0.5^2 = 0.25$.

This is illustrated in *Figure 4*. Here we are considering the half-sib progeny of a number of sires. With such a design we can only show that a QTL is linked to a particular marker by looking at mean differences between groups of progeny *within* a sire. Consider sires 1 and 2 in *Figure 4*. Each is heterozygous, both for the marker and for the QTL, but the linkage phase differs: marker allele 1 is associated with QTL allele 1 in the progeny of sire 1, but with QTL allele 2 performance in the progeny of sire 2. Thus there is no association between the marker and the QTL over the whole population: it is

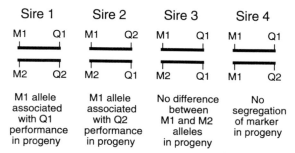

Figure 4. Marker and QTL association in a half-sib population. The association between a particular marker allele inherited from a sire and the performance of half-sib offspring depends upon the (a priori unknown) genotype of the sire for the QTL and its linkage relationship with the marker. Hence the association varies from sire to sire and will be absent from some sires (those homozygous for the QTL) and unobservable in others (those homozygous for the marker).

only seen *within* sires. The problem is further complicated by the fact that some sires may not be heterozygous for the QTL (sire 3 in *Figure 4*) and so no differences between progeny inheriting alternative marker alleles are seen. Some sires are homozygous for the marker (sire 4 in figure 4) and so it is not possible to divide their progeny into marker classes to look for trait differences. Thus the proportion of families in which a QTL is segregating *and can be seen to be segregating*, is greatly reduced compared to a cross between inbred lines. Furthermore, an analysis is needed which looks for differences within sire groups and which allows these to differ between families.

Some of the problems of non-informative markers can be overcome by simultaneous use of information from several markers—if one marker is not informative it can be replaced by one linked to it. The use of information from multiple markers simultaneously is much more valuable for outbred populations than it is for populations from inbred crosses: not only does it increase the power of studies, but it also helps avoid estimates of QTL position being biased towards the more informative markers (7). General methods for analysing data on multiple markers generated from QTL studies of outbred populations are not yet available. However, some methods are available for specific situations such as populations of half-sibs (8, 9). In studies of human populations much effort has been focused on sib-pair methods, and analyses incorporating multiple marker information for such methods have been developed (10). Such methods can be applied to plant and animal populations, but are probably not optimally efficient in populations with large families containing many pairs of sibs.

3. Populations for QTL mapping

The design of QTL studies and the choice of populations to be analysed depend on both philosophical and practical considerations. If the aim is to identify QTLs in a particular population (e.g. in order to develop the opportunity for marker-assisted selection within that population) the choice is rather limited. If the aim is to identify candidate loci for a particular trait, then there is the possibility of designing a population which maximizes the chance that such genes will be segregating. Detectable QTLs are more likely to segregate in a cross between two phenotypically divergent lines than they are within a population which has been under selection, in which individuals will tend to be homozygous for major QTLs. Where QTLs *are* segregating, their heterozygosity in the F_1 will be one if inbred lines are crossed, and may also be one if divergent outbred lines are crossed.

Various different types of population can be derived from a cross between divergent lines including F_2, single- or double-backcross, and recombinant inbred. If the phenotypic variation is known to be controlled by QTLs which all exhibit complete dominance in the same direction, then a backcross to the recessive parent results in the QTL segregating in two classes (heterozygote and recessive homozygotes) and this cross is a more powerful means of locating QTLs than an F_2. There is no information in this backcross on QTLs which exhibit complete directional dominance in the opposite direction to the majority of loci, as their effect in heterozygotes will be identical to that in recessive homozygotes. An F_2 is more powerful than either individual backcross for detecting QTLs of additive effect, and can also be used to estimate the degree of dominance for detected QTLs. In general, several traits may be of interest in any single cross and the level and direction of dominance (and hence the relative informativeness of either of the two backcrosses alone) will depend upon the trait. In this case the F_2, or a combination of the two backcrosses together, may be optimal both in terms of overall power and in terms of the ability to estimate the effects of detected QTLs. Recombinant inbred lines may also be powerful tools for QTL detection in some circumstances, but provide no information on dominance relationships at any QTL.

The lower heterozygosities of QTLs and markers (and the need to infer linkage phase between them for each founder animal) in outbreeding populations make these a less attractive prospect for the detection of QTLs. None the less, divergent lines may not always be available, or it may only be a particular population that is of interest to the investigator. In many such cases designed crosses will not be possible and the population as it stands will have to be used for QTL detection. In these instances, there may be strategies which improve our chances of detecting QTLs such as progeny testing or selective genotyping (see below).

4. Maps and markers for QTL analysis

Genetic maps based on DNA markers are available for many species. However, the density and type of marker varies from case to case. In man, mouse, and livestock short tandem repeats (STRs) or microsatellites are the markers of choice. At least 1000 of these are available in cattle and pigs, and several thousands in man and mice. Such markers are highly polymorphic, enabling experimenters in these species to choose a subset which are informative in their chosen population and which cover the genome to the desired resolution.

Until recently, plant geneticists have tended to focus on restriction fragment length polymorphisms (RFLPs) or random amplified polymorphic DNAs (RAPDs) as markers. In part this is because so many different species are of interest that the investment in developing STRs is less worthwhile. RFLPs are inherently less polymorphic than STRs, and RAPDs are not easily portable between populations. These factors mean that the experimenter often has less control over the density of the map used for QTL studies. Dominant markers such as RAPDs also suffer from not being completely informative in some types of cross, such as an F_2 where dominant homozygotes are indistinguishable from heterozygotes. The inherent low repeatability of RAPDs may also create a relatively high error rate in the genotyping. Amplified fragment length polymorphism (AFLP) is another very powerful method for generating large numbers of anonymous DNA markers using non-specific primers, and is widely used in plant genetics (11). This method appears more robust than the RAPD technique, but shares the limitation that the markers may not be easily portable between populations. (Further discussion of RAPDs and AFLPs will be found in Chapter 2.)

Where the experimenter can dictate his choice of map density through the selection of suitably spaced microsatellites, his choice should be guided by factors such as the aim of the study; the size and type of the experiment; and the relative cost of genotyping versus that of collecting and measuring traits on the study population.

For a cross between inbred lines analysed by interval mapping, the power to detect QTLs increases little with a map density of greater than one marker per 20 cM (6, 12, 13). In fact markers spaced 40 cM apart provide around 80% of the information obtained from 20 cM markers (6) with only half of the genotyping. (This is for the worst case, where the QTL is mid-way between markers; if it is fortuitously nearer to one marker there is even less loss of information in using the 40cM spacing.)

In contrast, doubling the population size doubles the information available. Thus if a study is set up to do a complete genome scan for QTLs, the balance may often be in favour of widely spaced markers (i.e. 30–50 cM apart) and a large population for initial studies (13). The density of markers can be increased once QTLs have been located, but it is seldom worth going below 20 cM spacing in such populations and, as noted below, the accuracy with which QTLs are located depends more on the number of meioses studied than on the marker density against which they are mapped.

The situation is slightly different in outbred populations or in crosses between outbred populations, because markers are not completely informative. Thus marker information content in the population under study needs to be taken into account both in selecting individual markers (to ensure that they are as informative as possible) and in deciding upon the marker spacing. With microsatellites, information content is 50–90% of that of a completely informative marker. At the lower end of this range one requires twice the density of markers to achieve equivalence with the situation of completely informative markers seen with inbred line crosses. In these situations, one might aim for markers spaced at 10 cM to approach maximum power for a given sample size.

5. Setting significance thresholds

Some of the major problems of QTL analysis stem from the fact that QTLs are detected in scans of the complete genome. Thus, many statistical tests for the presence of QTLs are performed. If marker-by-marker analysis is used, as many tests are performed per trait as there are markers (i.e. 100–200 in a typical study). If interval mapping is used, tests are performed every 1–2 cM (i.e. 1000–3000 tests in a typical genome). The problem is compounded by the fact that individual tests at adjacent positions are correlated—highly so in interval mapping. Thus if a standard significance level was used that was expected to give 5% false positives per test, we would be practically certain to find one or more false positives somewhere in the genome. Lander and Kruglyak (14) calculate that one would expect around 24 false positive results in the human genome if the standard 5% threshold were used. Thus we need to set a more stringent significance threshold, so that there is only a small (e.g. 5%) chance of detecting a single false positive *anywhere* in the genome. It is possible to use some theoretical approaches to derive this genome-wide significance threshold (6, 14, 15). However, such approaches depend upon underlying assumptions such as the normal distribution of the traits being studies, markers being equally spaced and equally informative, and so forth.

An alternative to theoretical methods is to derive the test threshold from a simulation study. Churchill and Doerge (16) have suggested using permutation analysis to do this. In this approach, the measured phenotypes are randomly scrambled with respect to the genotypes. Thus scrambled, no *real* QTL effects remain, and any that the analysis throws up must be 'false positives'. By analysing many such data sets, the statistical threshold can be found which gives false positives in an acceptably low number (e.g. 5%) of the scrambled data-sets analysed. This threshold can then be applied to the real (non-scrambled) data, with only a 5% risk of detecting a false positive.

Thresholds obtained either by theory or by simulation are typically equivalent to setting the significance threshold for a single test to $P = 10^{-4} - 10^{-5}$ (14). Such stringent thresholds mean that only the larger QTLs can be detected in most studies, but are certainly appropriate to avoid resources being wasted on attempting to clone or select for non-existent QTLs.

6. Predicting the power of an experiment to detect QTLs

An investigator should always have some idea of the size (i.e. degree of effect) of QTLs that he could hope to detect, before embarking on the time-consuming and costly study itself. A first consideration is whether there are likely to be QTLs segregating at all, and this has been discussed above. Many factors impact upon the ability of a particular study to detect QTLs that *are* segregating, such as the size and structure of the population and the markers used. One important fact is that detection of QTLs of a certain size is not guaranteed, but only carries a greater or lesser probability. QTLs of a larger effect will have a higher probability of detection than those of a smaller effect, with the probability of detection often being referred to as the 'power' of the experiment. Whether a particular QTL is actually detected will depend on chance events such as its distance from nearby markers, particular segregation patterns at those markers, the distribution of recombination events, and environmental influences found in that study.

Some of the problems of predicting the power of an experiment can be eased by considering the case of an infinitely dense map of fully informative markers (i.e. where the QTL is completely linked to one of the markers). In this case we can forget about the complexities of interval mapping, which would provide the same result as single marker analysis. With this assumption we can predict the power of an experiment by using a simple method such as a *t* test. Falconer and Mackay (1) give the number, n, required in each class of marker genotypes to detect a particular QTL as:

$$n \geq 2 \, (z_\alpha + z_{2\beta})^2 \, / \, (\delta/\sigma_w)^2 \tag{1}$$

where:

z_α is the ordinate of the normal distribution corresponding to the nominal significance, α, of a single test (i.e. the chosen significance threshold)

z_β is the ordinate of the normal distribution corresponding to the nominal significance of a single test of β,

β is one minus the desired probability of detecting a given effect in the study,

δ is the smallest trait difference between marker classes we can detect and

σ_w is the standard deviation of the trait within a single marker genotype class.

This formula is appropriate for comparing the two classes in a backcross, or for the two homozygous classes in an F_2 population. As already noted, one of the most difficult tasks is setting the nominal significance threshold so that a genome-wide 5% error rate is expected. A few examples of the sort of size of population required to detect QTLs of given effect are given in table 1. We have chosen to use nominal significance thresholds of 10^{-4} and 10^{-5} for these studies and compare these with the usual single test significance level of 5×10^{-2}. The results in *Table 1* emphasize that very large experiments are required to detect QTLs of modest size if the required stringent significance thresholds are to be met. Consequently, only QTLs of large effect are likely to be found in most studies. Increasing the probability of finding a QTL of given effect from a reasonable (50%) chance to a good chance (90%) also requires a substantial increase in the size of the experiment.

Monte-Carlo simulation can be used as a general tool to predict the power of a given experiment. It is valuable because simulations can be tailored to the exact proposed design and analysed using the proposed method. However, it can be very time-consuming to do sufficient simulations to get a realistic idea of power (the proportion of simulations that detect a given effect) over a range of QTL effects. Simulation can also be used once the experiment is complete to test the actual power of an experiment. The actual marker genotype data can be used and new phenotypes simulated which

Table 1. Example experiment sizes required to detect QTLs.

Effect of QTL[b]	Significance threshold	Power		
		50%	75%	90%
0.25	5×10^{-2}	123	221	336
	10^{-4}	484	665	855
	10^{-5}	625	829	1040
0.5	5×10^{-2}	31	56	84
	10^{-4}	121	166	214
	10^{-5}	156	207	260
1.0	5×10^{-2}	8	14	21
	10^{-4}	30	42	53
	10^{-5}	39	52	65

[a] The body of the table shows the minimum number of individuals in each marker genotype class needed to detect a given effect at the stated significance threshold with a given power. The total sample size needed would be twice the given value for a backcross (with two classes) or four times for an F_2 (where the two homozygous classes are compared). The calculations assume a high density map, with no recombination between the marker and QTL.
[b] The effect of the QTL is given in terms of the difference in standard deviations between the genotypic classes, i.e. (δ / σ_w) in *Equation 1*.

incorporate one or more QTLs at particular points amongst the markers. This approach can be useful for probing the effect of variation in marker information content and segregation distortion on the study's ability to locate QTLs.

7. Reducing background 'noise'

The effect of a QTL is seen against background of variation caused by other effects. When testing for the effect of a QTL, we compare the trait difference between marker genotype classes with the trait variation within those classes. Thus, anything that we can do to reduce the variation within classes will reduce the background 'noise' and so increase our ability to detect QTLs. Several ways to reduce this background noise or to minimize its effects are considered below.

7.1 Careful design

It should go without saying that careful design and execution of any experiment is important to optimize its chance of success. Thus, we can attempt to ensure that undesirable environmental sources of variation are minimized. Such effects might include temperature, nutrient, or water fluctuations between cultures of plants or animals. It is equally important to ensure the highest standards of accuracy in measuring and recording trait data. It is wise to implement systems to validate data where possible, such as double measurement and double data entry into computer databases.

The possibility of replicating measurements should not be overlooked. Repeated measures of a trait on the same individual are very valuable if the 'repeatability' of the trait (the correlation between successive measures) is low due to measurement errors or to uncontrollable environmental fluctuations which are peculiar to an individual (ref. 1, pp. 136–41). No more than two or three repeated measures of the same trait are usually needed and can be invaluable for traits with a low repeatability. If a trait is measured n times, and the mean of the measurements used, its apparent heritability will be $n/\{1 + r(n - 1)\}$ times greater than if single measurements were made (where r is the repeatability of the trait measurements). For example, the heritability of the mean of two measures of trait with a repeatability of 0.5 is $2/\{1 + 0.5(2 - 1)\} = 1.33$, i.e. 33% greater than the heritability of a singly—measured trait. Considering the mean of repeated measures can also be valuable when we are not considering strictly the same trait (e.g. growth rate in two seasons; yield in two lactations; fat depth at two sites in an animal). But here we must be more cautious and the judgement on the value of repeated measures should take account of the genetic correlation between the repeated measures which can not be considered identical.

The marker genotypes should be treated with the same care as phenotypic measurements and verified where possible. Not all genotype errors will reveal themselves as detectable Mendelian errors, so the data should be checked for diagnostic evidence of errors such as unlikely double recombinants or clustering of recombination events in particular families. Complete replicated genotyping is usually impractical and unnecessary, but a degree of replication is valuable to give an estimate of the level of genotyping errors.

7.2 Identifying and allowing for 'noise' in analysis

Some sources of 'noise' in the experimental data are inescapable. Environmental influences which affect the trait of interest, for example, may be unavoidable or may be imposed as part of the experimental design. Resource limitations may mean that an experiment has to be performed over several time periods (e.g. seasons) or locations (plots, fields, or animal facilities). The sex of an animal, although not strictly environmental, often has a big effect on the mean of a trait. Other environmental influences may be imposed by the experimenter to investigate their effect on QTLs (for example to look for genotype–environment interactions). Whilst it may not be possible to remove these sources of noise from the data, we can at least recognize their existence and allow for them in the subsequent analysis.

The best way to allow for these effects is to include them in the analysis, removing (and estimating) their effects simultaneously with detecting and estimating QTL effects. In principle this can be done with any form of analysis. In practise, computational restrictions mean that it is not practical with maximum likelihood analyses such as those implemented in *Mapmaker/*

QTL (see Section 11). In this case the data may be 'pre-corrected' for these effects, with the QTL analysis undertaken on the residual effects after correction. Such analysis is not optimal because it cannot easily incorporate interactions between QTLs and environmental effects, it does not take account of any chance correlations between QTLs and environmental effects, and it does not allow for the degrees of freedom lost when adjusting for environmental effects. Analyses based on least-squares can much more easily account for such effects and can explicitly test for QTL–environment interactions. Such analyses can be performed either when markers are looked at singly (17, 18) or in interval analysis (19).

Part of the 'noise' when looking at a particular QTL is segregation of other unlinked QTLs affecting the same trait. It is also possible to take account of these effects when analysing the data. Thus, a preliminary genome-wide screen may identify only QTLs of large effect against the background noise. Subsequently, these can be allowed for to reduce the background 'noise' when analysing for QTLs of smaller effect (6, 20–24). In principle and in simulation studies this method has been shown to be effective, but there are as yet few demonstrations of its efficacy in practise.

7.3 Progeny testing

The accuracy of estimating the genetic merit of an individual (and hence the effective heritability of the trait) can be increased by progeny testing. Instead of measuring the trait in the individual which has been genotyped, one instead measures that trait in a large number of its offspring. If sufficient offspring are measured, the effects of random environmental contributions can be almost eliminated; this more than outweighs the 'dilution' of the QTL in the second generation. It has the additional advantage that it can detect 'latent' QTLs which are carried by the genotyped individuals but only expressed in their offspring (for example, genes carried by bulls which affect the milk yield of their daughters).

The kind of progeny testing applicable, and its value, depend upon the breeding system. In plants, it is possible to produce an F_2 generation for genotyping, but to record phenotypes based on F_3 progeny. In livestock similar strategies can be employed. For example, widespread use of artificial insemination in dairy cattle produces sires with large groups (often 100 or more) of half-sib offspring. Thus Weller *et al.* (25) suggested the 'granddaughter' design for QTL detection, in which sires (and their sons) are genotyped. The value of the genes for milk traits being carried by the sons (but of course not expressed!) is then assessed from milking records of large numbers of their daughters, i.e. the granddaughters of the original bulls. Often, only a third as many sires need be genotyped to detect a given QTL if the granddaughter design is used (25). Milking records (i.e., trait measurements) are needed from many more females in the granddaughter design, but these are readily available as they are collected anyway in national breeding programmes.

7.4 Use of recombinant inbreds, dihaploids, and clones

Environmental noise can be virtually eliminated by measuring traits in large numbers of genetically identical individuals. Ideally, a line of clonal individuals is derived from each segregant, and each member of that line is scored for the relevant trait. The mean trait measurement within each clonal line thus substitutes for the trait measurement for each original segregant.

One method of achieving this is to inbreed from an F_2 cross to form a number of recombinant inbred lines. Alternatively, in some plant and fish species it is possible to produce genetically identical individuals directly as progeny from the F_2 through dihaploidization or gynogenesis. Such processes are slow, costly, and technically demanding in most species, and hence the number of clonal lines is usually limited. Furthermore, effects such as inbreeding depression and genetic-lethal mutations can cause lines to be lost and can produce major segregation distortions and biases in genome representation. In some species, direct cloning of individuals (e.g. from cuttings or potato tubers) can be used with ease and without any deleterious consequences of inbreeding. These methods can increase the effective heritability to unity, thus allowing better detection of QTLs. However, QTL effects will still be seen against a background variation caused by the segregation of other QTLs, and thus a reasonable number of distinct segregants and corresponding clonal or inbred lines is still needed.

8. Selective genotyping and DNA pooling

The use of 'selective genotyping' has been suggested as a means of making more effective use of a limited genotyping capability (6). In the context of a cross between divergent lines, Lander and Botstein (6) point out that most of the evidence on the existence of QTLs for a trait comes from the highest and lowest performing individuals. Therefore, if the cost of genotyping is limiting, it makes sense to genotype only the extreme individuals. For example, genotyping only 50% of the population (the top and bottom 25%) gives more than 90% of the information that would be obtained from typing all individuals (6). The optimum proportion to genotype will depend on the balance of costs of producing and measuring versus genotyping individuals (26). The cheaper it is to produce and measure individuals, or the more expensive it is to genotype them, the smaller the proportion which should be genotyped.

However, there are situations where selective genotyping is not appropriate. Often, QTL studies are set up to simultaneously study a number of traits. In these, selection of individuals for one trait will conflict with that for another (unless the traits are correlated), and is unlikely to be advantageous.

It should be noted that simple analysis of data from selective genotyping will give over-inflated estimates of QTL effects, because only extreme individuals will represent the QTL. Analysis by maximum likelihood can resolve this problem if the data upon which the selection was made is included in the analysis (even though individuals in the centre of the distribution have no marker genotype data). We should also note that selection which omits one tail of the trait distribution (as may happen in a breeding population if poor performing individuals are discarded before they can be genotyped) can not only bias QTL estimates but also reduce the power to detect QTLs (27).

An even more drastic way to reduce the work-load needed to carry out a genome scan is to perform genotyping on pooled DNA of samples selected from the upper extremes of the distribution and on pools from the lower end (28). The selective pooling is expected to cause a biased representation of alleles at linked marker loci and the method has successfully been used for mapping monogenic traits in several species. This method is expected to be most effective for QTLs with large effects and in crosses between inbred strains, since the detection of a representation bias will be more robust for biallelic systems than for the multiallelic systems typical of outbred populations. There are as yet few reports of successful application of this approach, however, for detection of QTLs.

Although the discussion here has applied largely to populations derived by crossing inbred lines, similar conclusions apply to other sorts of populations. For example, genotyping individuals with high and low trait values within human families can markedly increase the power of a study for a given number of individuals genotyped (29).

9. Pitfalls in interpretation

9.1 Overestimation of QTL effects

One potential problem in QTL analyses is the overestimation of the effect of QTLs caused by the selection of a few significant effects from statistical tests at a large number of positions. Occasionally we will detect a spurious QTL (a type I error) and this of course represents an overestimation of an effect that is actually zero! More commonly we may detect true QTLs but overestimate their effects. This arises simply because QTLs whose effects are over-estimated (for whatever reason) are more likely to be detected above the necessarily stringent threshold than are those whose effects are correctly esti-mated or underestimated. This problem is worst in cases where we are least likely to detect a QTL: then, the only QTLs we are likely to detect at all are those whose effect has been greatly overestimated (9). This problem has been recognized for some time, but there is as yet no easy solution. Lande and Thompson (30) suggest identifying significant effects in one study and then re-estimating the effects of significant QTLs in a second study. Although this is effective in ameliorating the problem, the cost of QTL studies makes it unattractive. A more practical solution may be to recognize the existence of the problem and try to design further studies (whether they are aimed at marker-assisted selection or towards high-resolution mapping of the QTL for positional cloning) so that revised estimates of the QTL effect can be obtained.

9.2 Linked QTLs

The standard QTL mapping techniques such as interval mapping look for the presence of a single QTL on a chromosome, and can be shown to work well when this is in fact the case. However, when there are several linked QTLs, we can be seriously misled by using methods designed to look for a single locus. For example, two QTLs at either end of a chromosome can be mis-interpreted as a single QTL in the middle (31, 32). Similarly, a large number of QTLs of small effect can look like a single QTL of large effect (12, 33). The complete solution to this problem has yet to be resolved, but it can at least be partially ameliorated by explicitly testing alternative models which assume multiple QTLs (33). Analysing for two or more QTLs simultaneously is possible, but can be cumbersome (31, 32). An alternative approach is to attempt to control for the effects of linked QTLs by fitting additional markers as cofactors whilst performing interval mapping (20–24). Such approaches are effective in some circumstances, but have yet to be fully explored.

9.3 Interaction between genotype and environment

The possibility of genotype–environment interactions should always be borne in mind. For example, a QTL allele that promotes growth rate through increased nutrient uptake when conditions are good may have little effect when conditions are poor and nutrients are limiting. The presence of such interaction may mean that a QTL seen in one study is not seen when a similar population is studied in another environment, or its effect may even be reversed.

Genotype–environment interaction has been cited in the past when replicate studies have failed to find the same QTLs. However, as noted above it is easy enough to test for these interactions (for example using least squares methods) and when this is done, there is often little evidence for their existence. For example, in the analyses of QTLs affecting growth rate and fatness in pigs, sex and feeding regime did indeed have significant effects on the mean performance, but did not appear to interact with QTLs. Most of the failures of reproducibility in QTL studies can be attributed to their relatively small size (and hence limited chances of detecting QTLs), such that QTLs which are just above the threshold for detection in one study happen to fall below it in another. Only QTLs of major effect would be detected reproducibly in such studies. This seems to be the case in studies of maize and tomato in different environments reported by Stuber *et al.* (17) and Paterson *et al.* (18), respectively.

10. Towards positional cloning of QTLs

If we are to clone a QTL starting from a knowledge of its map position, several steps are required:

(a) We should have good evidence (obtained, for example, by the application of some of the tests discussed above) that we are dealing with a single gene whose position can be pinpointed, rather than the agglomerated effects of several linked genes.

(b) The QTL must be mapped with sufficient resolution for there to be a manageable number of genes for further study in the target region.

(c) The candidate genes need to be identified in the target region. This can be done, for example, by alignment of the target region with the homologous region of better studied species. Candidate loci can then be identified based on comparative information. Alternatively, large-fragment clones coming from the target region can be isolated and examined for coding sequences.

We have already discussed the problem of mapping genetically linked QTLs. However, even when only a single QTL is working in a particular region, it is difficult to map it precisely. For example, the genomes of higher mammals contain up to 100000 genes. A linkage distance of 1 cM (around 1/3000 of the total genome) thus covers a region containing around 30 genes on average. Simulation studies have shown that even QTLs of relatively large effect mapped in large experimental populations have poorly estimated positions (12, 13). For example, a QTL causing 11% of the variation in an F_2 population of 1000 individuals and mapped using markers spaced at 10 cM intervals would have a position known only to within $\pm$ 4 cM (12). One would expect this 8 cM candidate region to contain around 240 genes, which would have to be sifted through to find the QTL.

Positional cloning will therefore usually require that the QTL is located to a region of 1 cM or less. A number of strategies could be applied here, such as selective genotyping (if the cost of genotyping is substantially more than the cost of measuring phenotypes). In general, the most important factor in increasing resolution is to increase the number of meioses in the study, as increasing the density of markers alone has little effect in most cases (13). An alternative to simply increasing the size of the population under study is to focus on recombinations within the target region. For example in organisms with a relatively short generation interval, individuals which are recombinant in the target region can be identified in the early generations of a cross between two lines. Such individuals can be multiplied and bred to homozygosity, for example by repeated backcrossing to one of the founder lines to create a series of 'near isogenic lines' which only differ in the target region (34). Analysis of these lines can then refine the QTL location. In addition, such a suite of lines will confirm that the effect is due to a single QTL, as effects due to linked groups of QTLs will tend to be dissipated by this process.

Another major problem with positional cloning of QTLs is that we often do not have a clear idea of the type of mutation we are looking for. Consider for instance a QTL allele increasing growth rate by 20% in the homozygous condition. Is this likely to be a regulatory or structural mutant in a gene controlling growth rate? Is it expected to influence expression patterns of the gene? Thus it will be hard to verify that an observed genetic polymorphism in the candidate region is really the causative mutation. This is in sharp contrast to positional cloning of a gene causing an inherited disorder where we expect to find a clear defect in the gene product or in its expression.

11. Software

There are a number of options for analysing data from a QTL study. For some designs it may be necessary for the investigator to write his own computer programs. However, a range of software exists for QTL analyses. For single marker analyses in simple population structures, standard statistical software such as *Genstat* (35) and *SAS* (36) can often be used. Analyses using information from two or more markers simultaneously require software designed especially for this case and several packages have been written, some of which are readily available. Brief details of some of the packages are given below. However, it should be noted that as the methods are refined, old software is modified and new software written to take account of these developments. In addition, no package can undertake all analyses that might be required. In particular, there are few packages capable of analysing data from outbred populations and some of these have been designed for human populations. Such software may be inefficient or incapable when faced with data from animal or plant populations with large sibships and many inbreeding loops in the pedigree.

(a) *Mapmaker/QTL* is the companion program to *Mapmaker* and is the original QTL interval mapping software. It is widely distributed, and runs on many types of computer. It will analyse F_2 or backcross data from crosses between inbred lines using a maximum likelihood implementation of standard interval mapping (6). The software is freely available, and further information and the software itself (and other useful software, including *Mapmaker/SIBS* discussed below) is available from the Whitehead Institute at (http://www-genome.wi.mit.edu/).

(b) *Mapmaker/SIBS* is designed to analyse sib-pair data from human populations (i.e. small family sizes) and uses data from multiple linked markers (10). Several alternative analyses are implemented, including the 'traditional' least squares Haseman and Elston (37) method, maximum likelihood methods (in which marker identity by descent and QTL variance are estimated) and a non-parametric method.

(c) *QTL Cartographer* is written for either UNIX, Macintosh, or Windows. It performs single-marker regression, interval mapping (6), and composite interval mapping, i.e. the inclusion of markers outside an interval in an attempt to control for background genetic influences and the effects of linked QTLs outside the interval being explored (23, 24). It permits analysis from F_2 or backcross populations derived by crossing inbred lines. Further information on the programs and availability can be obtained from http://www2.ncsu.edu/ncsu/CIL/stat_genetics/.

(d) *MapQTL* is part of the JoinMap suite of linkage analysis programs, which are only available on payment of a registration fee. The package runs on many computers including Macintosh, Windows, and UNIX systems. *MapQTL* conducts maximum likelihood versions of standard interval mapping (6), multiple QTL mapping (MQM) (20, 21), and non-parametric mapping (Kruskal–Wallis rank sum test). It can analyse a variety of pedigree types including some types of outbred pedigrees. Further information is available from j.w.vanooijen@cpro.agro.nl.

(e) *MQTL* analyses data from a cross between inbred lines producing recombinant inbred or doubled haploid lines or backcross progeny. It does not currently handle data from F_2 populations. Its main feature is the ability to analyse data from multiple environments: phenotypic data may represent observations from several or many different environments. Two types of analysis are performed: simple interval mapping, and a simplified form of composite interval mapping, both implemented via least squares. In each type of mapping, tests are performed for the presence of main effect and QTL-by-environment interaction. MQTL can also perform permutation tests to establish thresholds for control of type I error rate (38). Contact the authors (Tinker@gnome.agrenv. mcgill.ca), or obtain the software by anonymous ftp from gnome.agrenv.mcgill.ca/pub/Genetics/software/MQTL/.

(f) *LINKAGE* has been widely used for linkage analysis using markers alone, but can also analyse data with the inclusion of phenotypic information for association studies (39, 40). This can be obtained by anonymous ftp from linkage.cpmc.columbia.edu/software/ linkage.

In addition to the contacts listed, general information can be obtained from sites on the World Wide Web, for example:

- http://s27w007.pswfs.gov/qtl/software.html
- http://linkage.cpmc.columbia.edu/soft/

Further information on linkage mapping software can be found in Appendix 2.

12. Conclusions

In this chapter it has only been possible to give the briefest overview and flavour of the many considerations that go into designing and analysing a QTL mapping study. The design, performance and analysis of such a study have to be considered as a whole, and will be affected by factors such as the aims of the study and the populations and other resources available. There is thus no single protocol, and no two studies will be the same. Furthermore, the methodology and technology is still evolving rapidly and this will impact upon the detail of the design and analysis of studies.

However, the basics of QTL mapping are likely to remain more constant and there are a number of 'take-home messages' for the would-be investigator. First, one must be realistic about what is achievable. At present, only QTLs of moderate or large effect can be detected with manageable population sizes. To detect even these QTLs, studies must be designed, performed, and analysed carefully. There are a number of pitfalls in QTL analysis, such as setting significance thresholds too low or overestimating a QTL's effects, and many unresolved issues such as the optimum analysis of linked QTLs.

None the less, much is possible as is evidenced by the explosion in reported QTLs in plants and mice over the past few years and more recently in man and in livestock species. The future will see many more QTLs being mapped and studied, as well as their use in plant and animal breeding programmes and intensive efforts to positionally clone mapped QTLs in order to study their molecular function in detail. It is clear that a major challenge in current biology and human medicine is to unravel the molecular genetic basis for polygenic traits. The next eight years of QTL analysis are likely to be as active and exciting as the last.

References

1. Falconer, D. S. and Mackay, T. (1996). *Introduction to quantitative genetics, 4th edition.* Longman, Harlow, UK
2. Geldermann, H. (1975). *Theor. Appl. Genet.*, **46**, 319.
3. Hill, W. G. and Knott, S. A. (1990). In *Advances in statistical methods for the genetic improvement of livestock.* (ed. K. Hammond and D. Gianola), p. 477. Springer–Verlag, Berlin.
4. Hospital, F., Chevalet, C., and Mulsant, P. (1992). *Genetics*, **132**, 1199.
5. Meuwissen, T. H. E. and van Arendonk, J. A. M. (1992). *J. Dairy Sci.*, **75**, 1651.
6. Lander, E. S. and Botstein, D. B. (1989). *Genetics*, **121**, 185.
7. Haley, C. S., Knott, S. A., and Elsen, J. M. (1994). *Genetics*, **136**, 1195.
8. Knott, S. A., Elsen, J. M., and Haley, C. S. (1994). *Proc. 5th World Cong. Genet. Appl. Livest. Prod. (Guelph)*, **21**, 33.
9. Georges, M., Nielsen, D., Mackinnon, M., Mishra, A., Okimoto, R., Pasquino, A. T. *et al.* (1995). *Genetics*, **139**, 907.
10. Kruglyak, L. and Lander, E. S. (1995). *Am. J. Hum. Genet.*, **57**, 439.

11. Vos, P., Hogers, R., Bleeker, M., Reijans, M., van-de-Lee, T., Hornes, M. *et al.* M. (1995). *Nucleic Acids Res.,* **23**, 4407.
12. Knott, S. A. and Haley, C. S. (1992). *Genet. Res.,* **60**, 139.
13. Darvasi, A., Weinreb, A., Minke, V., Weller, J. I., and Soller, M. (1993). *Genetics,* **134**, 943.
14. Lander, E. and Kruglyak, L. (1995). *Nature Genet.,* **11**, 241.
15. Lander, E. S. and Schork, N. J. (1994). *Science,* **265**, 2037.
16. Churchill G. A. and Doerge, R. W. (1994). *Genetics,* **138**, 963.
17. Paterson, A. H., Damon, S., Hewitt, J. D., Zamir, D., and Rabinowitch, H. D. (1991). *Genetics,* **127**, 181.
18. Stuber C. W., Lincoln S. E., Wolff D. W., Helentjaris T., and Lander E. S. (1992). *Genetics,* **132**, 823.
19. Andersson L., Haley, C. S., Ellegren, H., Knott, S.A., Johansson, M., Andersson, K. *et al.* (1994). *Science,* **263**, 1771.
20. Jansen, R. C. (1993). *Genetics,* **135**, 205.
21. Jansen, R. C. (1994). *Genetics,* **138**, 871.
22. Jansen, R. C. and Stam, P. (1994). *Genetics* **136**, 1447.
23. Zeng, Z.-B. (1993). *Proc. Natl. Acad. Sci. USA,* **90**, 10972.
24. Zeng, Z.-B. (1994). *Genetics,* **136**, 1457.
25. Weller, J. L., Kashi, Y., and Soller, M. (1990). *J. Dairy Sci.,* **73**, 2525.
26. Darvasi, A. and Soller, M. (1992). *Theor. Appl. Genet.,* **85**, 353.
27. Mackinnon, M. J. and Georges, M. A. J. (1994). *Genetics,* **132**, 1177.
28. Darvasi, A. and Soller, M. (1994). *Genetics,* **138**, 1365.
29. Risch, N. and Zhang, H. (1995). *Science,* **268**, 1584.
30. Lande, R. and Thompson, R. (1990). *Genetics,* **124**, 743.
31. Haley, C. S. and Knott, S. A. (1992). *Heredity,* **69**, 315.
32. Martinez, O. and Curnow, R. N. (1992). *Theor. Appl. Genet.,* **85**, 480.
33. Visscher, P. M. and Haley, C. S. (1996). *Theor. Appl. Genet.,* **93**, 691.
34. Paterson, A.H., Deverna, J.W., Lanini, B. and Tanksley, S.D. (1990). *Genetics,* **124**, 735.
35. Genstat 5 Committee (1993) *Genstat 5 Release 3 Reference Manual.* Clarendon Press, Oxford.
36. Cary, N. C. (1989). *SAS/STAT User's Guide, Version 6* (4th edn). SAS Institute Inc.
37. Haseman, J. K. and Elston, R. C. (1972). *Behav. Genet.,* **2**, 3.
38. Tinker, N. A. and Mather, D.E. (1995). *J. QTL.,* **1**, http://probe.nalusda.gov:8000/otherdocs/jqtl/
39. Lathrop, G. M. and Lalouel, J. M. (1984). *Am. J. Hum. Genet.,* **36**, 460.
40. Lathrop, G. M., Lalouel, J. M., and White, R. L. (1986). *Genet. Epidem.,* **3**, 39.

4

Radiation hybrid mapping

ELIZABETH A. STEWART and DAVID R. COX

1. Introduction

Constructing a map of an area of interest has been one of the first endeavours of explorers throughout the ages. As the scientific community delves further into one of its newest areas of exploration, the human genome, fast and accurate methods of mapping are becoming increasingly important.

There are a variety of methods currently available to construct maps of the genome, each differing in efficiency, resolution, and accuracy. One of the earliest approaches—simple observation of the genome under a microscope —enabled scientists to determine the relationship of telomeres to centromeres, and gave insights into the function of each. *In situ* hybridization took this observational approach a step forward, allowing researchers to map fragments of interest rather relying on the inherent landmarks found on chromosomes. There has been recent renewed interest in *in situ* mapping with the advent of multicoloured fluorescent tagging of the probes, and newer methods of template preparation which allow higher resolution, sometimes down to 50 kb (Chapter 9). However, this method is labour-intensive and not well suited to automation.

Genetic mapping relies on recombination events within (preferably) large multi-generational families in order to track the inheritance of genetic markers among family members (Chapters 1–3). Of necessity, the markers must be at least biallelic in order to track them within diploid samples. The most common type of marker currently used is the microsatellite (also known as the short tandem repeat polymorphism, STRP), which is often multiallelic and highly polymorphic. Many pedigree members may need to be assayed to resolve two markers of interest since the timing and placement of the recombinations are under Nature's control. Genetic mapping relies on probability comparisons to determine the most likely order of the markers, usually with a resolution no better than 1 Mb.

Physical mapping relies on the detection of overlap between cloned DNA fragments. Markers lying on the same clone are presumed to lie near to one another, and different clones which share one or more markers are presumed to represent overlapping parts of the genome (Chapters 10, 11). The markers are similar to those used in genetic mapping, but need not be polymorphic. This method is highly dependent on the type of clone used. More distance can be covered with less effort if the clones are larger but, as many researchers have found, the largest clones currently available (yeast artificial chromosomes or YACs) have a high rate of chimerism and may thus give misleading results. Although physical mapping is amenable to automation, it requires a very large number of clones to cover a region, misses regions in which the DNA is unclonable, and gives only a very crude estimate of the distance between adjacent markers.

Radiation hybrid mapping combines aspects of both the physical and genetic approaches. This method is based on studies by Goss and Harris (1–3) in which lethally irradiated human diploid lymphoblast cells were fused with Chinese hamster cells and grown under conditions which selected for the retention by the hybrids of a human X-linked marker. Non-selected human chromosomal material was rapidly lost from such radiation hybrids. It was assumed that loci further from the selected locus would be more likely to have been broken away from the piece of DNA containing the selectable marker and would therefore be lost from the hybrids more readily. As a result, the order and distance of several human X-linked loci in relation to the locus under selection could be inferred by their loss rates within a panel of such hybrids.

More recent work (4, 5) showed that a hamster–human somatic cell hybrid containing only one human chromosome could be used as a donor in hybrid formation, and that the selectable marker could lie in the hamster portion of the irradiated donor genome. It was found that fragments of human DNA, though not directly selected for, were retained by the hybrids at high enough frequencies to allow for extensive map building. This is in contrast to the findings by Goss and Harris in which non-selected fragments were lost quite rapidly. Cox *et al.* (5) assumed that the further apart any two markers were on the donor chromosome (in this case, the human component of the hamster–human donor), the more likely that the irradiation would break them apart so that they would segregate to different hybrids. Markers are assayed for their presence or absence within each member of a panel of radiation hybrid cell lines, and the frequency with which they co-segregate reflects their proximity to one another and hence allows determination of their order. This method is similar to genetic mapping but the frequency of radiation-induced breaks, in contrast to that of recombination events, can be controlled by the radiation dose: an increase in the dose causes more breaks and therefore finer resolution. The frequency of radiation-induced breakage appears to be linearly related to physical distance and, unlike genetic recombination, there do not seem to be 'hot spots' or 'cold spots' for breakage along the chromosome (6). Several chromosome-specific regional maps have been constructed using this method (7–14).

A chromosome-specific radiation hybrid panel allows a map to be built of

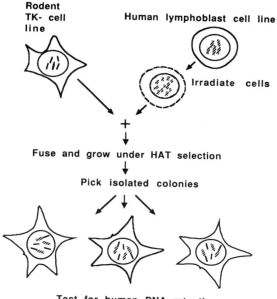

Rodent TK- cell line

Human lymphoblast cell line

Irradiate cells

+

Fuse and grow under HAT selection

Pick isolated colonies

Test for human DNA retention
Assemble into a mapping panel

Figure 1. Construction of radiation hybrid panel. Donor cells (such as human lympho-blasts) are irradiated and fused to recipient cells (such as a thymidine kinase-deficient hamster cell line), and the resultant cells are grown under selection (in this case hypo-xanthine/aminopterin/thymidine) to isolate hybrids. Colonies are picked and tested for retention of human DNA fragments (shaded), and assembled into a panel.

one chromosome at a time but, as a panel of roughly 100 different hybrids is needed for each chromosome, a genome-wide project becomes unwieldly (15). Most recently, two groups have gone back to the original work of Goss and Harris and have made radiation hybrid panels using diploid human cells as the donor (16; Stewart *et al.*, manuscript in preparation; http://www-shgc.stanford.edu; see *Figure 1*).

The advantage of these 'whole-genome' hybrids is that a single panel of such hybrids can be used to map the entire genome, provided that the donor cell was male. However, the panel as a whole must have sufficient retention of the unselected fragments to have the statistical power to differentiate between the patterns of retention across the panel. In general, it is necessary to have 80–100 hybrids, each retaining at least 12% of the human genome, in order to construct whole-genome RH maps. There are two caveats to this statement. The first is that, because the hybrids are selected by virtue of their retaining a specific human marker (such as HPRT or TK), all hybrids will tend to retain this region of the human genome, which may therefore be unmappable (though recent work by Foster *et al.* (17) suggests that this may

75

not be a limitation). Secondly, the sex chromosomes will be underrepresented in a male-derived panel, and hence may be difficult to map. Three whole-genome radiation hybrid maps have recently been constructed using this approach (18, 19, and Cox *et al.*, manuscript in preparation; see `http://www-shgc.stanford.edu`).

Radiation hybrids have also been used for purposes other than mapping. Those retaining a very small part of the donor genome—perhaps of just one fragment of human DNA—can be used as sources of DNA in an area of interest, or for assigning new probes to an area (20–22). However for the purposes of mapping it is the hybrids with high retention frequencies (each retaining 15–50% of the donor genome) which are most useful.

This chapter is intended to allow a research group to construct its own panel of radiation hybrids, to score it for STSs, and to begin the analysis of the data. The protocols within the chapter are tailored for the donor and recipient species mentioned, but limited experience suggests that the protocols are applicable to other mammalian species.

2. Preliminary considerations

There are several points to remember when starting a radiation hybrid mapping project. First, the type of donor and recipient cell line will depend on the goals of the project. If the goal is to map a single chromosome then the ideal donor will be a somatic cell hybrid containing only that chromosome (14). The recipient is typically a hamster cell line deficient in hypoxanthine phosphoribosyl transferase (HPRT) or deficient in thymidine kinase (TK). The hamster HPRT or TK gene in the donor contributes the selectable marker to the RH hybrids. Some combinations of donor and recipient lines yield better hybrids (i.e. better retention of donor fragments) than others; this is generally determined empirically. Extensive discussion of the formation of somatic cell hybrids, which has much in common to that of radiation hybrids, will be found in Chapter 6. The factors which influence the retention of non-selected donor fragments in the hybrid are not yet known, although it appears that radiation dose (and therefore the size of the fragments) plays a role. In general, the higher the dose of radiation, the lower the retention frequency of the subsequent radiation hybrids.

If the goal of the project is to build maps of the entire genome then it is best to start with a human lymphoblast cell line as the donor. The human HPRT or TK gene will complement the deficiency in the recipient rodent cell, allowing hybrids to be selected (16; Chapter 6). RH panels have also been constructed for mapping species other than human, thus illustrating the broad applicability of this method (23–25).

Unlike somatic cell hybrids, in which the goal is usually the development of a hybrid containing one intact human chromosome, radiation hybrids contain an unknown number of human chromosomal fragments within them. These fragments can be independent mini-chromosomes or (more often) are incorporated into the chromosomes of the recipient cell (26). This incorporation causes changes in the structure of the recipient chromosomes, so that not only does the recipient have to survive the influx of foreign DNA but also the disruption of its endogenous genome.

One final point to consider is that radiation hybrids are not completely stable. From one culture passage to the next, a particular cell line may lose fragments, or even appear to gain fragments when a subpopulation of cells overgrows the cell line. It is necessary, therefore, to isolate a large quantity of DNA from each cell line and use only that batch for the entire project.

3. Creation of radiation hybrid panels

Protocol 1 describes the construction of human radiation hybrids using a hamster recipient cell, though modifications of this protocol are likely to be applicable to other donor/host species. Many aspects of the protocol are similar to those used for the construction of somatic cell hybrids (Chapter 6), but the version given here has been optimized for making radiation hybrids. *Protocol 2* describes the isolation of DNA, used for the preliminary assessment of the hybrids, from the small samples of the clones grown in 6-well plates

---

Protocol 1. Construction of whole-genome radiation hybrid cell lines

Equipment and reagents

- Cell culture medium: DMEM containing 4.5 g/litre glucose + 582 mg/litre glutamine (Fisher [Cellgro])
- Foetal bovine calf serum (FBCS)
- Penicillin + streptomycin, 100 × stock: 10 000 mcg/ml streptomycin, 10 000 U/ml penicillin (Fisher)
- HAT supplement, 50 × stock: 0.68 g/litre hypoxanthine, 20 mg/litre aminopterin, 0.145 g/litre thymidine (Fisher)
- Selection medium: DMEM supplemented with 10% FBCS, 1 × penicillin + streptomycin, and 1 × HAT supplement

- Trypsin/EDTA solution: 0.05% trypsin, 0.53 mM EDTA (Fisher)
- Phosphate-buffered saline (PBS), Mg- and Ca-free (Fisher)
- Polyethylene glycol, 1500 M_r (PEG 1500), 50% (w/v)

All of the above solutions should be purchased sterile, or be filter-sterilized
- 100 mm tissue culture Petri dishes
- 24-well tissue culture plates
- 6-well tissue culture plates
- γ-irradiation source (e.g. ^{137}Cs source, J. L. Sheppard and Associates Inc.)

Method

1. Grow recipient cells (in this example, A23, a diploid Chinese hamster TK⁻ fibroblast cell line) (27) to confluency in a 100 mm tissue culture Petri dish. Split 1:4 or 1:6 (into 100 mm Petri dishes) four days before the fusion, changing the medium at least once before the actual fusion. The aim is to have the recipient just reach confluency on the day of the fusion.

Protocol 1. *Construction*

2. Feed donor cells (in this example RM, a diploid male human lympho-blast cell line) on the day before the fusion; cells should be confluent on the day of fusion.

3. Harvest 9×10^7 donor cells into 15 ml or 50 ml centrifuge tubes and pellet in a bench-top centrifuge at 1000 r.p.m. ($\sim$ 200 g) for 3 min. Resuspend each pellet in 15 ml of DMEM (containing no serum, antibiotics, or selective supplement) and pellet as before.

4. Resuspend each pellet in a small volume of DMEM and combine the suspensions in a single centrifuge tube. Add additional DMEM to 15 ml total and pellet as before.

5. Resuspend in 15 ml DMEM, pellet, and resuspend in 15 ml DMEM. Place on ice in readiness for irradiation.

6. Irradiate the donor cells, following the instructions appropriate to the irradiator. The optimal dose will need to be determined, as described in section 4.

7. After irradiation, place the cells on ice only if transit to the tissue culture facility will be prolonged. Rinse the cells three times, each time pelleting at 200 g for 3 min and resuspending in 15 ml of DMEM (no serum, antibiotics or selective agents).

8. Pellet the irradiated donor cells at 200 g for 3 min. Aspirate off 10 ml of the DMEM, resuspend the cells in the remaining 5 ml, and layer the suspension over the recipient cells in the Petri dish. Incubate (without agitation) for 1 h at 37 °C to allow the irradiated donor cells to attach to the recipient monolayer.

9. Gently aspirate the liquid from the recipient plate. Avoid disturbing the monolayer, as this may dislodge attached (but as yet unfused) donor cells.

10. Take up 2 ml of PEG solution at room temperature in a 2 ml pipette. Touch the edge of the dish with the pipette, and slowly add the solution whilst rotating the dish with the other hand. When the dish has rotated a quarter turn about 0.5 ml PEG should have been released; it should take about 1 min to add all the PEG solution in one full rotation. Gently tilt the dish back and forth just enough to distribute the PEG over the centre. Leave undisturbed for 1 min.

11. Use one of the other Petri dishes of confluent recipient cells as a negative control. Aspirate off the medium and add PEG solution as in step 10. In subsequent steps, fusion and control dishes should be processed in parallel.

12. Slowly add 8 ml DMEM (without serum, antibiotics or selective agents) by running it down the edge of the Petri dish while rotating

the dish slowly. Aspirate the liquid and repeat. Aspirate again, and add 10 ml of DMEM. Incubate at 37°C for 30 min.

13. Aspirate off the medium and add 10 ml DMEM supplemented with 10% FBCS and 1 × penicillin + streptomycin. Incubate overnight at 37°C.

14. Aspirate off the medium, add 10 ml of PBS, and aspirate. Add 2 ml of trypsin solution.

15. Pipette the liquid up and down with a Pasteur pipette to dislodge the cells, then add 8 ml of selection medium.

16. Add 0.5 ml of the cell suspension (above) to each of 20 100 mm Petri dishes, each containing 10 ml of selection medium.

17. Incubate the dishes at 37°C, replacing the medium with fresh selection medium on the third and tenth days.

18. Colonies should begin to appear in the fusion dishes during this period. Few if any colonies should appear in the control dishes; if the 20 control dishes contain more than eight colonies in total, this indicates that the recipient cell line includes a number of hamster cells that have 'reverted' at the selectable locus. In this case, the fusion should be discarded and a new vial of recipient cells, free of revertants, should be used.

19. Colonies (typically 15–20) should be visible to the naked eye by day 10–14. Use a marker pen to circle the larger 'parent' colonies; 'satellite' colonies (derived from cells which have broken away from the parent colonies) will be numerous by day 14, but are easily distinguished by their smaller size.

20. Choose only well-isolated, large colonies for picking. Avoid those which are surrounded by many satellite colonies, and pick only two or three clones from each plate to reduce the risk of picking 'sister clones'. Typically 50% of clones picked will grow well and retain useful quantities ($> 12\%$) of the donor genome, so 200 clones will need to be picked for a typical panel of 100 members. Several fusions may be performed simultaneously, but it is preferable to do several fusions sequentially to avoid processing large numbers of clones in parallel.

21. To pick the chosen clones, aspirate the medium from the Petri dish and rinse with 10 ml of PBS. Aspirate the PBS thoroughly, tilting the dish to enable as much liquid as possible to be removed.

22. Drop 5 μl of trypsin solution onto each chosen colony and leave for 1 min.

23. Set an adjustable pipette to 10 μl, but draw only 2–3 μl of fresh trypsin solution into the end of the tip. Use this to 'prime' the pool of

Protocol 1. *Construction*

trypsin on the Petri dish so that all of the liquid can be drawn up into the pipette. (Simply trying to draw the liquid from the dish into a dry pipette tip often fails to work, due to surface tension.)

24. Eject the liquid into one well of a 24-well dish containing 20 μl of trypsin solution. When all colonies have been picked, add 1 ml of selection medium to each well.

25. Incubate at 37°C. If the colony is healthy, it should be confluent within five days.

26. Add 200 μl of trypsin solution to each healthy clone and pump up and down with a Pasteur pipette to dislodge the cells. Transfer 120 μl, 40 μl and 40 μl respectively to wells of three 6-well tissue culture plates. Incubate at 37°C

27. The cells in the first plate (which received 120 μl of cells) should be confluent within five days, and may be harvested for DNA isolation for an initial assessment of the clone (*Protocol 2*). Those in the other two plates should be grown until confluent, then harvested and frozen as in ref. 28.

Protocol 2. Small scale DNA isolation from hybrids in 6-well plates

Equipment and reagents

- Lysis mix (the following is sufficient for 20 purifications): 5 ml TEN (10 mM Tris–HCl pH 8.0, 1 mM EDTA, 400 mM NaCl), 205 μl 20% SDS, 165 μl Proteinase K (10 mg/ml stock in 50 mM Tris–HCl pH 8.0, 1.5 mM CaCl$_2$), 15 μl 0.5 M EDTA, 615 μl H$_2$O
- Trypsin solution (*Protocol 1*)

- TE/RNase solution: add 2 μl of 10 mg/ml DNase-free RNase (Sigma) to 1 ml of TE (10 mM Tris–HCl pH 8.0, 1 mM EDTA)
- Saturated NaCl solution (~ 5–6 M)
- Ethanol, 100% and 70% (v/v)

Method

(Modified from ref. 29.)

1. Add 300 μl of trypsin solution to the well containing the hybrid. Leave for 1 min at room temperature, then pump the liquid up and down with a Pasteur pipette to dislodge the cells.

2. Transfer the cell suspension to a microcentrifuge tube and pellet at 13 000 r.p.m. for 2 min.

3. Remove the supernatant and resuspend the pellet in 300 μl of lysis mix. Incubate at 55°C overnight. If clumps of undigested material remain, vortex gently and incubate at 55°C for a further hour.

4. Add 100 μl of saturated NaCl and shake to mix well.

5. Spin for 7 min in a microcentrifuge at 13 000 r.p.m. at room temperature. Rotate the tube 180 degrees and spin for another 5 min; a hard white pellet should form and the supernatant should be relatively free of debris.

6. Transfer supernatant to a clean microcentrifuge tube and add 1 ml of 100% EtOH. Mix by gentle shaking and leave at room temperature for 5 min.

7. Pellet the DNA in a microcentrifuge at 13 000 r.p.m. for 5 min. The DNA should form a visible pellet.

8. Remove the supernatant, taking care not to disturb the pellet. Add 200 μl of 70% ethanol, spin for 30 sec at 13 000 r.p.m. in a microcentrifuge, and remove the supernatant. Add a further 200 μl of 70% ethanol and repeat. After the ethanol has been removed, spin briefly and remove as much residual ethanol as possible.

9. Leave the tubes uncapped at room temperature for 5 min to allow residual ethanol to evaporate.

10. Redissolve the pellet in 200 μl TE/RNase solution.

11. Incubate at 55°C for ~ 1 h, then cool to 37°C to dissolve pellet. It is best to shake the DNA into solution, by holding the tube and flicking it with a finger.

Protocol 2 will yield enough DNA (0.5–1 μg) for evaluation of the hybrid. However, once a hybrid has been determined to be useful it is best, as mentioned previously, to isolate sufficient DNA for all foreseeable requirements in a single large scale preparation to avoid the inevitable change in fragment retention from one culture passage to the next. Cells are grown in selective medium, and DNA is isolated using any standard procedure (see, for example, ref. 30). For PCR screening, 50 ng of hybrid DNA per (duplicated) reaction should be allowed, with a safety margin of at least twofold to allow for failed reactions or wasted material. It is always advisable to overestimate the number of PCR screenings which will be performed on the hybrid panel, and to prepare appropriate amounts of DNA. The extra effort involved is infinitely preferable to running out of hybrid DNA part of the way through a large-scale mapping project!

4. Preliminary evaluation of panels and determination of optimal radiation dose

The radiation dose required in *Protocol 1* will vary depending on the cells involved, on the fragment size required in the panel, and on other factors, and must be determined empirically. Preliminary typing of small numbers of hybrids and analysis of the data will be required in order to do this. As an example, however, the Stanford Human Genome Center G3 human radiation hybrid panel was created with a dose of 10 000 rads; this panel has a typical map resolution of 500 kb, and 1 cR (a probability of 0.01 of breakage between two markers) corresponds to 30 kb (see Section 8). The TNG panel, also created at SHGC, was made using a dosage of 50 000 rads; in this panel, 1 cR corresponds to 4 kb.

Protocol 3 gives the steps required for the preliminary assessment of hybrids produced by varying doses of radiation. It gives not only the mean size of fragments present in the hybrids but also their retention frequency of donor fragments. These results can be used to guide the choice of radiation dose for full-scale hybrid panel production. Most of the analytical procedures are condensed versions of those used in the subsequent mapping, and are therefore detailed in later parts of this chapter.

Protocol 3. Estimating the optimal radiation dose

Equipment and reagents
- See protocols 1, 2 and 4

Method
1. Perform a series of fusions (*Protocol 1*) whose only variable is the radiation dose given to the donor cell line. Prepare 'sample panels' of 20–30 hybrids per radiation dose.
2. Purify DNA from these hybrids (*Protocol 2*).
3. To determine retention frequency, select a set of STSs which have been previously mapped to disparate areas of the genome (to avoid biases caused by correlation of retention patterns between closely linked STSs). If possible, two STSs should be chosen from each chromosome. Score these STSs against each of the hybrids (*Protocol 4*).
4. The retention frequency of a specific hybrid is simply the proportion of donor STSs which it retains. Hybrids with retention frequencies of < 12% or > 50% should be rejected at this stage.
5. For accurately determining the breakage frequency in the sample panels (and hence identifying the optimal radiation dose), a set of reference STS markers is required which represent a range of known intermarker distances. This range should span the range and resolution ultimately required of the mapping panel, and should include approximately 50 STSs.
6. These STSs should be typed on all of the sample panels (*Protocol 4*) and the data analysed as described in later sections of this chapter. This analysis will determine which of the hybrid panels gives the desired range and resolution of mapping, and hence which radiation dose should be used in constructing the full-scale mapping panel of 80–100 hybrids.

---

5. Existing human radiation hybrid panels

Researchers involved in human genome analysis can make use of several existing human radiation hybrid panels. Indeed, for most purposes it is preferable to use existing, well-characterized panels in order that newly placed markers can be integrated directly with previous ones.

Three whole-genome human radiation hybrid panels are currently available commercially from Research Genetics (http://www.resgen.com; e-mail info@resgen.com; telephone [1] 800-533-4363), as detailed below.

5.1 G3 panel

The G3 panel was constructed at the Stanford Human Genome Center using as a donor a male lymphoblast cell line irradiated with 10 000 rads, fused to the TK⁻ hamster cell line A23. There are 83 hybrids in the panel, plus one positive and one negative control. In this panel, 1 cR (a probability of 0.01 of breakage between two markers) corresponds to 30 kb. The average resolution of maps built with this panel is 500 kb (Cox, unpublished data, http://www-shgc.stanford.edu).

5.2 TNG panel

The TNG panel was constructed at the Stanford Human Genome Center using a radiation dose of 50 000 rads and, like the G3 panel, used a male lymphoblast donor and hamster A3 host. It consists of 90 hybrids, plus one positive and one negative control. 1 cR corresponds to 4 kb, allowing closely-spaced markers to be resolved.

5.3 Genebridge4 panel

The Genebridge4 (or GB4) panel was developed by a collaboration between the laboratories of Peter Goodfellow and Jean Weissenbach, and contains 93 hybrids plus one positive and one negative control. The lower radiation dose used in the construction of this panel makes it more useful for longer-range mapping and for providing better coverage of the genome, with 1 cR corresponding to 300 kb.

6. Selection of STSs for radiation hybrid mapping

Whether using one's own radiation hybrid panel or a pre-existing one, suitable markers must be chosen for mapmaking. These can in principle be any type of sequence-tagged sites (STSs)—short regions of sequence, typically 100–300 bp long, defined by PCR primers which can be used to amplify them. The STSs used in RH mapping need not be polymorphic, and can be derived from any unique sequence including coding regions. Currently there are several large efforts underway by industrial and academic research centres to sequence large numbers of cDNAs or cDNA fragments. For example, the Merck/Washington University collaboration deposits all of its sequences with Genbank as well as maintaining a public database which is an excellent source of sequences from which to design STSs. Several of the genome centres are co-ordinating efforts to use these sequences, develop STSs from them, and deposit STS-specific information into Genbank's dbSTS database (see Appendix 2).

There are several criteria which an STS must meet if it is to be suitable for use in radiation hybrid mapping:

(a) It must not amplify sequences from the genome of the host species used in making the radiation hybrids. Coding regions are the most likely to be conserved between donor and host. The recipient cell line should be used as a negative control to identify and eliminate such markers.

(b) Amplification should be robust. Because radiation hybrids may occasionally lose fragments during culture, a particular STS may not be present in all members of a particular hybrid line. Amplification of the STS must therefore be efficient enough to detect, on ethidium-stained agarose gels, sequences present in less than equimolar ratios (22).

(c) It must previously be assigned to a single chromosome. It is not currently feasible to construct genome-wide maps using radiation hybrids, without prior chromosomal assignment of the STSs. This assignment is commonly done by initially typing candidate STSs against a somatic cell hybrid panel (Chapter 6). This process also helps to weed out any repetitive markers (unless all copies lie on the same chromosome), and serves as a preliminary test of the robustness of amplification.

(d) If the project is a large one and availability of STSs is not a limiting factor, it is wise to standardize the PCR conditions which will be used. Primers can be designed or chosen to amplify under these standard conditions, and any STS which fails to amplify robustly under this regime is abandoned. If the project is small or if certain STSs are extremely important then a wider range of conditions can be examined to optimize amplification of a greater variety of STSs.

7. PCR typing of radiation hybrid panels

Once suitable STSs have been identified, they are scored or typed against the panel of radiation hybrids. This is done using conventional PCR, the main requirements being that the protocol should be as simple as possible and capable of processing large numbers of samples under (preferably) uniform conditions.

Protocol 4 describes the typing procedure which we routinely use.

Protocol 4. Radiation hybrid typing procedure

Typing should be carried out in duplicate for the radiation hybrid panels.

Equipment and reagents

- PCR buffer C (10 ×): 25 mM MgCl$_2$, 500 mM KCl, 200 mM Tris pH 8.3
- dNTP mix: 2.5 mM each dATP, dCTP, dGTP, dTTP
- Oligonucleotide primer pair (10 μM each of left and right primers)
- *Taq* polymerase (5 U/μl)
- Thermocycler (Perkin-Elmer 9600)
- 96-well thermocycler plates (Robbins Scientific)
- Foil plate sealers (Beckman Instruments)
- Micro-Amp full plate covers (Applied Biosystems)

- 3 × gel loading dye: 10% (w/v) Ficoll 400, 0.1 M EDTA pH 8.0, 0.025% (w/v) bromophenol blue
- Agarose
- TBE electrophoresis buffer (10 × stock): 108 g Tris base, 55 g boric acid, 9.3 g Na$_2$EDTA; make to 1 litre with deionized water
- Ethidium bromide, 10 mg/ml stock
- Agarose gel electrophoresis apparatus
- Gel imaging system (see text)
- Radiation hybrid DNAs at 5 ng/μl
- DNA size standards (e.g. 100 bp ladder, Pharmacia)

Method

1. Prepare a master PCR mix. The recipe below is for one reaction; approximately 10% excess should be prepared to allow for dispensing using multichannel pipettes:

   - 1 μl 10 × PCR buffer C
   - 0.8 μl dNTP mixture
   - 0.8 μl primer pair
   - 0.07 μl (0.35 U) *Taq* polymerase
   - 2.33 μl water

2. Aliquot 5 μl of the hybrid DNAs into each well of the thermocycler plate and add 5 μl of master mix.

3. Seal the plate with a foil sealer and the full plate cover. Place in the thermocycler and cycle as follows:

   - 94°C × 1 min 15 sec

   followed by 30 cycles of:

   - 94°C × 15 sec
   - annealing temperature × 23 sec (as appropriate to the primers being used)
   - 72°C × 1 min 30 sec

   followed by:

   - 4°C (indefinite).

4. Add 5 μl of 3 × gel loading dye to the side of each well and tap the plate to make the dye fall to the bottom of the wells.

5. Analyse 12 μl samples on a 3% agarose gel in 1 × TBE buffer (8 V/cm, 45 min). Load DNA size standards in the outermost lanes of the gel.

Protocol 4. *Continued*

6. Stain the gel in ethidium bromide solution (0.9 mg/litre in 1 × TBE) for 15 min, and destain in 1 × TBE for 30 min.

7. The gel is now ready for imaging and analysis (see text).

We image the gels using a UV transilluminator and a 640 × 480 pixel CCD camera system which captures images in TIFF format on a UNIX work-station and provides a hard copy of the gel image. These are then analysed using a modified version of the *DNA/GUI* software (31) and samples are scored for the presence or absence of PCR products of the expected size in a semi-automated way. However, any suitable imaging system can be used and results can be scored manually if necessary.

It should be emphasized that errors or ambiguities in typing can greatly decrease the chances of successfully detecting linkage between markers on the G3 (or other) panels. We strongly recommend the use of duplicate typings; if more than seven discrepancies occur between the duplicates, the assaying conditions should be optimized or the PCR primers redesigned.

Scoring is normally performed in duplicate, and an STS must be seen in each duplicate for it to be scored as positive. Bands which appear in only one of the two duplicates are scored as ambiguous. The scoring results may be stored in any suitable electronic format. It is usually most effective to store the results in a file in which each entry is represented by the STS name followed by a string of 1s, 0s and Rs (for positive, negative, and ambiguous results, respectively) representing the typing results for that STS. This simple file format can then be edited as necessary to provide input data for various analytical programs.

8. Analysis of radiation hybrid data

Due to the fact that the number and composition of human DNA fragments within a hybrid is unknown it is not possible to directly ascertain the relative positions and distances within a set of markers by direct analysis, as it is in the case of physical mapping (Chapters 10, 11). It is therefore necessary to use statistical methods for determining the order and distance between any two markers. There are many methods, or combinations of methods which can be used to construct maps with radiation hybrid data. There is a trade-off, however, in that exhaustive searches for the correct order of a large number of markers are not yet computationally feasible, yet less rigorous methods which can handle many markers at once in a reasonable amount of time can fail to examine all possibilities and may therefore miss the best order.

All analytical approaches, however, have certain features in common. The most fundamental step is the initial analysis of the data to estimate the distance between two chosen markers. This estimate takes the form of a 'two-point lod score', reflecting the likelihood (actually the logarithm of the odds, hence 'lod') that the two markers are in fact linked; a high lod score reflects tight linkage and hence short distance between two markers. The lod score is in turn derived from analysis of the typing results: markers with very similar typing patterns (i.e. those which co-segregate amongst the panel of hybrids) have the highest lod scores.

The two-point lod scores, therefore, allow estimation of the distance between any two markers in the data-set. Subsequent analysis—the most demanding part of the procedure—then aims to find that order and spacing of all of the markers along a linear chromosome which maximizes the *total* lod score, and hence best fits the data.

8.1 Calculation of two-point lod scores

Before analysis, it is assumed that the STSs and their associated data have been divided into separate chromosomal groups. Analysis is then performed independently for each chromosome, even if they were typed on a whole-genome panel of hybrids. After typing, the results are in the form of a table, in which each STS is associated with a series of scores —1, 0, or R—indicating its presence, absence, or ambiguous presence in each member of the hybrid panel. Conversely, the data can be considered as a table listing the presence, absence or ambiguous presence of each STS in each hybrid (*Table 1*).

Consider two STSs, A and B, within one of the chromosomal groups. The first step is to calculate the frequency of breakage (θ) between two STSs. This is done on the basis of a model (5) which assumes:

(a) That radiation-induced breaks occur at random along the chromosome.

(b) That physically independent fragments are retained independently within the recipient cell (i.e., the retention by the hybrid of one fragment does not affect the probability of another fragment being retained).

In the first instance, we will consider the situation where the donor of the hybrids contained only a single copy of the chromosome in question (i.e. the

Table 1. Example of scoring results for ten members of a radiation hybrid panel.[a]

	Hybrid									
STS	1	2	3	4	5	6	7	8	9	10
1876	1	0	0	0	0	R	1	1	0	0
1882	1	1	1	1	0	0	0	0	0	0
1885	0	0	0	0	0	R	1	1	0	0
1887	1	1	1	1	0	0	0	0	1	1
1891	1	0	0	0	0	0	1	0	0	1
1893	1	R	0	1	1	1	1	0	0	0

[a] The presence or absence of each STS (numbered at left) in each of the hybrids (1–10) is indicated by a 1 or 0 respectively; R indicates an ambiguous result, in which duplicate typings were in disagreement.

'haploid' case); this is true when the donor is itself a somatic cell hybrid, or when looking at X or Y chromosomes in a panel of male-derived whole-genome hybrids. On the basis of the typing results for the two STSs, the hybrids may be divided into four categories: those which contain A but not B; those which contain B but not A; those which contain both A and B; and those which contain neither. (Hybrids for which the typing result for either A or B was ambiguous are discarded from the analysis at this stage, as they cannot be assigned to any of these four categories.)

A hybrid containing both A and B may do so for either of two reasons:

(a) A and B are present on the same (retained) chromosomal fragment.

(b) A and B are present on independent fragments which have both been retained.

By the same token, a hybrid containing neither A nor B may do so either because:

(a) A and B lie on the same fragment, which was not retained.

(b) A and B lie on independent fragments, neither of which was retained.

First, let:

$o(A+ B-)$ = the number of hybrids observed to retain A but not B.
$o(A- B+)$ = the number of hybrids retaining B but not A.
$o(A+ B+)$ = the number of hybrids which retain both A and B.
$o(A- B-)$ = the number of hybrids which retain neither A nor B.
T = the total number of hybrids giving valid results (i.e. no ambiguous typings) for A and B.

If we assume A and B to be retained with equal frequency amongst the hybrids (the 'equal retention model') then:

R = mean retention frequency of A and B
L = mean frequency of loss = $1-R$

According to the equal retention model, $e(A+ B-) = e(A- B+)$, where 'e' denotes the expected number in each class. Considering only marker A, it can be seen that

$$[e(A- B+)/T)] + [e(A- B-)/T] = L \qquad [1]$$

since marker A has been lost from classes $(A- B+)$ and from $(A+ B-)$. Similarly,

$$[e(A+ B+)/T)] + [e(A+ B-)/T] = 1-L = R \qquad [2]$$

as marker A is retained in both of these classes. We can then use the observed class $o(A- B-)$ to derive the expectations for all of the other classes:

$$o(A- B-)/T = \text{observed fraction of } (A- B-) \qquad [3]$$

Rearranging *Equation 1* and substituting the observed data for $(A- B-)$ gives:

$$L - [o(A- B-)/T] = e(A- B+)/T \qquad [4]$$

and as $e(A- B+) = e(A+ B-)$ (according to the equal retention model), it follows that:

$$e(A+ B-)/T = e(A- B+)/T \qquad [5]$$

and, from *Equation 2*:

$$(1-L) - e(A+ B-)/T = e(A+ B+)/T \qquad [6]$$

Finally,

$$[e(A+ B-)/T] + [e(A- B+)/T] + [e(A+ B+)/T] + [e(A- B-)/T] = 1 \qquad [7]$$

each of the four terms of this equation can be also be expressed in terms of θ, the probability of radiation-induced breakage between A and B:

$$e(A+ B-)/T = e(A- B+)/T = \theta L(1 - L) \qquad [8a]$$
$$e(A+ B+)/T = (1 - \theta)(1 - L) + \theta(1 - L)^2 \qquad [8b]$$
$$e(A- B-)/T = \theta L^2 + L(1 - \theta) \qquad [8c]$$

Equation 8c rearranges to give:

$$\theta = \left[1 - \left(e\frac{(A- B-)}{T}\Big/ L\right)\right](1 - L) \qquad [9]$$

Substituting the observed result $o(A- B-)$ into *Equation 9* gives:

$$\theta = \left[1 - \left(o\frac{(A- B-)}{T}\Big/ L\right)\right](1 - L) \qquad [10]$$

thus allowing θ to be calculated from the observed data. For any chosen value of θ, the relative likelihood (Lk_θ) of obtaining the observed data-set (i.e. the observed numbers of hybrids in each of the four possible classes) can be calculated as:

$$Lk_\theta = [e(A+ B-)/T]^{o(A+ B-)} \times [e(A- B+)/T]^{o(A- B+)} \times$$
$$[e(A+ B+)/T]^{o(A+ B+)} \times [e(A- B-)/T]^{o(A- B-)} \qquad [11]$$

where $e(A+ B-)/T$ etc. are as in *Equations 8a–8c*. The linkage between markers A and B is then simply the log of the ratio of the likelihood for the chosen θ to that for $\theta = 1$, i.e.

$$Lod_{AB} = \log_{10}[Lk_{(AB,\theta)}/Lk_{(AB,\theta = 1)}] \qquad [12]$$

The foregoing considerations apply, as stated, to the haploid case in which only one copy of each marker can be contributed to the hybrid by the donor cell. In the diploid case, the donor cell contains two copies of each marker, each of which may be either retained or lost by the hybrid. Retention of two

89

copies of a marker in the hybrid (e.g. A++) is indistinguishable from retention of only one copy (A+− or A−+), and hence the only class of hybrids which can be unequivocally observed is (A− B−). The absence of a given marker in the hybrid is the product of two events, the loss of copy one and the loss of copy two. Therefore an observed diploid loss (*Ld*) is the square of the haploid loss (*L*) which was used in the earlier equations. Hence, the expected proportion of (A− B−) diploid hybrids is simply the square of the expected proportion of (A− B−) in the haploid case with the square root of the observed diploid loss:

$$(A- B-)/T = [\theta(\sqrt{Ld})^2 + \sqrt{Ld}(1 - \theta)]^2 \qquad [13]$$

which can be solved for θ to give:

$$\theta = [\sqrt{Ld} - \sqrt{(o(A- B-)/\underline{T})}]/[\sqrt{(Ld)}(1 - \sqrt{(Ld)})]. \qquad [14]$$

Hence θ, the probability of breakage between the two markers in question can be determined in either the haploid or diploid case. This probability of breakage, in turn, is related to the distance between the two markers: the greater the distance, the greater the probability of radiation-induced breakage. If breaks occur at random along the chromosome, then:

$$D = -\ln(1 - \theta) \qquad [15]$$

where *D* is the distance in Rays. In practice, distances are normally expressed in centiRays (cR), where 1 cR = 0.01 Rays. One centiRay is thus the distance between markers at which there is a 1% probability of breakage. The value of 1 cR in kilobases depends upon the radiation dose used to construct the panel and must be calibrated against markers at known distances (Section 4).

8.2 Map building

For any proposed order of markers, the two-point lod scores derived in Section 8.1 can be summed to give the overall likelihood of that order. The best estimate of the true order is then deemed to be that which has the highest likelihood. This approach is strictly valid only when all markers have been typed on all members of the panel, and have equal retention frequencies. If inconsistencies between duplicate marker typings cause some results to be discarded, the approximation is no longer completely valid. In addition, retention frequencies are not normally identical for all markers, again complicating the analysis. There are computer programs available which allow the user to test different models of retention frequency, which can have an influence on the accuracy of the distance measurements in particular (32).

Many of the current programs for map building (such as *RHMap*2.01) (33) are unable to analyse more than about 20 markers in a useful time, though software is now becoming available which alleviates this problem, such as '*Mapper*' from MIT (18) and '*SAMapper*' from Stanford (manuscript in preparation; available from `ftp://shgc.stanford.edu`). These pro-

Odds					Position						
	1	2	3	4	5	6	7	8	9	10	11
1:1	A	B	C	D	E	F	G	H	I	J	K
1:10	A	C	B	D	E	F	H	G	I	J	K
1:100	A	C	B	D	F	E	H	G	I	K	J
1:1000	C	A	B	F	E	D	H	G	I	K	J

Figure 2. Placement of markers into high confidence 'bins'. The likeliest order of the 11 markers (A–K) is shown in the top line. Possible alternative maps with progressively lower relative likelihoods are shown on the following lines; for example, maps in which B is interchanged with C (or G with H) are tenfold less likely than the best order. By comparison of these maps, the markers can be divided into high confidence bins divided by vertical lines.

grams allow the optimal order (that with the highest likelihood) to be determined for large numbers of markers. However, there may be a number of other possible marker orders with likelihoods almost as great as that of the most likely, reflecting ambiguities in the map. *SAMapper* therefore examines not only the best order, but also all other orders whose likelihood is at least 1/1000th as great as the best order. By comparing all of these suboptimal orders, it constructs a map in which markers are placed into 'high confidence bins'—groups within which the marker order is less certain (75% confidence), but between which the order is more certain (98%; *Figure 2*).

This brief overview is by no means an exhaustive study of the statistical methods used in radiation hybrid mapping. For further information and discussion of radiation hybrid analysis, including polyploid analysis, the authors refer the reader to several articles on the subject (32–34).

8.3 Resources for the analysis of radiation hybrid data

The Stanford Human Genome Center supports an automated form-based web or email server for providing radiation hybrid localization information of anonymously submitted markers (http://www-shgc.stanford.edu/rhserver/info.html or send email with subject 'info' to rhserver @shgc.stanford.edu). This server accepts the results of an assay against the Stanford G3 whole-genome radiation hybrid panel (Section 5.1), analyses the data, and attempts to localize the marker on the whole-genome maps which have been built by SHGC with the G3 panel. The server will also make use of chromosomal assignment information if available. If chromosomal information is not available, the rhserver will return the identity of the marker with the highest lod score to the submitted marker, if the lod score is six or greater. If the chromosomal assignment is submitted, the server will return the identity of the marker with the highest lod score with the submitted marker, if the lod score is three or greater and the marker is on the designated chromosome. If there is a lod score of six or greater to a marker on a chromosome other than the designated one this information is returned also.

A high lod to a chromosome other than the designated one may indicate an incorrect chromosomal assignment. Based on the scoring of 322 randomly distributed ESTs, we find that 75% of markers can be linked with a lod of six or greater to the G3 maps. If the user knows the chromosomal assignment of the marker, the probability of detecting linkage with a lod of three or greater increases to 90%. For those wishing to perform more detailed analysis of the G3 maps, the raw typing data are available from the Stanford Human Genome Center's FTP site: `ftp://shgc.stanford.edu`.

References

1. Goss, S. J. and Harris, H. (1975). *Nature*, **255**, 680.
2. Goss, S. J. and Harris, H. (1977). *J. Cell Sci.*, **25**, 17.
3. Goss, S. J. and Harris, H. (1977). *J. Cell Sci.*, **25**, 39.
4. Benham, F., Hart, K., Crolla, J., Bobrow, M., Francavilla, M., and Goodfellow, P. N. (1989). *Genomics*, **4**, 509.
5. Cox, D. R., Burmeister, M., Price, E. R., Kim, S., and Myers, R. M. (1990). *Science*, **250**, 245.
6. Cox, D. R. and Myers, R. M. (1992). *Curr. Biol.*, **2**, 338.
7. Gorski, J. L., Boehnke, M., Reyner, E. L., and Burright, E. N. (1992). *Genomics*, **14**, 657.
8. Tamari, M., Hamaguchi, M., Shimizu, M., Oshimura, M., Takayama, H., Kohno, T., *et al.* (1992). *Genomics*, **13**, 705.
9. Frazer, K. A., Boehnke, M., Budarf, M. L., Wolff, R. K., Emanuel, B. S., Myers, R. M., *et al.* (1992). *Genomics*, **14**, 574.
10. Richard, C. W., Boehnke, M., Berg, D. J., Lichy, J. H., Meeker, T. C., Hauser, E., *et al.* (1993). *Am. J. Hum. Genet.*, **52**, 915.
11. Warrington, J. A., Bailey, S. K., Armstrong, E., Aprelikova, O., Alitalo, K., Dolganov, G .M., *et al.* (1992). *Genomics*, **13**, 803.
12. Burmeister, M., Kim, Suwon, Price, E. R., de Lange, T., Tantravahi, U., Myers, R. M., *et al.* (1991). *Genomics*, **9**, 19.
13. Shaw, S. H., Farr, J. E. W., Thiel, B. A., Matise, T. C., Weissenbach, J., Chakaravarti, A., *et al.* (1995). *Genomics*, **27**, 502.
14. James, M. R., Richard, C. W., Schott, J, J., Yousry, C., Clark, K., Bell, J., *et al.* (1994). *Nature Genet.*, **8**, 70.
15. Barrett, J. H. (1992). *Genomics*, **13**, 95.
16. Walter, M. A., Spillett, D. J., Thomas, P., Weissenbach, J., and Goodfellow, P. N. (1994). *Nature Genet.*, **7**, 22.
17. Foster, J. W., Schafer, A. J., Critcher, R., Spillett, D. J., Feakes, R. W., Walter, M. A., *et al.* (1996). *Genomics*, **33**, 185.
18. Hudson, T. J., Stein, L. D., Gerety, S. S., Ma, J., Castle, A. B., Silva, J., *et al.* (1995). *Science*, **270**, 1945.
19. Gyapay, G., Schmitt, K., Fizames, C., Jones, H., Vega-Czarny, N., Spillett, D., *et al.* (1996). *Hum. Mol. Genet.*, **5**, 339.
20. Cox, D. R., Pritchard, C. A., Uglum, E., Casher, D., Kobori, J. and Myers, R. M (1989). *Genomics*, **4**, 397.

21. Pritchard, C. A., Casher, D., Uglum, E., Cox, D. R., and Myers, R. M., (1989). *Genomics*, **4**, 408.
22. Goodfellow, P. J., Povey, S., Nevanlinna, H. A., and Goodfellow, P. N. (1990). *Som. Cell Mol. Genet.*, **16**, 163.
23. Hunter, K., Housman, D., and Hopkins, N. (1991). *Som. Cell Mol. Genet.*, **17**, 169.
24. Sefton, L., Arnaud, D., Goodfellow, P. N., Simmler, M. C., and Avner, P. (1992). *Mamm. Genome*, **2**, 21.
25. Schmitt, K., Foster, J. W., Feakes, R. W., Knights, C., Davis, M. E., Spillett, D. J., *et al.* (1996). *Genomics*, **34**, 193.
26. Walter, M. A. and Goodfellow, P. N. (1993). *Trends Genet.*, **9**, 352.
27. Westerveld, A., Visser, R. P., Meera-Khan, P., and Bootsma, D. (1971). *Nature New Biol.*, **234**, 20.
28. Ramivez-Solis, R. (1993). In *Methods in Enzymology* (ed. Paul M. Wasserman and Melvin L. Depamphilis), Vol. 225, p. 855. Academic Press, San Diego.
29. Miller, S. A., Dykes, D. D., and Polesky, H. F. (1988). *Nucleic Acids Res.*, **16**, 1215.
30. Ausubel, F. M., Brent, R., Kingston, R. E., Moore, D. D., Seidman, J. G., Smith, J. A., *et al.*. (1995). *Current protocols in molecular biology*. John Wiley & Sons, Inc., New York.
31. Drury, H. A., Clark, K. W., Hermes, R. E., Feser, J. M., Thomas, L. J. Jr., and Donis-Keller, H. (1992). *Biotechniques*, **12**, 892.
32. Boehnke, M., Lange, K., and Cox, D. R. (1991). *Am. J. Hum. Genet.*, **49**, 1174.
33. Lange, K., Boehnke, M., Cox, D. R., and Lunetta, K. L. (1995). *Genome Res.*, **5**,136.
34. Lunetta, K. L., Boehnke, M., Lange, K., and Cox, D. R. (1995).*Genome Res.*, **5**, 151.

5

HAPPY mapping

PAUL H. DEAR

1. Introduction

Most methods for genome mapping rely on some form of cloning to isolate a subfraction of the genome for analysis, whether into yeast or bacterial hosts (as in physical mapping, Chapters 10 and 11), into hybrid cells (as in radiation hybrid mapping, Chapter 4), or as offspring amongst which polymorphic markers segregate (as in linkage mapping, Chapters 1–3).

In all of these methods, drawbacks arise from the way in which the DNA is cloned. Yeast or bacterial clones do not faithfully represent the genome, leading to errors and gaps in contig maps. Radiation hybrid panels require considerable effort to produce even from mammalian species, and complications can arise because the 'donor' fragments are not biologically inert. Linkage mapping is confined to the analysis of polymorphic markers and is limited in its resolution: 1 Mb is the practical limit in man, and the vagaries of meiotic recombination mean that the genetic distances which it reveals may not accurately reflect physical distances.

HAPPY mapping relies on the analysis of minute (roughly haploid) quantities of DNA by the polymerase chain reaction (PCR) (1). It involves no cloning, no hybrid cells, and no reliance on meiotic recombination. By stripping away all of the biological paraphernalia we are left with an *in vitro* method which is efficient, accurate, and versatile.

Figure 1 explains how HAPPY mapping works. Genomic DNA is broken at random and the fragments are dispensed into (typically) 96 aliquots. Each aliquot contains *only a minuscule sample* of DNA—less than a single genome's worth of fragments (1–2 pg in the case of mammalian genomes).

Each of these aliquots is the *in vitro* analogue of a radiation hybrid cell: it contains a large portion of the genome in the form of many fragments. Two DNA markers (sequence-tagged sites, or STSs) may occur together in the same aliquot, or *co-segregate*, either by chance (as, for example, A and Z in the third aliquot shown in *Figure 1*) *or* because they lie on the same DNA fragment (such as A and B in *Figure 1*). However, only co-segregation of the latter type will occur consistently: the linked markers A and B co-segregate

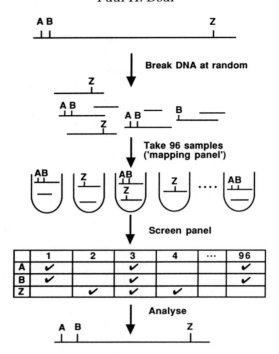

Figure 1. Principle of HAPPY mapping. The example shows the analysis of three markers, though in practice many hundreds can be mapped. Genomic DNA (top) is broken at random, and 96 samples (the 'mapping panel') are taken from this pool of fragments, each sample containing less than one genome's worth of DNA. When these samples are screened by PCR to determine their marker content (table), A and B are found to occur together (i.e. co-segregate) in many of the samples, whereas Z does not co-segregate with either A or B. Analysis of the co-segregation frequency of any two markers allows the distance between them to be determined, leading to a map.

more often than A and Z. Hence, statistical analysis of the co-segregation frequency of any two markers will reflect the distance between them.

If we screen the panel of aliquots for many STSs (up to several thousands) using PCR, we can estimate all of the distances between them and construct a map, free of the distortions which can afflict biological methods: there is no opportunity for chromosome structure or biological activity of the DNA to affect the segregation of markers amongst the aliquots.

HAPPY mapping panels are easy to construct regardless of species and, suitably amplified by PCR, can be used repeatedly as a mapping resource. They give distance estimates which accurately reflect physical distances. The resolution of HAPPY mapping can be varied according to need: a mapping panel made from large fragments will reveal the order and distances of widely spaced markers, whilst one made of smaller fragments will resolve local clusters of markers at the expense of long-range continuity. A HAPPY

reference sequences known to lie within a few kilobases of one another. Unless the DNA is very badly fragmented, these two markers should almost always be detected jointly (since any piece of DNA carrying one must also carry the other); if they are seen to occur independently, it can be deduced that something is amiss with the PCR used to detect them.

In the case of the human genome, we use two nearby sequences in the Duchenne locus as references (2); the primer sequences are given in *Protocol 3*. (As these lie on the X chromosome, they will underestimate by exactly twofold the DNA concentration of male-derived samples.) Reference primers for other species should be designed according to the following guidelines:

(a) The two amplified sequences should be 2–5 kb apart.

(b) Primers should be 18–24 bases long, with GC contents of approximately 50%; they should if possible have one or two Gs or Cs at their 3′ ends.

(c) For each sequence, left and right 'external' PCR primers should be 200–500 bp apart; internal PCR primers should lie within them and be 150–400 bp apart. The lengths of the two internal amplimers should be sufficiently different to be resolved on agarose minigels.

(d) All four external primers should have similar annealing temperatures; there should be no regions of complementarity of > 6 bp between or within any of them. The same applies to the four internal primers.

Protocol 3.   Quantitation 'test panel' of DNA samples by nested PCR

Equipment and reagents

- Thermocycler and compatible microtitre plates and lids.
- 5M0 (5 × PCR buffer, no magnesium): 250 mM KCl (purchased as a saturated 4.02 M solution; Aldrich), 50 mM Tris–HCl pH 8.3 (prepared from pH-adjusted powder; Fluka), 1 mM each dATP, dCTP, dGTP, dTTP (Pharmacia)—prepare under clean conditions and store in 240 μl aliquots for single use, at −20°C (< 1 month) or −70°C.
- *Taq* polymerase, 5 U/μl
- Light mineral oil (Sigma)
- Reference sequence external primers (see text), combined at 20 μM each, as follows:
  D31extL: 5′AGG AAG CTA CAT GGT AGA GG3′
  D31extR: 5′CAA GTT GTC CAA TAT AGA CTG G3′
  D32extL: 5′CAA GAT GTC TCC ATG AAG TTT CG3′
  D32extR: 5′GAA AGT CAA GGG GCA CCT GC3′

- Reference sequence internal primers (see text), combined at 20 μM each, as follows:
  D31intL: 5′GAG GAG AGT TTC TGA ATT TCG3′
  D31intR: 5′GTA TAA TGC CCA ACG AAA ACA CG3′
  D32intL: 5′CCA GCC AAT TTT GAG CAG CG3′
  D32intR: 5′TTT GCC ACC AGA AAT ACA TAC C3′
- MgCl₂ solution, 25 mM (prepare from 1.0 M stock; Sigma)
- Horizontal agarose gel electrophoresis apparatus; we use the Electro-4 system (Hybaid; six 18-well combs can be used in an 8 × 12 cm gel)
- NuSieve 3 : 1 agarose (FMC Bioproducts)
- TBE buffer: (10 × stock is 108 g Tris base, 55 g boric acid, 9.3 g Na₂EDTA in 1 litre)
- SyBr loading dyes: 15% (w/v) Ficoll 400, 0.15 mg/ml bromophenol blue, 4 × SyBr GreenI (FMC Bioproducts)

Protocol 3. *Continued*

- Size standards: ΦX174 RF DNA digested with *Hae*III (HT Biotech), 1 μg/μl in SyBr loading dyes
- UV transilluminator and gel photography system

A. *Phase 1 PCR*

Steps A1–A2 must be performed under clean conditions (Section 2.1).

1. Prepare a reaction master mix as follows:
   - 120 μl 5M0
   - 96 μl 25 mM $MgCl_2$[a]
   - 30 μl external primer mix
   - 6 μl *Taq* polymerase (5 U/μl)
   - 228 μl water

2. Dispense 4 μl of the master mix into each well of the test plate (*Protocol 2*). The same pipette tip(s) may be used throughout, provided that the master mix is added to the negative controls before the remaining (positive) wells. Place the lid on the microtitre tray, and centrifuge briefly at ~ 200 *g*.

3. Thermocycle as follows:
   - 93°C × 3 min

   followed by 25 cycles of:
   - 94°C × 20 sec
   - 50°C × 30 sec[a]
   - 72°C × 60 sec

4. Add 45 μl of water to each well (using a clean pipette tip for each). This should *not* be done in the clean area, but should be performed with care using pipette tips containing an aerosol barrier.

B. *Phase 2 PCR*

1. Prepare a master-mix as follows:
   - 240 μl 5M0
   - 96 μl 25 mM $MgCl_2$[a]
   - 60 μl internal primer mix
   - 12 μl *Taq* polymerase (5 U/μl)
   - 312 μl water

2. Place one drop (~ 30 μl) of mineral oil and 6 μl of master mix in each well of a microtitre plate.

3. Transfer 4 μl of the diluted aqueous phase from each well of the phase 1 PCR plate (step A4) to the corresponding well of the new (phase 2) plate.

4. Place the lid on the plate and thermocycle as follows:
   - 93°C × 3 min

followed by 33 cycles of:

- 94°C 20 sec
- 57°C × 30 sec[a]
- 72°C × 60 sec

5. Add 10 µl of SyBr loading dyes to each sample and centrifuge the plate briefly at ~ 200 g.

6. Ensure that the loading dyes are well mixed with the PCR products before analysing: either leave the plate for > 1h to mix by diffusion, or mix by pipetting when loading the samples.

7. Analyse 10–15 µl samples of each PCR product by electrophoresis through 3% NuSieve 3 : 1 agarose gels in TBE, and photograph the gel using UV transillumination.

C. *Interpretation*

1. None of the negative controls should contain PCR products.[b] Some of the positive samples should contain both reference markers; very few should contain either marker alone. All bands should be strong and unambiguous (see *Figure 2*). The 80 positive samples should contain a total of 20–140 bands.

2. If this is the case, the concentration of the DNA, in genomes per aliquot, is given by $G = -\log_e([2T - N]/2T)$ where T is the number of samples analysed (in this case 80) and N is the *total* number of PCR products seen amongst the samples.

3. If the total number of PCR products in the positive samples is < 20 or > 140, the DNA concentration is too low or too high to estimate reliably, and outside of the suitable range for use as a mapping panel.

4. If the number of samples containing only a single reference band is more than 5% of the number of samples containing both reference bands, and if there is no contamination (step C1) then the PCR is not performing efficiently. Optimize the conditions (try altering the concentration of $MgCl_2$ over the range 1.5–3 mM, or the annealing temperature over the range 50–60°C, for the first and second phase PCRs).[c]

[a] Annealing temperatures and magnesium concentrations may need to be altered for primers other than those shown. Nested PCR can normally be performed under lower stringency than a conventional single-phase PCR, as the use of nested primers eliminates spurious PCR products arising from mis-priming in either phase. Reducing the stringency ensures high efficiency of detection.

[b] Contamination of the first-phase PCRs can be distinguished from contamination of the second phase by repeating the second-phase PCR from the same first-phase plate. Contamination arising in the first phase will recur in the same samples whereas contamination of the second phase will not.

[c] Note that several second-phase PCRs can be performed from one first-phase PCR if necessary, to optimize PCR conditions for the second phase.

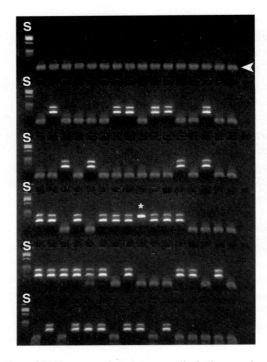

Figure 2. Quantitation of DNA content in a 'test panel' of aliquots. A panel of 16 negative controls (top row) and 80 DNA samples (bottom five rows) was screened by multiplex nested PCR for the two closely linked 'reference' sequences D31 and D32 (see text). These appear as a doublet of bands (D31 upper; D32 lower) in many of the samples. The negative controls contain no PCR products, indicating no contamination of the PCRs. Only one of the samples contains a single reference sequence (asterisk), indicating that the PCR is performing efficiently. Based on a total of 73 bands amongst 80 samples, the mean DNA content can be estimated as 0.61 genomes per sample (see text). The diffuse bands (arrowed) are primer dimer. S: DNA size standards (ΦX174 DNA digested with *Hae*III).

2.3.3 Preparation of large-fragment panels using pulsed-field gels

The larger fragments required for long-range mapping are prepared by irradiation of genomic DNA in agarose strings (to break it at random), followed by size selection on a pulsed-field gel (PFG). The aliquots of DNA are taken directly from the appropriate region of the PFG using a glass capillary to punch out small cylinders of agarose containing the fragments, and amplified directly by PCR.

Very small quantities of DNA—too small to be seen by ethidium fluorescence—are used. Therefore, lanes on either side of the gel are loaded with more concentrated irradiated DNA and with suitable size standards. The sides of the gel are then excised, stained and photographed, and used as a

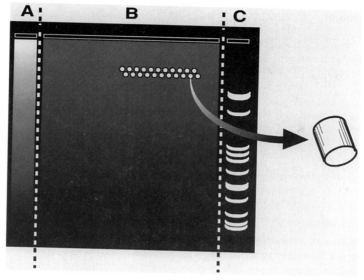

Figure 3. Preparation of large-fragment aliquots using pulsed-field gels. Samples containing (A) high and (B) low concentrations of irradiated genomic DNA, and (C) yeast chromosomal size standards are resolved by pulsed-field gel electrophoresis. The irradiated samples give smears of DNA (shaded) covering a broad size range. The sides of the gel are cut off (dotted lines), stained, and photographed. Using a glass capillary, 96 'plugs' of agarose (right) are punched out of the central part of gel; these plugs, containing DNA fragments of the chosen size, are the aliquots which constitute the mapping panel.

guide for taking the samples from the central, unstained portion of the gel (*Figure 3*). As with the sheared DNA panel, test samples are first evaluated for their DNA content. If suitable, further samples are taken from the gel to make the mapping panel (*Protocol 4*).

Aliquots of DNA prepared in this way typically contain fragments of up to 1.5 Mb. Panels made from such aliquots give a resolution of 50–100 kb, and can detect linkage between markers several hundred kilobases apart (2). We have used such a panel to produce a map of human chromosome 14, with a mean marker spacing of 100 kb (in preparation).

Protocol 4.   Preparation of large-fragment panels using PFGE

Equipment and reagents

- Pulsed-field gel system (Gene Navigator, Pharmacia) and preparative (single-well) comb.
- HCl, approximately 2 M
- Agarose strings containing genomic DNA (*Protocol 1*)
- Yeast chromosomal DNA size standards (Bio-Rad)
- 0.5 × TBE buffer (*Protocol 3*; prepare using HPLC grade water)
- Bio-Rad chromosomal-grade agarose
- Plastic Petri dishes, 35 mm diameter

Protocol 4. *Continued*

- Gamma irradiator (e.g. Gravatom RX30/ 55M ^{137}Cs source)
- SaranWrap food film
- Ethidium bromide solution (10 mg/ml stock)
- TE: 10 mM Tris–HCl pH 7.5, 1 mM EDTA, prepared with HPLC grade water
- UV transilluminator and gel photography system

- Transparent plastic sandwich box (> 15 cm × 12 cm base area)[a]
- Silicone tubing, ~ 60 cm length, ~ 4 mm bore
- Glass capillaries ('Microcaps', Drummond Scientific), 7 µl, 10 µl, 20 µl, and 30 µl volumes
- Reagents and equipment as for *Protocol 3*

A. *Gel preparation and running*

1. To destroy any DNA present in the gel apparatus, place the gel tray, gel former, gel comb, supporting clips, and electrode array in the gel tank. Fill with 2M HCl, run the recirculating pump for a few seconds and leave for 45 min–1 h.[b]

2. Draw one complete, unbroken agarose string containing 10^5 cells/ml into a 1ml disposable pipette. Expel the string (plus lysis mix) into a 35 mm plastic Petri dish and remove the surplus lysis mix. Place the lid on the Petri dish. Transfer one string containing 10^6 cells/ml (only 1–2 cm will be required) into a second dish.

3. Irradiate the strings (with the lids on the Petri dishes) with 30–40 J/kg of gamma rays. Store at 4°C until required.

4. When the tank has been decontaminated (step 1), remove the HCl solution. Fill the tank (containing electrode, gel tray, etc.) with tissue culture grade water and run the recirculating pump for a few seconds.

5. Remove the water, and fill the tank with 1 litre of 0.5 × TBE. Run the recirculating pump for a few seconds before discarding the liquid. Remove the gel tray, former, comb and supporting clips and place them on a clean sheet of SaranWrap. Fill the tank with 3 litres of 0.5 × TBE and replace the lid.

6. Dry the gel tray, former, comb, and clips with clean tissues. Prepare 120 ml of a 1% agarose solution in 0.5 × TBE and cast the gel using 90 ml. Cover with SaranWrap (forming a 'tent' over the gel comb) while the gel sets. Place remaining agarose in an incubator at 70°C.

7. When the gel has set, remove the comb and gel former. Using disposable pipette tips, carefully nudge the low DNA-content irradiated agarose string from the Petri dish onto the gel surface. Taking great care not to break it, use pipette tips to straighten it and bring it adjacent to the sample well, centred from left to right, then nudge it into the well.

8. In the same way, load a ~ 1 cm segment of the high DNA-content irradiated string into the left-hand side of the sample well, and yeast

chromosomal size standards into the right-hand side. Ensure that no fragments of agarose from the samples remain on the gel.

9. Seal the samples into the well with agarose retained from step 7 and leave for 3–5 min.

10. Place the gel in the tank and run for 14 h with a pulse time of 100 sec at 180 volts ($\approx$ 7 V/cm) at 12–17 °C.

B. *Gel staining and preparation of test plate*

1. Using a sterile scalpel blade, cut off the sides of the gel and stain them in ethidium bromide solution (0.6 μg/ml) for 1 h, then destain for 1 h.

2. Place the remainder of the gel in a transparent plastic sandwich box and cover with 1–2 cm TE; replace the lid and leave at 4 °C for at least 1 h. Pour off the liquid, and replace with 0.1 × TE. Leave at 4 °C for at least 1 h, or preferably overnight to remove residual TBE buffer.

3. Photograph the stained parts of the gel using UV illumination; include a size scale in the photograph. The yeast chromosomes should be well resolved; refer to the supplier's data sheet for sizes. The high DNA-content irradiated sample should show a smear covering a broad size range up to and beyond the largest yeast chromosome.

4. Pour off the 0.1 × TE from the unstained part of the gel. Draw a straight line across the underside of the transparent box, parallel with the sample well and approximately midway between the well and the position of the largest yeast chromosome (as judged from the gel photograph).

5. Working in the clean area, fit two aerosol-barrier pipette tips into either end of the length of silicone tubing. Insert one of them point first (to act as a mouthpiece), the other base first. Trim a short section from the end of the second tip, so that a 10 μl Microcap can *just* be pushed into it leaving 10–15 mm of the capillary protruding.

6. Stab the capillary into the gel in the region above the sample well (this will serve as a negative control). Drag the capillary slightly to one side to break the airlock and remove it; it should contain an agarose 'plug' equal in length to the thickness of the gel. Blow gently into the mouthpiece to expel the plug into one well of a thermocycler-compatible microtitre plate.

7. Remove further plugs from above the sample well to give 24 negative controls (three columns in the microtitre plate). Take 32 similar plugs from the DNA-containing part of the gel, working horizontally along the marked line. This should be done within 5–10 min, before the first plugs start to dry out. Place a lid on the plate and store at

Protocol 4. *Continued*

    4°C. Place the lid on the box containing the gel, seal with tape, and store at 4°C.

8. Prepare a master mix of:
   - 70 µl 5M0
   - 56 µl 25 mM $MgCl_2$
   - 17.5 µl external primer mix
   - 3.5 µl *Taq* polymerase (5 U/µl)
   - 168 µl water

9. Add 4.5 µl of this to each well containing an agarose plug, adding it to the negative controls first. Add one drop of mineral oil to each well.

10. Perform thermocycling[c], second-phase PCR, gel electrophoresis and analysis as in *Protocol 3* (steps A3 onwards); in step C2, substitute $T = 32$ in the calculation.

C. *Preparation of mapping panel*

1. If the negative controls are satisfactory and the positive samples are found to contain 0.3–0.7 genomes of DNA per sample, take a further 96 similar samples from the same region of the gel. Overlay each with a drop of mineral oil. Store at −20°C or proceed directly to pre-amplification and screening (section 3).

2. If the samples contain far too little or far too much DNA, another gel must be run, using an irradiated agarose string containing more or fewer cells.

3. If the samples are within a factor of two of the correct DNA content, samples for the mapping panel may still be taken using a capillary with an appropriately larger or smaller bore area.[d]

[a] Newly purchased boxes, supplied with the lids fitted, can generally be regarded as free of contamination. Otherwise, decontaminate before use by filling with 2 M HCl and leaving for >1 h, then rinse thoroughly with double-distilled water.
[b] This treatment is not kind to the gel tank, and should not be prolonged. If it is repeated several times, check the condition of the contacts between the tank lid, electrode connection bars, and electrode, and clean with fine abrasive paper if necessary.
[c] As this first-phase PCR contains dilute agarose, it may set if kept at 4°C. Either add the 45 µl of water (*Protocol 3*, step A4) as soon as the last cycle of first-phase PCR is complete; or add the water to the set PCRs and leave for > 30 min to allow PCR products to diffuse out of the agarose.
[d] Bore area in mm² = capillary volume in microlitres divided by capillary length in millimetres.

3. Panel pre-amplification and screening

The mapping panels prepared in *Protocols 2* or *4* must now be screened for marker sequences by PCR. This is done in two stages. The first (pre-

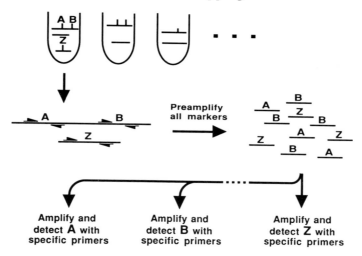

Figure 4. General procedure for screening aliquots. Each aliquot is first *pre-amplified* to give many copies of each marker which it contains (only one aliquot is shown). This is done either using a cocktail of primers for all markers ('nested' PCR approach) or with repeat element primers which amplify many parts of the genome ('IRS-PCR'). This pre-amplified material can then be screened in successive PCRs for each individual marker in turn, using marker-specific primers.

amplification) amplifies *all* marker sequences present in the aliquots. The second (screening) amplifies chosen markers sufficiently to detect their presence on agarose gels (*Figure 4*).

We have successfully applied two different approaches to perform these steps. The first (Section 3.1) is used for mapping a relatively small number of known markers; pre-amplification is done using a multiplex PCR with a cocktail of primers for all the markers, and subsequent screening is performed by PCR with 'nested' primers internal to those used in the pre-amplification (2, 3). The second approach (Section 3.2) uses interspersed repeated sequence-mediated PCR (IRS-PCR) to pre-amplify many sequences throughout the genome; screening is then done using sequence-specific PCR primers for individual markers.

A third approach, using a 'whole-genome' PCR to pre-amplify *all* sequences, is discussed in Section 3.3.

3.1 Pre-amplification and screening using nested PCR

In this approach, the pre-amplification and marker screening stages are similar to the first- and second-phase PCRs used in the quantitation of the test panels (*Protocol 3*). Four primers are required for each marker: two external primers for pre-amplification and two internal (nested) primers for screening. All external primers are used simultaneously in a multiplex first-phase PCR for the pre-amplification (*Protocol 5*). At least nine external primer pairs can be used in combination without any loss in PCR efficiency (2). External primers may include degenerate regions, or may be targeted against conserved regions of members of gene families, allowing many family members to be pre-amplified (3). The products of this amplification are then re-amplified to detectable levels using the internal primers, either separately or again in a multiplex format (*Protocol 6*). The design of external and internal primers should follow the guidelines in Section 2.3.2.

Protocol 5. First-phase amplification for screening by nested PCR

Equipment and reagents
- Thermocycler and compatible microtitre plates and lids.
- 5M0 (*Protocol 3*)
- MgCl$_2$ solution (*Protocol 3*)
- *Taq* polymerase, 5 U/μl
- Light mineral oil (Sigma)
- External primers, as primer pairs containing 20 μM of each primer

Method

Steps 1 and 2 should be performed under clean conditions (Section 2.1).

1. Prepare a reaction mix of:
   - 120 μl 5M0
   - 96 μl MgCl$_2$ solution[a]
   - 30 μl of each external primer pair[b]
   - 6 μl *Taq* polymerase

   Add water to a total volume of 480 μl (for amplification of 1 μl aliquots of sheared DNA from *Protocol 2*) or 540 μl (for amplification of agarose plugs from *Protocol 4*).

2. Dispense 4 μl of the mixture into each well containing 1 μl aliquots of sheared DNA, or 4.5 μl into each well containing agarose plugs. Place a lid on the plate and centrifuge briefly at ~ 200 *g*.

3. Transfer to the thermocycler and cycle as in *Protocol 3*, step A3.[a]

4. Add 45 μl of water to each well, using clean tips for each. The plate may be stored at $-20\,^\circ$C.

---

[a] These are low stringency annealing conditions to maximize detection efficiency (see footnote *a* of *Protocol 3*). The magnesium concentration and/or annealing temperature may need to be optimized for some primer combinations.

[b] If more than eight external primer pairs are used, the concentrations of the primer stocks will need to be higher, and smaller volumes of each primer added. If this is not possible, scale up the volume of the reaction mix. The final concentrations in the PCR (including the 1 μl or 0.5 μl volume of the sheared DNA aliquot or agarose plug) should be 0.2 $\times$ 5M0, 4 mM MgCl$_2$, 0.05 U/μl *Taq* polymerase, and 1 μM of each primer.

Protocol 6. Second-phase amplification for screening by nested PCR

Equipment and reagents

- As for *Protocol 3*, except that the only primers required will be the internal primers for the relevant markers, combined as pairs containing 20 μM each primer.

Method

Some or all of the second-phase PCRs may be multiplexed if preferred. Internal primers to be multiplexed must operate under identical PCR conditions and give products resolvable on gels. Alternatively, each marker may be detected by an independent second-phase PCR from the same first-phase plate.

1. Prepare a reaction mix of:
   - 240 μl 5M0
   - 96 μl MgCl$_2$ solution[a]
   - 60 μl internal primer pair (or 60 μl each pair, for multiplexed screening)
   - 12 μl *Taq* polymerase

   Adjust to a total volume of 720 μl with water.

2. Place one drop of mineral oil and 6 μl of reaction mix in each well of a thermocycler-compatible microtitre plate (the second-phase plate).

3. Transfer 4 μl of the diluted first-phase PCR products (*Protocol 5*, step 4) to the corresponding wells of the second-phase plate, using a clean pipette tip for each. Put the lid on the plate and centrifuge briefly at ~ 200 *g*.

4. Perform thermocycling[a] and gel electrophoresis as in *Protocol 3*, steps B4–B7.

[a]See footnote *a*, *Protocol 3*.

3.2 Pre-amplification and screening using IRS-PCR followed by sequence-specific PCR

IRS-PCR uses primers complementary to the ends of repetitive sequence motifs to amplify regions of unique sequence which are flanked by repeat elements. *Alu* element-mediated PCR (4) can be used to pre-amplify aliquots containing human DNA, which are then screened using PCR with specific primers directed against individual inter-*Alu* sequences. As *Alu* PCR pre-amplifies several tens of thousands of distinct inter-*Alu* sequences by a factor of > 10^7-fold, very large numbers of markers may be mapped.

IRS-PCR suffers from the limitation that only those sequences flanked closely by repeat elements will be pre-amplified and can hence be mapped.

Although this is adequate to give dense map coverage (amplifying one sequence every few tens of kilobases on average), it does mean that any *arbitrarily* chosen sequence is unlikely to be represented, and hence cannot be mapped. Our solution to this has been to choose marker sequences from *Alu*-amplified human DNA so that, by definition, all markers are inter-*Alu* sequences and will be represented in the *Alu* PCR pre-amplified aliquots.

The following sections deal with the design of suitable STS markers for mapping by this approach; the *Alu*-mediated pre-amplification; and the screening procedure.

3.2.1 Selection of STSs for mapping with *Alu* pre-amplified panels

As mentioned, STSs for mapping by this route must be derived from inter-*Alu* regions of the human genome. The most convenient way to ensure this is to use PCR to amplify human inter-*Alu* sequences from genomic DNA, large-insert clones, or somatic cell hybrids (*Protocol 7*), and to clone the products. STSs are then derived by sequencing randomly chosen clones from this 'inter-*Alu*' library. We have used this approach only for human, but there is no reason why analogous procedures, using appropriate species-specific repeat motifs, should not be applied to other species.

Protocol 7. Production of inter-*Alu* libraries as sources of human STSs

Equipment and reagents

- Thermocycler and compatible microtitre plates or tubes
- 5M0 (*Protocol 3*)
- MgCl$_2$ solution (*Protocol 3*)
- *Taq* polymerase, 5 U/μl
- *Alu* primer mix: 20 μM each of Alu1 and Alu2 (4):
  Alu1: 5'GGA TTA CAG GYR TGA GCC A3'
  Alu 2: 5'RCC AYT GCA CTC CAG CCT G3'
  (Y = C/T; R = G/A)

- Template DNA and appropriate negative control (see steps 2, 3 below)
- Light mineral oil (Sigma)
- SyBr loading dyes (*Protocol 3*)
- Horizontal agarose gel electrophoresis system, agarose, and TBE gel buffer (*Protocol 3*)
- UV transilluminator and gel photography system

Method

1. Prepare two reaction mixtures as follows:
   - 10 μl 5M0
   - 2 μl MgCl$_2$ solution
   - 2.5 μl *Alu* primer mix
   - 0.25 μl *Taq* polymerase

2. To one mixture, either:
   (a) Add 60 ng human genomic DNA (for isolating genome-wide inter-*Alu* sequences).

(b) Add 60 ng DNA from a hybrid cell line containing human chromo-some(s) or fragment(s) (for isolation of regional inter-*Alu* sequences).[a]

(c) Add 1000 human flow-sorted chromosomes (FSCs; for isolation of chromosome-specific inter-*Alu* sequences).[b]

(d) Touch a sterile wooden toothpick onto a bacterial or yeast colony containing a large-insert human clone, then swirl the end of the toothpick in the mixture (for isolation of inter-*Alu* sequences from clones).

3. Use the second reaction mixture as a negative control, substituting either water (as a control for genomic DNA, flow-sorted chromo-somes, or clones) or 60 ng of the host species DNA (as a control for hybrid cell DNA) in place of the template.

4. Adjust the final volumes of each reaction to 50 μl with water and over-lay with one drop of mineral oil.

5. Thermocycle as follows:

   • 93°C × 5 min

   followed by 22 (for human or hybrid DNA) or 33 (for FSCs or clones) cycles of:

   • 94°C × 20 sec
   • 65°C × 30 sec
   • 72°C × 2 min

   followed by:

   • 72°C × 4 min

6. Remove a 5 μl sample from each reaction and add 5 μl of SyBr load-ing dyes. Analyse by electrophoresis in 3% agarose in 0.5 × TBE. Store the remainder of the PCR product at −20°C for later cloning (see text).

7. The negative control should contain no PCR products. Complex sources (human DNA, FSCs, or hybrids containing large pieces of human chromosomes) should give a smear of PCR products from < 200 to > 1500 base pairs. Clones containing human DNA will give a variable number of bands (usually 0–3) depending on their *Alu* con-tent. Those giving no products of between 100 and 2000 base pairs should be rejected.

---

[a] Chapter 6 gives details of sources of many suitable hybrid DNAs.
[b] Flow-sorted chromosomes are seldom more than 95% pure (less in many cases); hence, a proportion of markers derived from these sources will lie on other chromosomes. It is prefer-able to have the chromosomes sorted directly into the PCR reaction vessel, as they may adhere to tube walls.

Products from these PCRs may be cloned in any convenient system (see, for example, ref. 5). We routinely blunt-end clone into M13-derived vectors. Randomly selected clones from these libraries should then be sequenced using any conventional approach; we use dye-primer or dye-terminator cycle sequencing followed by analysis on Applied Biosystems 373A automated sequencers, according to protocols recommended by ABI. PCR primers are then designed from these sequences according to the following guidelines:

(a) Avoid sequences with internal *Alu* primers; these are often the result of co-ligation.

(b) Primers should amplify regions of 100–300 bp. Avoid placing primers within ~ 50 bp of the start of the sequence or (if the full length of the sequence has been determined) within 50bp of its end. (Primers lying in these weakly-conserved regions may cross-react with other inter-*Alu* sequences.)

(c) Primers should be 18–24 bases long with GC contents approximating 50%, and preferably with one or two GC bases at their 3' ends.

(d) Avoid runs of > 5 of any one base or > 4 dinucleotide repeats.

(e) Avoid complementarity of > 6 bases within or between primers, particularly at the 3' end.

3.2.2 *Alu* pre-amplification of mapping panels

The *Alu* PCR used to pre-amplify the HAPPY mapping panel (*Protocol 8*) is closely similar to that used for making inter-*Alu* libraries in Section 3.2.1. To achieve as great as possible a degree of pre-amplification, *two* rounds of *Alu* PCR are used. The aliquots are *Alu* amplified, and the products are then re-amplified in a secondary *Alu* PCR; the products of this reaction are themselves diluted further for use as templates in the screening reactions. This gives enough material for the subsequent screening of $> 10^4$ markers.

Protocol 8. *Alu* preamplification of aliquots

Equipment and reagents
- Thermocycler and compatible microtitre plates
- 5M0 (*Protocol 3*)
- MgCl₂ solution (*Protocol 3*)
- *Alu* primer mix (*Protocol 7*)
- *Taq* polymerase, 5 U/μl
- Light mineral oil (Sigma)
- Rigid microtitre plates and lids (for storage of aliquots; e.g. Corning, Cat. No. 25850)

A. *Primary* Alu *PCR*

Steps 1 and 2 should be performed under clean conditions (Section 2.1).

1. Prepare a reaction mixture of:
   - 120 μl 5M0

- 24 μl MgCl$_2$ solution
- 30 μl *Alu* primer mix
- 6 μl *Taq* polymerase

Add 300 μl of water (for pre-amplification of 1 μl aliquots of sheared DNA from *Protocol 2*) or 360 μl of water (for pre-amplification of DNA in agarose plugs from *Protocol 4*).

2. Dispense either 4 μl (for amplification of sheared DNA aliquots) or 4.5 μl (for amplification of DNA in agarose plugs) of the mixture into each well of the microtitre plate from *Protocol 2* or *4*.

3. Place a lid on the plate and centrifuge briefly at ~ 200 *g*.

4. Thermocycle as follows:

- 93°C × 5 min

followed by 22 cycles of:

- 94°C × 20 sec
- 65°C × 30 sec
- 72°C × 2 min

followed by:

- 72°C × 4 min

5. If amplifying DNA from agarose plugs, proceed *immediately* to part B, as the melted agarose in the PCR reaction may set if left at 4°C. If amplifying from liquid (sheared DNA) samples, the reaction may be left overnight at 4°C if necessary.

B. *Dilution and storage of primary* Alu *PCR products*

1. Transfer the microtitre plate from step A4 to the clean work area, placing it on two sheets of absorbent paper towel. Remove the lid and dispose of it immediately, to minimize the risk of contaminating the clean area. Change gloves immediately, and after any other occasion on which you touch the microtitre plate.

2. Using fresh tips for each well, add 100 μl of water per well. Mix gently by pipetting up and down once, without introducing air bubbles.

3. Transfer 20 μl of the aqueous phase into each of five Corning microtitre plates, to which 80 μl of water per well has been added. Overlay each well with one drop of mineral oil (to prevent sublimation after freezing), and store at −70°C.

4. Dispose of the primary *Alu* PCR plate and absorbent paper and change gloves.

C. *Secondary* Alu *PCR*

All steps except step 4 should be performed under 'clean' conditions.

Protocol 8. *Continued*

1. Prepare a reaction mixture of:
   - 240 μl 5M0
   - 48 μl MgCl$_2$ solution
   - 60 μl *Alu* primer mix
   - 6 μl *Taq* polymerase (5 U/μl
   - 366 μl water

2. Place one drop of mineral oil and 6 μl of reaction mixture in each well of a thermocycler-compatible microtitre plate (the reaction plate).

3. Transfer 4 μl of the diluted primary *Alu* PCR products (step B3) from one of the replica plates into the corresponding wells of the reaction plate. Use fresh tips for each well. Place the lid on the plate and centrifuge briefly at ~ 200 *g*.

4. Thermocycle as for step A4, but for 25 rather than 22 cycles.

5. Place the plate on two layers of absorbent paper in the clean area. Remove and discard the lid, and change gloves.

6. Add 90 μl of water to each well, mixing as in step B2.

7. Transfer 30 μl of the aqueous phase into each of three Corning microtitre plates, containing 170 μl of water per well.

8. Each of these three plates now contains sufficient template (*Alu* pre-amplified aliquots) for ~ 50 screening reactions (section 3.2.3). Store at −70 °C (for long-term storage) or at −20 °C.[a]

[a] Further secondary *Alu* PCRs are performed as necessary, on further 4 μl aliquots of the diluted primary *Alu* PCR products (step B3). To avoid repeated thawing/freezing of the primary *Alu* PCR products, split one of the five replica plates (step B3) into further subfractions, so that each is only thawed < 5 times.

3.2.3 Screening of STSs against *Alu* pre-amplified panels

Screening of *Alu* pre-amplified panels with STSs designed as in Section 3.2.1 is done using straightforward PCR under routine laboratory conditions (*Protocol 9*). We do not multiplex the screening reactions as it precludes the use of the small, high sample-density agarose minigels, since these cannot resolve multiple STSs reliably.

Protocol 9. Screening of *Alu* pre-amplified mapping panels

Equipment and reagents

- Thermocycler and compatible microtitre plates
- 5M0 (*Protocol 3*)
- MgCl$_2$ solution (*Protocol 3*)
- *Taq* polymerase, 5 U/μl
- Marker-specific primers (see text) prepared as primer pairs containing 20 μM each oligo
- *Alu* pre-amplified template (from *Protocol 8*)
- Light mineral oil (Sigma)

- SyBr loading dyes (*Protocol 3*)
- Size standards (*Protocol 3*)
- UV transilluminator and gel photography system

- Horizontal agarose gel electrophoresis apparatus, agarose, and TBE gel buffer (*Protocol 3*)

Method

1. Prepare a reaction mix of:
   - 240 μl 5M0
   - 48 μl MgCl$_2$ solution[a]
   - 60 μl marker-specific primer pair
   - 6 μl *Taq* polymerase
   - 365 μl water

2. Add one drop of mineral oil and 6 μl of reaction mix to each well of a microtitre plate (the 'screening plate').

3. Centrifuge the template plate briefly at ∼ 200 *g* to dislodge air bubbles which tend to adhere to the bottoms of the wells (they can result in pipetting errors).

4. Transfer 4 μl of the template into the corresponding wells of the screening plate.[b]

5. Place the lid on the screening plate and centrifuge briefly at ∼ 200g. Re-freeze the template plate and store at -20°C[c]

6. Transfer the screening plate to the thermocycler and cycle as follows:
   - 93°C × 3 min

   followed by 33 cycles of:
   - 94°C × 20 sec
   - 60°C × 30 sec[a]
   - 72°C x 1 min

7. Add 8 μl of SyBr loading dyes to each well of the plate.

8. Centrifuge the plate briefly at ∼ 200 *g*.

9. Ensuring that the samples and loading dyes are well mixed, analyse 10–15 μl of the products on 3% agarose minigels in 0.5 × TBE. Load 5 μl of size-standards in some wells (see *Figure 5*). Photograph using UV transillumination.

[a] These conditions are optimal for primers designed as in Section 3.2.1, but may need to be modified for primers of different compositions.

[b] If several screening reactions (i.e. for several different markers) are being prepared in parallel, use a multistep eight-channel pipette to dispense into each plate in turn: take up liquid from the first column of the template plate, then dispense into the first column of each screening plate. Then change tips, and repeat for the second column, etc. Use of a robotic workstation (e.g. Biomek 1000, Beckman Instruments) for adding template greatly accelerates screening and reduces the risk of error.

[c] To refreeze the template plate, place it in a non-airtight plastic box containing a little dry ice, and place the box in the freezer. This ensures that the plate freezes rapidly, and reduces evaporation/sublimation.

14h1717   **14h1728**   **14h2100**

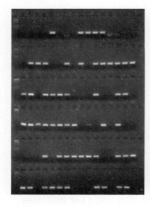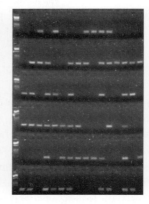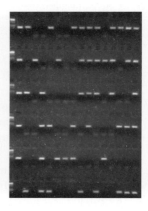

Figure 5. Examples of marker screening. A panel consisting of 96 samples of large DNA fragments, pre-amplified using *Alu* PCR, was screened for the presence of three STSs (14h1717, 14h1728 and 14h2100) as described in the text. Each gel shows the corresponding 96 PCR products, in six rows of 16; the leftmost lane in each row contains DNA size standards (ΦX174 DNA digested with *Hae*III). The distributions of STSs 14h1717 and 14h1728 are very similar, reflecting almost complete co-segregation and hence tight linkage between these two markers. STS 14h2100 gives a very different distribution: it does not co-segregate with either of the first two any more than would be expected by chance, and hence must be remote from them.

3.3 Pre-amplification using 'whole-genome' PCR

Pre-amplification using a 'whole-genome PCR' (WGPCR), capable of amplifying any and all sequences in the aliquots, would make it possible to map *any* chosen sequence, rather than those defined by specific external primers (Section 3.1) or those flanked by repeat elements (Section 3.2). However, none of the existing WGPCR protocols is entirely satisfactory for this purpose. In our hands, those which rely upon degenerate primers (6–8) either amplify only a subset of sequences, or amplify all sequences but only by a factor of a few hundredfold or less.

We have used WGPCR based on a wholly degenerate primer (7) to amplify the total sequence content of HAPPY mapping aliquots by > 1000-fold. However, this still does not provide sufficient material for more than a very few marker screenings, unless one then uses nested PCR to detect small numbers of marker copies (2). We hope soon to announce a more powerful and efficient form of WGPCR for use in HAPPY mapping but, at present, IRS-PCR provides the best compromise between the need to pre-amplify many sequences and the requirement for their efficient amplification.

4. Data entry

Data from the gel photographs produced by *Protocols* 6 or 9 are entered into the computer to give a record of the presence or absence (or ambiguous status) of each STS in each of the aliquots. We use our own software (*Gelenter*, running on Macintosh computers; P.H.D. unpublished) to perform data entry.

Screening of a marker should be repeated (possibly with altered PCR conditions) if it gives more than six ambiguous bands, or the marker rejected. (Rejected markers may still give enough data to allow them to be placed on the map, but should be excluded from the initial mapmaking so that they cannot disrupt the placement of other, reliably typed, markers.)

The number of positive aliquots should be similar for all markers; those with unusually high numbers of positives probably represent multicopy sequences and should be rejected. Very low numbers may be due to inefficient detection, or to a polymorphic marker of which only one allele is detected. Markers should be accepted only if the number of positives is within the range $p \pm 2\sqrt{(p - [p^2/n])}$, where p is the mean number of positives per marker (averaged over all markers), and n is the number of aliquots (usually 96) in the panel.

5. Data analysis and mapmaking

Analysis of HAPPY mapping data relies on the same principle as that of conventional linkage mapping or radiation hybrid data:

(a) For any two markers, the distance between them is estimated from their degree of co-segregation amongst the aliquots of the mapping panel: closely linked markers will co-segregate often and give very similar screening patterns; distant markers will not co-segregate and will have very different patterns (*Figure 5*).

(b) When all pairwise distances (that is, the distances between each marker and each other marker in the data-set) have been estimated, a map is constructed. As with all maps based on segregation analysis, marker order and spacing is determined not absolutely, but within known confidence limits.

What follows is only an outline of the mapmaking process. We have a number of computer programs to aid this process, which we will make available to interested users (contact phd@mrc-lmb.cam.ac.uk).

5.1 Pairwise distance estimates

The distance between any two markers is expressed as a value known as θ (theta) between 0 and 1, representing the probability that a radiation- or shearing-induced break will occur between the two markers in any one instance. Associated with the value of θ is a 'lod' (log of odds) score, reflecting the certainty of this estimate of the distance. Two closely-linked markers will have a small θ (they are rarely broken apart) and a high lod score (we are very sure that they are linked). Two widely-spaced markers will have a θ close to 1 (they are almost always broken apart) and a near-zero lod score (there is little or no evidence that they are linked). As an example, the lod score between markers 14h1717 and 14h1728 of *Figure 5* is 20 (θ = 0.07), meaning that the observed pattern of co-segregation is 10^{20}-fold more likely to be due to the markers being linked (at a distance of θ = 0.07) than due to their being unlinked.

Lod scores are invariably calculated by computer programs rather than manually. We have several programs running on Macintosh computers to perform lod score calculations between pairs of markers or between the members of large groups of markers. A good introductory discussion of lod scores will be found in ref. 9.

5.2 Map construction

5.2.1 Division of markers into linkage groups

Before mapmaking begins, it is essential to ensure that all markers are indeed linked to one another: no two consecutive markers must lie so far apart that they are unlinked. As a first step, therefore, the markers in the data-set are analysed into 'linkage groups'. This is done using our own software (*HAPlg*; P.H.D. unpublished). One marker is chosen at random as a 'founder' for the linkage group, and all markers which are linked to it (with a lod score above a certain threshold) are found and added to the group. When this has been done, all markers linked to these newly added markers are also found and added to the linkage group. This process is continued until no more markers can be found which link to any member of the linkage group. If any markers remain, they are then put into separate linkage groups by the same process.

It is important, when using *HAPlg*, to choose the correct threshold lod score for deciding whether two markers are truly linked. Too low a threshold will mean that markers which are in reality unlinked will be included in the same linkage group, confusing the mapmaking process. In general, the threshold lod score should be equal to the log of the square of the number of markers in the data-set. For example, if the data-set comprises 1000 markers, then a lod of 6 should be the minimum for declaring that two markers belong to the same linkage group. This is because if N markers are analysed, then there will be approximately N^2 (actually $[N^2 - N]/2$) pairwise distance estimates between them. Hence, the probability of *wrongly* declaring linkage between any pair of markers must be less than 1 in N^2; a lod score of $\log(N^2)$ ensures that this is the case.

Once linkage between all markers has been verified in this way (or the markers have been divided into two or more linkage groups) a map is built for each linkage group as described in the following section.

5.2.2 Map building

In an ideal world, construction of the map from the complete set of pairwise distance estimates between linked markers (calculated as in Section 5.1) would be straightforward. For example, if the distances between markers A, B, and C were A − B = 0.3, B − C = 0.4, and A − C = 0.7, then the correct marker order (ABC) and distances could be unambiguously determined.

As with all forms of segregation analysis, however, two complications arise. The first is that the distances represented by θ values are not linear distance measures, but are probabilities of breakage; hence, they may not be simply summed. (For example, θ can never exceed 1, even between remote markers.) The second is that the θ values are themselves only *estimates*. Hence, for a large data-set is it unlikely that the correct marker order and spacing will agree perfectly with all of the estimated intermarker distances. Instead, the 'correct' map is deemed to be that which best agrees with the total set of intermarker distance estimates.

For a small number of markers, the best marker order and spacing can be found by exhaustive searching of all possibilities. This approach, however, becomes impractical for large numbers of markers: for 100 markers, $\sim 5 \times 10^{157}$ candidate orders would have to be evaluated.

A number of programs have been devised to address this problem in the closely related contexts of meiotic linkage and radiation hybrid mapping (see Chapters 1 and 4, and Appendix 2). We favour the approach known as *distance geometry* for the solution of HAPPY maps containing many markers. The program which we use (*DGmap*) (10) is not generally available, but its author has agreed to make copies available to interested users (contact wnewell@oxmol.co.uk). Distance geometry works by first placing the markers in multidimensional space (which can accommodate all of the pairwise distance estimates, regardless of their errors), and then 'squashing' this arrangement progressively down to a single dimension (a line), maintaining the optimal placement of markers as it does so. Because the map is derived from the distance estimates (rather than candidate maps being tested against the distance estimates), the algorithm is extremely fast and can solve maps involving large numbers of loci.

For *DGmap* (or any other map-solving program) to yield the correct order and spacing, it must be given information on the relationship between the θ values (in which the pairwise distances are expressed) and physical distances. This relationship, known as the 'mapping function', depends upon the distribution of fragment sizes in the original mapping panel and is influenced by factors such as whether the fragments were size selected (for instance by PFG) or not. Fortunately, this problem is ameliorated in densely populated maps, since small values of θ are linearly proportional to physical distances. We find that *DGmap* yields similar results regardless of whether the 'linear' or 'RH' (logarithmic) mapping functions are used. Our own software reformats the pairwise distance estimates into a form suitable for use by the *DGmap* program, which runs under Unix on a DEC Alpha workstation.

DGmap gives both a graphical output and a file giving the calculated marker order and spacing. In addition, it gives confidence values for the placement of each marker, and can draw attention to any markers whose placement is serious doubt and which should be re-examined.

5.2.3 Conversion of map distances to physical distances

The output of *DGmap* is a map in which intermarker distances are expressed in units reflecting the probability of breakage between consecutive markers. As mentioned in Section 5.2.2, the relationship between these units and physical distance (kilobases) is known as the mapping function, and depends upon the mean size and distribution of the fragments from which the panel was prepared.

For a densely populated map, in which intermarker distances are a small fraction (less than about 0.4) of the mean fragment size, intermarker distances as indicated by *DGmap* are linearly proportional to physical distances. In this case, all that is required is a scaling factor between map units and kilobases. This is provided by including two or more markers, whose separation is known, in the analysis. It is also possible to calculate the scaling factor from the known size distribution of the fragments used in making the panel, and to correct for the departures from linearity which arise at very small or large intermarker distances (2).

6. Conclusions and future developments

In its present form, HAPPY mapping allows rapid distortion-free mapping either of small clusters of markers (using the nested PCR approach) or of unlimited numbers of repeat element-flanked markers (using the IRS-PCR approach) at any chosen level of resolution from 10 kb to > 1 Mb. Mapping panels are easily prepared, making comparative mapping straightforward. The use of an efficient 'whole-genome' PCR approach will shortly make it possible to map unlimited numbers of STSs of any type.

As in many STS-based mapping strategies, a limiting step is often the design and synthesis of suitable PCR primers for each marker. We have recently found that *Alu* amplified HAPPY mapping aliquots may be screened by hybridization (rather than by PCR) with *Alu* PCR products derived from large-insert bacterial clones (M. B. Piper, unpublished). This opens the possibility of placing such clones directly on a HAPPY map, without the need to obtain sequence information and design STSs from them.

Acknowledgements

I would like to thank the many colleagues who have given advice and encouragement. This work was supported by the Medical Research Council. Additional funding was provided by Techne (Cambridge) Limited.

References

1. Saiki, R. K., Scharf, S., Faloona, F., Mullis, K. B., Horn, G. T., Erlich, H .A,.*et al.* (1985). *Science*, **230**, 1350.
2. Dear, P. H. and Cook, P. R. (1993). *Nucleic Acids Res.*, **21**, 13.
3. Walter, G., Tomlinson, I. M., Cook, G. P., Winter, G., Rabbitts, T. H., and Dear, P. H. (1993). *Nucleic Acids Res.*, **21**, 4524.
4. Liu, P., Siciliano, J., Seong, D., Craig, J., Zhao, Y., de Jong, P. J., *et al.*. (1993). *Cancer Genet. Cytogenet.*, **65**, 93.
5. Messing, J. and Bankier, A. T. (1989). In *Nucleic acids sequencing: a practical approach* (ed. C. J. Howe and E. S. Ward), p. 1. IRL Press, Oxford.
6. Telenius, H., Carter, N. P., Bebb, C. E., Nordenskjold, M., Ponder, B. A., and Tunnacliffe, A. (1992). *Genomics*, **13**, 718.
7. Zhang, L., Cui, X., Schmitt, K., Hubert, R., Navidi, W., and Arnheim, N. (1992). *Proc. Natl. Acad. Sci. USA*, **89**, 5847.
8. Bohlander, S. K., Espinosa, R. 3rd, le Beau, M. M., Rowley, R. D., and Diaz, M. O. (1992). *Genomics*, **13**, 1322.
9. Ott, J. (1986). In *Human genetic diseases: a practical approach* (ed. K. E. Davies), p. 19. IRL Press, Oxford.
10. Newell, W. R., Mott, R., Beck, S., and Lehrach, H. (1995). *Genomics*, **30**, 59.

6

Construction and use of somatic cell hybrids

SUSAN L. NAYLOR

1. Somatic cell hybrids and genome research

Somatic cell hybrids provided the first parasexual means for studying human genes (1, 2). Interspecific hybrids were found to lose the chromosomes of one species, allowing genetic characters to be assigned to particular chromosomes by following their segregation among hybrid clones (1). Since this tenuous beginning in 1967, thousands of genes and DNA markers have been located on human chromosomes using somatic cell hybrids. The same hybrid technology is also used in the formation of hybridomas to make monoclonal antibodies (3) and to map genes in other species (4).

The production of 'whole-cell' somatic cell hybrids is illustrated in *Figure 1*. The cytoplasmic membranes of two parental cell lines are fused using a fusogen such as polyethylene glycol to produce a cell with two or more nuclei. Cells with one or more nuclei from each parent are termed heterokaryons. Approximately one in a hundred such cells will undergo nuclear fusion and survive as a proliferating hybrid. Successful nuclear fusion depends upon a number of factors, including the relative phases of the cell cycle in each parental cell: fusion of cells at different points in the cycle can lead to premature chromosome condensation (5) and pulverization of one of the chromosome complements.

In the case of the fusion of a normal diploid cell to a transformed or immortalized cell, the chromosomes from the diploid parent are generally lost at random during subsequent proliferation. Consequently, each clone will have a different complement of chromosomes derived from the diploid donor. This spectrum of chromosomes can be used to assign genes to a particular chromosome or, if the donor has a chromosome translocation or deletion, to a region of the chromosome.

Somatic cell hybrids containing small numbers of donor chromosomes are particularly effective tools for genome studies (2, 5, 7, 8). Microcell hybrids were devised to obtain a relatively pure population of cells having only one to five chromosomes of the donor species (9, 10) . This is achieved by first

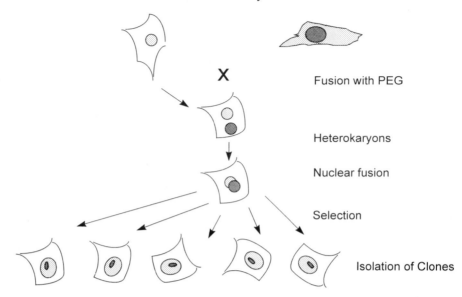

Figure 1. Principle of construction of whole-cell somatic hybrids. The cytoplasmic membranes of two parental cells are induced to fuse using a fusogen such as polyethylene glycol (PEG), forming a binucleate heterokaryon. If the two parental cells were at similar points in the cell cycle, the heterokaryon will then undergo nuclear fusion to form a true hybrid. During subsequent growth (under selection to eliminate unfused parental cells), hybrids lose the majority of one parent's genome, leading to a population of clones containing various chromosomes or chromosome fragments from the donor.

treating the donor cells with colcemid and cytochalasin B, causing them to form microcells containing one or a few chromosomes. These microcells are then fused with the recipient cells and selected in essentially the same way as for whole-cell hybrids (*Figure 2*). Panels of microcell hybrids are now available in which each of the human chromosomes is represented, intact and free of other human material, and carrying a selectable marker. Microcell hybrids have become extremely useful not only as mapping reagents (9) but also as sources for single donor chromosomes, and have been used extensively in the analysis of tumour suppression (11, 12).

Beyond their application as mapping reagents, somatic cell hybrids are key tools in positional cloning, the hybrids being used to define the region containing the gene being sought. Often, somatic cell hybrids with small amounts of donor chromosomal material can be used to isolate regional DNA to facilitate contig assembly. There have also been developed regional mapping panels—collections of hybrid lines carrying various fragments of particular human chromosomes.

The availability of 'mapping panels' of well-characterized hybrids (Section 8), and of PCR assays for DNA content have brought this approach within the

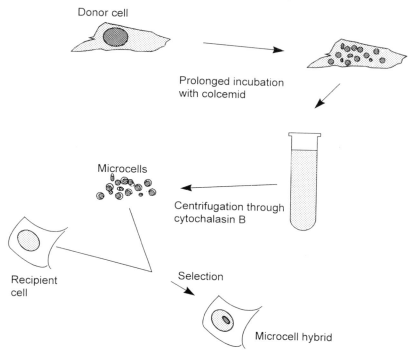

Figure 2. Production of microcell hybrids. Prolonged incubation of a donor cell with colcemid induces individual chromosomes (or small groups of chromosomes) to condense as micronuclei. Centrifugation of the cells in the presence of cytochalasin B results in the formation of microcells, each containing one micronucleus. These microcells are then fused to a recipient cell to form a hybrid.

reach of any laboratory. The protocols and information given in this chapter should allow any molecular biology laboratory to both make and use somatic cell hybrids effectively.

2. Choice of cell lines and selection systems

Interspecific somatic cell hybrids lose chromosomes from one of the parental cells (1), which is referred to as the 'donor'. Most often, chromosomes are lost from the parent which grows less well in culture. For mapping human genes, hybrids are usually made using a diploid human cell as the donor and a transformed rodent cell line as the recipient. Such hybrids will lose human chromosomes and retain those of the rodent parent. In contrast, *intra*specific hybrids usually retain both chromosomes from both parental lines (13). Somatic cell hybrids segregating any species' chromosomes can be formulated by fusing a diploid cell from the 'donor' species to an immortalized,

Table 1. Examples of suitable diploid donor cells and their sources

Cell line or origin	Cell type	Source[a]
JMR-191	Human diploid fibroblast	CCR
WI-38	Human diploid fibroblast	ATCC
Hs27	Human diploid fibroblast	ATCC
Foreskin fibroblasts	Human diploid fibroblast	(Primary culture)
Lymphocytes	Diploid lymphocyte	Blood or EBV-transformed cells
Spleen	Mouse lymphocyte	Mouse spleen

[a]CCR = Coriell Cell Repositories; ATCC = American Type Culture Collection.

transformed host cell. Most of the description here will apply to hybrids in which the donor is human but the principle and the majority of protocols are applicable to other species.

2.1 Donor cells

Sources of human chromosomes are usually fibroblasts or lymphocytes. There are commercially available diploid fibroblasts (*Table 1*) such as WI-38 and Hs27. Alternatively, lymphocytes isolated from blood or from EBV-transformed lymphocytes can be used as the diploid donor. It is essential that the hybrids be made from an individual with normal chromosomes if they are to be used for accurate mapping. However, donor cells with chromosomal aberrations such as translocations and deletions can provide material for regional mapping panels. There are many established lines from patient material with translocations (see Appendix 1); for example, the Coriell Cell Repositories in Camden, NJ list thousands of cells that can be used as starting material for hybrids with fragments of chromosomes.

Interspecific hybrids can be formed using other species, again with fibroblasts or lymphocytes as the donor parent. In addition, we have used mouse spleen as a source of diploid cells for interspecific hybrids.

It is often advantageous if one chromosome of the donor parent is marked with a selectable marker such as *neo*, *gpt*, *hisD*, or *hyg* (14–18). Microcell hybrids are often constructed using such marked chromosomes. All the human chromosomes have been marked and are available from a number of sources. These marked chromosomes are on a rodent background and sources have been listed in the section on existing hybrid panels (Section 7).

2.2 Recipient cells

There are many cell lines that have been used as recipients in constructing somatic cell hybrids; *Table 2* lists some of those which have been used successfully in the formation of hybrids. For whole-cell hybrids, a rodent parent is used which is deficient in a selectable marker (19, 20). The choice of rodent parent usually takes into account several factors. Mouse L cells (e.g. A9 ,

Table 2. Examples of suitable rodent host cells

Cell line	Species	Selectable marker(s)[a]	Human chromosome(s)[b]	Reference
A9	*Mus musculus*	HPRT, APRT	X, 16	19
La⁻t⁻	*Mus musculus*	APRT, TK	16, 17	20
LMTK-Cl1D	*Mus musculus*	TK	17	20, 22
RJK-36	Chinese hamster	HPRT	X	23
GM459	Chinese hamster	HPRT	X	24
FTO-2B	Rat	TK	17	25

[a]HPRT = hypoxanthine phosphoribosyl transferase; APRT = adenosine phosphoribosyl transferase;
TK = thymidine kinase),
[b]Human chromosomes which complement these deficiencies.

La⁻t⁻, LMTK, Cl1D) have been developed that are deficient in hypo-xanthine phosphoribosyl transferase, adenine phosphoribosyl transferase, or thymidine kinase. L cell hybrids are formed easily and tend to retain a few human chromosomes. In contrast, hybrids made with the mouse renal adenocarcinoma line, RAG (21), tend to retain a larger number of human chromosomes.

Chinese hamster cells are favoured as recipient cells for two reasons. First, the fact that they are functionally hemizygous makes it easy to produce mutants which facilitate the subsequent selection of hybrids (26, 27); many auxotrophic Chinese hamster lines have been produced, such as the CHO (Chinese hamster ovary) line (28, 29). Secondly, Chinese hamster cells pro-liferate well in culture. Although some researchers feel that Chinese hamster cells tend to fragment human chromosomes, they have been used as recipient cells in many stable hybrids.

2.3 Selection systems

Suitable selection systems enable the hybrids to be isolated from the parental cells and the choice of selection system, like that of the parental cells, should be made at the outset. There are several selection strategies which can be used to isolate the appropriate hybrid.

Some hybrids are formed using parental cells which both have a selectable marker: the hybrid will have both chromosome complements and be able to survive the double selection medium. Other schemes take advantage of cell adhesion. For example, if human lymphocytes are fused to a rodent cell with a selectable marker, the selection medium will destroy the rodent parent, while the non-adherent human parent can be washed away. Hybrids can also be identified by their growth characteristics. If a fibroblastic line is fused to a transformed line carrying a selectable marker, the transformed parent will be destroyed by the selection medium, while the slow-growing fibroblastic parent will be outgrown by the hybrid cell.

Table 3. Selection systems commonly used in the isolation of somatic cell hybrids

Gene	Chrom. location[a]	Type of selection[b]	Selective medium	Counterselect. medium	Reference
Hypoxanthine phosphoribosyl transferase (HPRT)	X	C	HAT	6-thioguanine	30
Thymidine kinase (TK)	17	C	HAT	BrdU	30
Adenosine phosphoribosyl transferase (APRT)	16	C	AAT	fluoroadenine	31, 32
Bacterial neomycin resistance (*neo*)	Int	D	Genetecin (G418)	–	33
Bacterial guanine phosphoribosyl transferase (*gpt*)	Int	D/C	HAT or 6-thioxanthine	mycophenolic acid	34
Hygromycin resistance (*hyg*)	Int	D	Hygromycin	–	35

[a]The chromosomal locations given are those in the human genome; Int indicates that the selectable gene is not endogenous, and must be integrated into the chosen donor chromosome prior to hybrid formation.
[b]C: donor gene complements host deficiency; D: donor gene is dominant.

The following sections describe the most commonly used selection systems. *Table 3* summarizes these systems, while *Table 4* gives the compositions of the relevant selection media.

2.3.1 HAT and AAT selection systems

The HAT selection system was developed by Szybalski and his co-workers (36) to study melanoma variants, and has been applied most successfully to the formation of somatic cell hybrids (30). HAT medium inhibits *de novo* purine synthesis and forces the cell to use alternative 'salvage' pathways. Two key enzymes in these pathways are hypoxanthine phosphoribosyl transferase (HPRT) and thymidine kinase (TK). HPRT⁻ mutants can be isolated on medium containing 6-thioguanine (36). Since the HPRT gene is located on the X chromosome, mutants are quite frequent and relatively easy to isolate. The majority of existing hybrids have been made using this selectable marker in the donor cells. One advantage of using HPRT selection is that it is possible to force the segregation of the complementing X chromosome by selecting the hybrid on 6-thioguanine. A functional TK gene is also needed for growth on HAT medium. Although TK⁻ mutants are much more difficult to

Table 4. Selective agents used in the isolation of somatic cell hybrids

Name	Preparation and use
HAT (100 × stock)	Add 136 mg hypoxanthine, 1.91 mg aminopterin, 38.6 mg thymidine and 0.5 ml phenol red to 90 ml ddH$_2$O. Add sufficient 10 M NaOH to solubilize; adjust to pH 9.5–9.7 with HCl and make to 100 ml. Filter sterilize, aliquot, and store at −20°C. Use 1 × as selective supplement to normal medium.
AAT (100 × stock)	Add 100 mg adenine hydrochloride, 2.5 mg aminopterin, 100 mg thymidine and 0.5 ml phenol red to 90 ml of ddH$_2$O. Add sufficient 10 M NaOH to solubilize; adjust to pH 9.5–9.7 with HCl and make to 100 ml. Filter sterilize, aliquot and store at −20°C. Use 1 × as selective supplement to normal medium.
Genetecin (G418)	Prepare a stock solution of 75 mg/ml in saline, based on the effective concentration of genetecin supplied by the vendor, rather than the mass of the substance supplied. Filter sterilize and store at −20°C. The final concentration required to completely kill the parental cells will depend on the cell line and must be titrated. For most L-cell hybrids, 500 μg/ml is used for selection, and 250 μg/ml for maintenance. Excessive concentrations will kill even 'resistant' hybrids.
Mycophenolic acid	Solution I (100 × stock): add 25 mg xanthine, 1.5 mg hypoxanthine, 0.2 mg aminopterin, 1 mg thymidine and 0.5 ml phenol red to 90 ml ddH$_2$O. Add sufficient 10 M NaOH to solubilize; adjust to pH 9.5–9.7 with HCl and make to 100ml. Filter sterilize, aliquot and store at −20°C. Solution II (100 × stock): dissolve 250 mg of mycophenolic acid in 90 ml 100 mM NaOH, adjust to pH 7.4 with 0.1 M HCl and adjust to 100 ml with ddH$_2$O. Store at −20°C. Solution III (100 × stock): L-glutamine, 15 mg/ml. Store at −20°C. Supplement normal medium with solutions I, II and III for *gpt* selection of HPRT$^+$ cells. For *gpt* selection of HPRT$^-$ cells, supplement medium with solution I only.
Ouabain (100 × stock)	Prepare 3mM solution in ddH$_2$O, filter sterilize, aliquot and store at −20°C. Supplement normal medium (1 × final) to kill parental human cells.
Hygromycin	Dilute the liquid supplied (CalBiochem) to 100 mg/ml; store at −20°C. The effective concentration for *hyg* selection must be titrated for each cell type, but is typically ~ 400 μg/ml.

obtain, both mouse (20, 22) and rat (25) TK$^-$ lines are available for use as recipient cells in hybrid formation.

A closely related selection system is AAT selection (31). This medium is used to select for a functional APRT (adenine phosphoribosyl transferase) gene, the human form of which is located on chromosome 16 (32). The mouse A9 cell line, deficient in APRT (19), can be used as the recipient when using AAT selection.

The list of rodent lines which have been used as recipient cells (*Table 2*) indicates which selectable markers they carry.

Table 5. Chinese hamster cell lines carrying auxotrophic and temperature-sensitive mutations, and the human chromosomes which carry complementing genes

Cell line	Selectable gene	Selection method	Human chromosome	Reference
Urd-A	CAD	Uridine⁻ medium	2	37
Urd-C	UMPS	Uridine⁻ medium	3	38
UCW56	LARS	Temperature	5	40
glyA	SHMT2	Glycine⁻ medium	12	39
UCW206	NARS	Temperature	18	41
adeB	GART	Hypoxanthine– F12 medium	21	42, 43

2.3.2 Auxotrophic selectable markers

A large number of auxotrophic mutant cells have been isolated (24, 37–39), particularly in Chinese hamster cells because they are functionally hemizygous (26, 27). Mutants have been made which require amino acids or nucleosides among others (see *Table 5*). This allows selection using media which lack this normally non-essential nutrient. Production and characterization of these mutants is painstaking: transport variants are often difficult to distinguish from mutants in an enzymatic pathway. Consequently, only a few auxotrophic mutants have been used widely in the production of hybrids.

2.3.3 Temperature-sensitive selectable markers

Another class of mutants that have been used in the isolation of hybrids carry temperature sensitive mutations (29, 40, 41), which render the cells non-viable above the permissive temperature. The defect is complemented in the hybrid cell, which therefore grows at selective temperatures. Table 5 includes several such mutants which have been used to select for hybrid cells.

2.3.4 Ouabain selection

Often, there will be no specific selective pressure that can be applied against the human parental line. Ouabain, a sodium/potassium ATPase inhibitor (44), has been used to overcome this obstacle. Human cells are normally more sensitive to ouabain than are mouse cells: a concentration of 3×10^{-5} M will kill human cells but will have little or no effect on mouse cells or on human-mouse hybrids (45). Thus, low concentrations of ouabain can provide one side of a double selection on any mouse–human hybrid.

3. Production of somatic cell hybrids

3.1 Production of whole-cell hybrids

Through the years several different methods have been used to produce whole-cell hybrids. Cell fusion was first observed with Sendai virus (46), and several investigators adapted the virus for making hybrids (47, 48). Modifications such as inactivation of the virus with ultraviolet light (48) or with propiolactone (47) were tried, but the fusion results were variable. Other methods for cell fusion include using lipid micelles (lysolecithin) (49). But by far the most successful method has been with polyethylene glycol (PEG; 50, 51). PEG was first used in fusion experiments with plant spheroplasts (52) and then, by Pontecorvo (50) and Davidson (51), in the fusion of animal cells. Once the toxicity of the compound was understood, PEG became the standard fusogen for the production of hybrids.

3.1.1 Whole-cell fusion techniques

Protocols 1, 2, and *3* respectively are methods for making whole-cell hybrids by fusing cells in monolayer; with an adherent recipient parent and a donor in suspension; or with both parents in suspension. The fusion agent in each case is polyethylene glycol (PEG 1000).

Note that standard tissue culture media appropriate to the parental lines, both serum-free and supplemented with appropriate serum (e.g. foetal calf serum) will be required for these protocols, as will the appropriate selection medium (Section 2.3). Where the two parental cell lines grow best in different media, each should of course be maintained in its own medium until the two are combined. Subsequent steps should be performed in the medium best suited to the recipient line, appropriately supplemented with selective agents where specified. Incubations should, unless stated otherwise, be performed at the appropriate temperature for the cell lines in question (and, during selection only, at the selective temperature if temperature-sensitive parents are used).

The selection method used depends on the parents used in the formation of the hybrid, and should be determined with reference to the foregoing sections. For example, HAT selection medium would be used to isolate hybrids made with an HPRT-deficient cell line. As indicated in the discussion of selection media, ouabain can be added to other selection medium to kill human cells.

Protocol 1.   Monolayer fusion

Equipment and reagents
- PEG (polyethylene glycol) 1000 M_r, 50% (w/v) in serum-free medium
- T25 tissue culture flasks

Method

1. Plate two T25 flasks (one control, one fusion) with 5×10^6 cells (2.5×10^6 each of donor and recipient), or at a density to insure cell to cell contact, in serum-supplemented medium. Incubate to allow attachment.

Protocol 1. *Continued*

2. Rinse the monolayers three times with serum-free medium. Leave the control flask whilst performing steps 3 and 4 on the fusion flask.

3. Add 1 ml PEG to the fusion flask and rock flask gently for 1 min.[a]

4. Quickly aspirate off PEG solution and rinse the monolayer three times with serum-free medium, diverting the stream away from the mono-layer to avoid dislodging it.

5. Feed the flasks with 5 ml of serum-supplemented medium and incubate for 24 h.

6. After 24 h, subculture the fusion and control flasks (1 : 20) and plate in selection medium in T25 flasks.

7. Feed flasks at four days with selection medium to remove dying cells and then once a week until the clones are picked. Clones should arise within 5–14 days in the fusion flasks. No cells should grow in the control flask.

8. Pick clones as soon as they have grown to approximately 50 cells, as described in *Protocol 5.*[b]

[a] Less PEG—down to 44%—can be used if 50% proves to be too toxic.
[b] One clonal line may give rise to several colonies in a flask; to avoid picking such duplicates, select only one clone from each flask.

Protocol 2. Suspension/monolayer fusion

Equipment and reagents
- PEG solution (*Protocol 1*)
- T25 tissue culture flasks
- PHA-P (phytohaemagglutinin-P), 100 μg/ml in serum-free medium

Method

1. Plate adherent recipient cells in two T25 flasks (fusion and control) 4–24 h before fusion, at a density such that they are 70–80% confluent at the time of fusion.

2. For each flask of recipient cells, pellet 10^6 donor cells (growing in suspension) at 1000 *g* for 10 min. Resuspend the cells in 4 ml of PHA-P solution. Disperse cells by gentle pipetting.

3. Remove the medium from the recipient monolayer cells and add the cell suspension to each flask immediately. Incubate until the suspension cells have adhered to the monolayer (10–20 min).

4. Carefully aspirate the medium. Add 0.5 ml PEG solution to the fusion flask and gently rock for 1 min.

5. Remove the PEG solution from the fusion flask. Rinse the monolayer in each flask three times with serum-free medium. Remove the medium, replace with serum-supplemented medium and incubate both flasks for two days.

6. Add selection medium and incubate. Feed flasks after four days with fresh selection medium to remove dying cells and then once a week until the clones are picked. Clones should arise within 5–14 days in the fusion flasks; no cells should grow in the control flasks.

7. Pick clones as soon as they have grown to approximately 50 cells, as described in *Protocol 5*.[a]

[a] One clonal line may give rise to several colonies in a flask; to avoid picking such duplicates, select only one clone from each flask.

Protocol 3. Suspension fusion

Equipment and reagents
- T25 tissue culture flasks
- PEG solution (*Protocol 1*)

Method

1. Combine 10^6 cells of each parental line in each of two 50 ml or 15 ml centrifuge tubes (fusion plus control). Pellet at 2000 *g* for 15 min at 37°C and resuspend in serum-free medium.

2. Decant the medium and resuspend the cell pellets in the small amount of residual medium. Over a period of 1 min, add 0.5 ml PEG solution to the fusion flask drop by drop, whilst mixing gently.

3. Immediately add 9.5 ml of serum-free medium to each tube over 1 min whilst mixing gently.

4. If the hybrid grows as a monolayer,[a] plate 20 T25 flasks each with the fusion mixture and with the control mixture in medium supplemented with serum. Incubate for 48 h, then proceed as for *Protocol 2*, steps 6 and 7.

5. If the hybrid grows in suspension,[a] proceed to clone selection by limiting dilution (Section 3.3)

6. Treat the control cells in the same way as the fusion. No cells should grow under selective conditions.

[a] In general, hybrids formed from two non-adherent parents will grow in suspension; those formed from two adherent parents will grow as monolayers. Fusion of an adherent transformed parent to a non-adherent parent generally produces adherent hybrids. However, there are exceptions to these rules which must be determined empirically.

3.2 Microcell-mediated chromosome transfer

Hybrid cells with reduced number of chromosomes are made using a special technique termed microcell-mediated chromosome transfer (MMCT) (9, 10, 53). During exposure of the donor cells to colcemid for periods of up to three days, metaphase chromosomes accumulate and the nuclear membrane re-forms around single chromosomes or small groups of chromosomes, forming 'micronuclei'. Centrifuging the cells in the presence of cytochalasin B forms microcells which are then fused to an appropriate recipient cell. To select microcells containing only one or two chromosomes, the microcells can be filtered through a Nucleopore filter before fusion, to eliminate those containing multiple chromosomes.

The MMCT protocol given here (*Protocol 4*) has been optimized for hybrid lines in which the donor is a mouse cell; making microcells from human diploid fibroblast donors is much more difficult (54) as these cells do not form micronuclei very well. Killary and Fournier (10) have a thorough discussion of the parameters affecting microcell production. Because of the prolonged exposure to colcemid in the formation of micronuclei, some microcell hybrids may contain chromosomal fragments (55) and several independent clones should be isolated and characterized. As with the protocols for whole-cell hybrid formation (Section 3.1), the appropriate media (serum-free, serum-supplemented, and selective) will be required, and the same comments on incubation temperatures apply. It is recommended that media be supplemented with antibiotics for this procedure. The choice of antibiotic will be dictated by the cells used and by personal preference; I prefer to use Gentamycin at 50 µg/ml.

Protocol 4. Microcell-mediated chromosome transfer

Equipment and reagents

- 150 mm tissue culture plates
- 50 ml sterile snap-cap tubes
- Colcemid 100 × stock solution (6 µg/ml in isotonic saline)
- Cytochalasin B 100 × stock (1 mg/ml in DMSO; protect from light)
- Sterile, plugged Pasteur pipettes
- PHA-P solution (*Protocol 2*)
- PEG solution (*Protocol 1*)
- Absolute ethanol
- Microscope (×10 objective; phase-contrast)
- Haemocytometer (optional)
- Aceto-orcein stain (optional): 0.5% orcein (Sigma) in 50% (v/v) acetic acid
- Nucleopore filters (5 µm and 8 µm) and sterile holders (optional)

Method

1. Cut 16 plastic 'bullets' from 150 mm tissue culture plates. The bullets should have the same shape as the profile of a 50 ml centrifuge tube. Two bullets should fit, back to back, into a 50 ml tube so that they effectively divide it longitudinally in half.

2. Sterilize bullets in absolute ethanol for 48 h, changing the ethanol once after 24 h.

3. Place four bullets in each of four 150 mm culture dishes. Allow the ethanol to evaporate completely, and add 48 ml of serum-supplemented medium per dish. Harvest one T150 flask of donor cells and resuspend in 8 ml of medium. Drip 0.5 ml of the suspension onto each bullet and incubate for 3–5 h to allow cell attachment.

4. Overlay the bullets with serum-supplemented medium containing 1 × colcemid (0.06 μg/ml) and continue incubation. Monitor micro-nucleation under a microscope: the formation of micronuclei is very apparent. The optimal time needed for micronucleation will need to be determined empirically; in the case of A9 cells it is 48 h.

5. Plate the recipient cells in a T25 flask to give 90% confluency (3–4 × 10^6 cells per flask for most cell lines). Incubate for 3–4 h or overnight to allow the cells to attach and spread.

6. After the donor shows micronuclei and the recipient is attached to the flask surface, the enucleation process can be started. Place 35 ml medium without serum, supplemented with 1 × (i.e. 10 μg/ml) cytochalasin B[a] into each of eight 50 ml sterile snap-cap tubes. Add two bullets per tube, back to back (cell-carrying surfaces outward). Centrifuge for 50 min at 27 000 g at 28–32 °C.

7. Remove bullets with flamed forceps and discard the cytochalasin supernatant (hazardous waste), without disturbing the pellet. Add approximately 1 ml of serum-free medium to each tube. Using a sterile, plugged Pasteur pipette, resuspend the pellet vigorously and pool all suspensions in a 15 ml conical centrifuge tube.

8. (Optional.) Remove a small aliquot for counting using a haemo-cytometer. Whole cells can readily be distinguished from smaller particles. Cytoplasmic particles can be distinguished from micro-nuclei by staining the sample with an equal volume of aceto-orcein stain before observing under bright-field illumination (micronuclei will be stained). There should be relatively few whole cells, com-pared to the numbers of micronuclei.[b]

9. Pellet the microcells at 300 g for 10 min at room temperature. Decant the supernatant, resuspend the pellet in the residual liquid by agitat-ing the tube manually, and add 2.0 ml PHA-P solution.

10. Wash the monolayer of recipient cells with serum-free medium. Add the microcells in PHA-P to the flask. Incubate for 10 min at 37 °C to attach microcells to the recipient monolayer.

11. Aspirate PHA-P solution and add 1.0 ml of PEG solution. Rock the flask gently for 1 min. Quickly aspirate off the liquid and wash the monolayer three times with serum-free medium, taking care not to dislodge the monolayer.

12. Proceed as in *Protocol 1*, steps 5–8.

---

[a] Medium must be pre-warmed before adding cytochalasin B or it may precipitate.
[b] If desired, filtration may be used at this stage to remove whole cells and the larger micro-nuclei containing several chromosomes. Filtration is performed under gravity or very light positive pressure through a 5 μm Nucleopore filter in a sterile holder (e.g. Swinnex, 25 mm). If there are many whole cells in the preparation, they should first be removed by passing though an 8 μm filter.

3.3 Hybrid clone isolation

Each of the foregoing protocols should give rise to a population of hybrid cells growing in selection medium either as a monolayer or as a suspension. To be of use, however, individual clones must be isolated from these populations using standard mammalian cell cloning procedures. *Protocol 5* describes one means of isolating clones of adherent cells.

Protocol 5. Subcloning of adherent hybrids

Equipment and reagents

- Glass cylinders, 6 × 8 mm (Bellco Glass Inc.)
- High-vacuum grease (Dow Corning): autoclave before use
- Soldering iron
- Trypsin solution, 0.25% (w/v) in sterile saline

Method

1. Once the clones have grown to > 50 cells, mark their location on the outer surface of the flask.
2. Using the soldering iron, cut out a circle of plastic from the flask, carrying the clone.
3. Dip one end of a glass cylinder in sterile vacuum grease, and press this end onto the plastic circle to form a well containing the clone.
4. Add 2–4 ml of trypsin solution to the well and leave at room temperature for until the cells are seen to 'round up' and lift off.
5. Using a Pasteur pipette, pump the liquid up and down until the cells are dislodged from the plastic.
6. Transfer the liquid to a fresh T25 flask containing selective medium, and continue growth.

Clones of suspension cells are most easily isolated by limiting dilution. Aliquots of the culture are grown in selective medium in 96-well tissue culture plates, at a dilution which ensures that only 20% or fewer of the wells give rise to viable cultures. At this dilution, it may be assumed that the majority of such cultures are derived from single founder cells (i.e. they are clonal). These cultures are then transferred to flasks and grown as usual in selective medium. The optimal dilution will vary from case to case. However, if the suspension cells have not been subjected to prior selection, only a minority of them will be viable; accordingly, a dilution giving approximately 1000 cells per well is a good starting point for experimentation.

4. Verification of chromosome content

Each hybrid must be assayed for its donor chromosome content. A quick survey of the cells can be done by hybridizing biotinylated human DNA to metaphase spreads (56) or by differential staining with G-11 (57). Both methods give a scan of the human material in the hybrid, but do not allow identification of specific chromosomes. Karyotyping the hybrids gives a much more detailed account of the chromosomes and any large translocations they may contain. However, this is a very labour-intensive assay and is difficult to do for large numbers of clones.

Assay of chromosomally assigned markers gives a quick assessment of the chromosomes in a hybrid. If a specific chromosome is sought, a PCR reaction can be performed on DNA isolated from the mitotic cells from a T25 flask (58). Genes and DNA markers can be assayed by Southern hybridization (59) or by PCR (60). Isozyme analysis, an old standard, still allows the rapid and inexpensive screening of large numbers of hybrids.

4.1 Karyotyping of human–rodent somatic cell hybrids

Metaphase spreads prepared from human–rodent hybrids (*Protocol 6*) can be karyotyped (*Protocol 7*) and G-banded (*Protocol 8*) (61) to identify their human chromosomal content. Mouse chromosomes (*Mus musculus*) are acrocentric (2N = 40), but many of the mouse cell lines which are used in the formation of hybrids are aneuploid. Consequently, the identification of human chromosomes can be complicated by the presence of unusual murine marker chromosomes. In addition, human chromosomes can become fragmented in the production of the hybrid, further complicating their identification.

Chinese hamster chromosomes are distinguished from human chromosomes by their banding patterns (62). Because of the small chromosome number (2N = 22) and the range sizes of metacentric chromosomes in this organism, it has been the recipient of choice for many laboratories. However, we normally avoid using Chinese hamster cells as recipients because, as noted previously, there are indications that they can cause fragmentation of human chromosomes.

Geimsa, pH 11 (G-11) differentially stains human and mouse chromosomes (57) (*Protocol 8*). With G-11 staining mouse chromosomes appear magenta while human chromosomes are light blue, giving a quick assessment of the number of human chromosomes present in a hybrid cell. However, the banding pattern of human chromosomes is very indistinct with this method.

Giemsa staining also works (with less efficiency) on Chinese hamster–human hybrids. As documented by Cuthbert *et al.* (17), fragments of human chromosomes are easily overlooked following G-11 staining. Because of these difficulties, it is important to further analyse the hybrids for human DNA or specific markers by another method.

Protocol 6. Preparation of metaphase chromosome spreads

Equipment and reagents

- 0.75 M KCl
- Fixative: methanol/acetic acid 3:1 (make fresh each time)
- Colcemid 100 × stock solution (6 μg/ml in isotonic saline)
- Microscope slides, washed and stored in 95% ethanol
- Double distilled water at 4°C
- Microscope with phase-contrast facility

Method

1. Grow adherent cells in a T75 or T150 flasks until 80% confluent, or harvest non-adherent cells in exponential growth.

2. Add colcemid to a final concentration of 0.04–0.1 μg/ml and incubate at 37°C for between 30 min and 4 h, monitoring the cells at intervals under the microscope.[a] The shortest incubation time that yields metaphases is the best, as the chromosomes will be optimally extended.

3. Remove all but ~ 7 ml of the medium. Perform a mitotic shake-off by rapping the flask sharply with your palm.

4. Collect the liquid and centrifuge at 1000 *g* for 10 min to pellet the mitotic cells.

5. Resuspend the cells in 5–10 ml of KCl solution and incubate for 20–30 min at room temperature to allow swelling.[b]

6. Pellet the cells at 1000 *g* for 10 min, decant the supernatant and resuspend the cells in the residual KCl solution. Add 0.5–1 ml of fixative one drop at a time whilst gently shaking the tube, then add further fixative to a volume of 7–10 ml, and incubate at room temperature for 10 min.

7. Pellet the cells at 1000 *g* for 10 min, decant the supernatant, and resuspend the cells in 5 ml of fresh fixative.

8. Pellet as before, decant the supernatant, and resuspend the cells in 0.2–1 ml of fresh fixative. Cells can be frozen at −20°C at this stage if required.

9. Dry the ethanol-washed slides and allow residual ethanol to evaporate. For preparation of spreads for subsequent Giemsa staining, rinse the slides in cold, double distilled water and use wet. For preparation of slides for G-11 staining, use dry slides.

10. Either:

(a) Drop one or two drops of cell suspension onto the slide from a height of approximately three feet.

(b) Drop the cells onto the slide at close range, then blow from one side to spread the cells.

(c) Drop the cells onto the slide at close range, then place the slide on a hotplate at 37 °C to spread the cells until the fixative has evaporated.

Check the density of the cells and the intactness of the metaphase spreads under phase-contrast (no staining is necessary) before preparing further slides.

11. Slides for Giemsa staining (*Protocol 7*) should be aged at room temperature in air for seven days. Those for G-11 staining (*Protocol 8*) need only be aged in this way for three to four days, or by soaking in five to six changes of double distilled water over 1–2 h.

[a] The optimal concentration of colcemid varies from case to case, and should be determined empirically; 0.06 μg/ml is often optimal. The time required is also variable. When working with a new hybrid for the first time, it is advisable to prepare several batches of cells under different conditions to establish the best conditions.
[b] If the cells break when dropping slides (step 10), a shorter incubation time should be used.

Protocol 7. Giemsa staining

Equipment and reagents

- Trypsin, 0.25% (w/v) stock in saline
- 50 mM Tris-HCl pH 7.5 at 4 °C
- Giemsa stain solution: 20% (v/v) R66 Giemsa (GURR), 80% (v/v) 50 mM Tris–HCl pH 7.5
- Glass Coplin jars
- Light microscope with ×100 objective
- Metaphase spreads, prepared for Giemsa staining (*Protocol 6*)
- Ethanol, 95%

Method

1. Stain one slide (as for step 3 below) and examine to make sure the chromosomes are intact before proceeding with the remaining slides.

2. Prepare 1:10, 1:20, 1:40, and 1:200 dilutions of the trypsin stock in cold 50 mM Tris–HCl. Immerse slides in each solution for periods of 15–90 sec, to determine the optimal trypsin concentration and digestion time. Stop each digestion by plunging the slide into 95% ethanol, then dry using a gentle stream of compressed air.

3. Immerse slides in staining solution for 5 min.

4. Rinse the slides in water and allow to air dry before observing under ×100 objective.

5. Once the optimal conditions for banding have been determined, process the remainder of the slides.

Protocol 8. G-11 staining

Equipment and reagents

- G-11 staining solution: 0.8 ml R66 Giemsa (Gurr), 50 ml 50 mM Na_2HPO_4; adjust to pH 11.35 with 1 M NaOH—prepare immediately before use, and do not use for more than 10 min.
- Glass Coplin jars
- Metaphase spreads, prepared for G-11 staining (*Protocol 6*)
- Light microscope with ×100 objective

Method

1. Pre-warm the G-11 staining solution in a Coplin jar to 37°C using a water-bath.

2. Immerse slides in the staining solution for periods of 3–5 min, rinsing in water after immersion, and air dry.

3. Examine the slides to determine the optimal staining period, before proceeding with the remaining slides. Maximal colour difference is achieved by staining for 30 sec longer than the time needed to produce a magenta colour in the rodent chromosomes.

4.2 FISH with total human DNA

To identify small fragments of human chromosomes as well as intact chromosomes, an assessment can be done by hybridizing fluorescently labelled human DNA to a metaphase spread of a hybrid cell line. Fluorescent *in situ* hybridization (FISH) (56), has been described in detail by many researchers and there are now kits available for researchers new to this technique. A complete description lies beyond the scope of this chapter, but Chapter 9 of this book describes basic FISH techniques for the hybridization of specific probes; Chapter 8 gives protocols for the labelling and FISH of DOP-PCR products, which can be modified for use with labelled total human DNA. Many somatic cell hybrids contain fragments of human chromosomes, often translocated to the rodent complement. FISH with total human DNA is very sensitive and will identify these small fragments which can otherwise complicate mapping.

4.3 Chromosome painting

Another type of analysis which is often used for characterizing monochromosomal hybrids is chromosome painting, in which a probe prepared from the hybrid cell is hybridized to metaphase spreads of normal cells. Inter-*Alu* PCR products are made from the hybrid using a human-specific primer to the *Alu* repeat (63, 64). These products, representing only the human component of the hybrid, are then fluorescently labelled and hybridized to a normal human metaphase spread (56, 65), with a large excess of unlabelled repetitive (Cot-1) human DNA blocking ubiquitous hybridization to human repeat elements. Such painting will quickly indicate which chromosomes or chromosomal fragments are represented in the hybrid cell, and will reveal any deletions larger than a band. It should be noted, however, that the centromeric regions of human chromosomes are devoid of *Alu* sequences, and are hence not represented in the hybrid-derived paint. Nor does painting of an entire chromosome mean that the corresponding DNA is contiguous in the hybrid. Also, the paint will not indicate the copy number of the human components of the hybrid. If only a few assays are needed, it is expedient to obtain paints from commercial vendors such as Oncor or Vysis, because of the complex nature of optimizing the best paint for each chromosome.

4.4 PCR karyotyping

PCR karyotyping, or PCR fingerprinting, is especially useful for analysing monochromosomal hybrids or those containing a fragment of a chromosome. Ledbetter *et al.* (58) have developed a method using the LINE repetitive element to give a characteristic 'fingerprint' for each human chromosome. The hybrid DNA is amplified using a primer specific for the human LINE element (L1Hs) (58), in a manner similar to inter-*Alu* PCR. The products are analysed by gel electrophoresis to give a 'fingerprint' characteristic of each chromosome. Examples of PCR karyotyping by this approach are given in refs 64 and 66, whilst ref. 17 gives the fingerprint pattern of each human chromosome. *Protocol 9* gives a simple means of isolating DNA suitable for PCR amplification; the LINE amplification is described in *Protocol 10*. In our experience, the relative intensity of each band can vary with minor changes in PCR conditions.

Protocol 9.   'Mitotic shake-off' method to prepare DNA for PCR amplification

Equipment and reagents

- PBS (Dulbecco's phosphate-buffered saline): 8 g NaCl, 0.2 g KCl, 1.15 g $Na_2HPO_4.2H_2O$, 0.2 g KH_2PO_4, 0.1 g $MgCl_2.6H_2O$, 0.1 g $CaCl_2$—make to 1 litre with distilled water, filter sterilize and store at 4°C.
- Proteinase K solution (make fresh each time): 1 mg Proteinase K, 100 μl 10 mM Tris pH 8.0
- Lysis buffer (store at 4°C): 1 ml 10 × PCR buffer (10 × is 100 mM Tris–HCl pH 8.3, 500 mM KCl, 15 mM $MgCl_2$, 0.1% gelatin), 45 μl NP-40, 45 μl Tween 20, 8.9 ml HPLC quality water (sterilized)
- TE buffer: 10 mM Tris–HCl pH 7.4, 1 mM EDTA

Method

1. Rap a T25 flask of 80% confluent cells very sharply on the side to dislodge mitotic cells.[a]

2. Transfer to 15 ml conical centrifuge tube and spin for 5 min at 200 *g*.

3. Decant the supernatant and resuspend the cells in 1 ml PBS. Transfer to a 1.5 ml microcentrifuge tube.

4. Spin in a bench-top microcentrifuge for 1 min to pellet cells. Carefully remove the supernatant.

5. Add 20 μl lysis buffer and 1 μl Proteinase K solution.

6. Incubate at 55°C for 1 h, then heat to 95°C for 1 min to inactivate the Proteinase K.

7. Add 32 μl TE. Store at 4°C.

[a] This protocol can also be applied to isolate DNA from ~ 10^5 suspension cells, by following steps 2–7.

Protocol 10. LINE-PCR fingerprinting of human hybrid DNA

Equipment and reagents

- L1Hs primer, 60 μg/ml: 5'CATGGCACATG-TACATATGTAAC(A/T)AACC3'
- DNA template (*Protocol 9*; 40 μg/ml)
- 10 × PCR buffer: 100 mM Tris–HCl pH 8.3, 500 mM KCl, 15 mM MgCl$_2$, 0.1% (w/v) gelatin; prepare using HPLC grade water and filter sterilize
- dNTP solution: 1.25 mM each of dATP, dCTP, dGTP, dTTP
- *Taq* DNA polymerase, 5 U/μl (Amplitaq, Perkin-Elmer)
- Thermocycler and compatible reaction tubes or microtitre plates

- Light mineral oil (Sigma)
- Horizontal electrophoresis apparatus
- Agarose (SeaKem LE, FMC)
- 10 × TAE electrophoresis buffer: 242 g Tris base, 100 ml 0.5 M EDTA pH 8.0, 57.1ml glacial acetic acid; make to 1 litre with distilled water
- Loading buffer: 50 mg Orange G, 2 g Ficoll 400, 2.4 g urea, 1 ml 0.5 M EDTA pH 8.0; adjust to 10 ml with double distilled water
- Ethidium bromide, 10 mg/ml stock
- DNA size standards (e.g. 100 bp ladder, Pharmacia)

Method

1. Prepare the PCR reaction in a reaction tube or microtitre well as follows:

   - 5 μl DNA
   - 5 μl 10 × PCR buffer
   - 8 μl dNTP solution
   - 5 μl L1Hs primer
   - 0.5 μl *Taq* DNA polymerase
   - 26.5 μl H$_2$O

   Overlay with one drop of mineral oil.

2. Transfer to a thermocycler and cycle as follows:

   - 95°C × 5 min

   followed by 30 cycles of:

   - 94°C × 1 min

- 67°C × 2 min
- 72°C × 3 min

followed by a final extension of:

- 72°C × 10 min

3. Add 3 μl of loading buffer to a 20 μl aliquot of the reaction products, and analyse by electrophoresis on a 2% agarose gel in TAE buffer with size standards. Stain the gel with ethidium bromide (20 μg/ml) and photograph on a UV transilluminator.

4.5 PCR analysis of known markers

Although PCR has revolutionized the analysis of somatic cell hybrids, its exquisite sensitivity can also be a disadvantage due to the heterogeneous

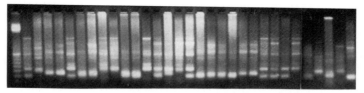

L 1 2 3 4 5 6 7 8 9 10 11 12 13 14 15 16 17 18 19 20 21 22 23 24 25 26 27 C M H

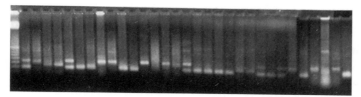

L 1 2 3 4 5 6 7 8 9 10 11 12 13 14 15 16 17 18 19 20 21 22 23 24 25 26 27 C M H

Figure 3. PCR amplification of donor sequences from human–mouse and human–Chinese hamster hybrid DNAs. Top: PCR primers directed against a non-coding human sequence (AFM a2857b5) were used to amplify DNA from human–mouse hybrids (1–27), and from pure Chinese hamster (C), mouse (M), and human (H) DNA. All three single-species DNAs give rise to amplification products, making it impossible to reliably type the hybrids. Bottom: the same amplifications were repeated at an annealing temperature 2°C higher than that used previously. Non-specific amplification products have been reduced, and the human and mouse PCR products can easily be distinguished in the hybrids. Note that the mouse sequence (upper of the two bands in the hybrids) amplifies less strongly, or not at all, when the human sequence is also present, presumably due to similar sequence in Chinese hamster. The use of *Taq* Gold polymerase (Perkin-Elmer) can eliminate most of these artefacts. L indicates DNA size standards (100 bp ladder).

nature of these hybrids. Because one parental cell's chromosomes are progressively lost during growth, the hybrid is a mixed population of cells, and only the average chromosome content is being examined at any one time. PCR is a very sensitive technique, and it will detect chromosomes present in only a minority of cells in the population. In our experience, PCR assays are normally self-consistent, but can conflict with the results of less sensitive techniques.

There are many PCR-formatted markers which have been mapped to specific regions of each chromosome (e.g. 67–69). We have previously published a list of primers for each chromosome (70) based on PCR analysis of specific genes. Others (17) have chosen markers near each of the telomeres to quickly assay for chromosomes. With the large number of markers available, primers specific for any chosen chromosome can easily be gleaned from GDB (the Genome Data Base; `www.gdb.org`).

Standard PCR protocols can be used to amplify from hybrid-derived DNA (*Protocol 9*); a typical procedure is given in *Protocol 11*. It must be cautioned that PCR amplification can detect sequences in the host as well as in the donor genome; this is particularly true for coding sequences, but non-specific host sequence amplification can also occur when amplifying non-coding targets (*Figure 3*). Thus, PCR conditions must be carefully established to determine those which selectively amplify the correct product from the donor species, whilst giving little or no amplification of the host DNA. The conditions of the reaction can be modified by altering the annealing temperature or the magnesium concentration to reduce the non-specific priming. Higher annealing temperatures or lower magnesium concentrations (typically in the range of 1.5–4 mM) will reduce non-specific priming. Generally, Chinese hamster DNA yields more host-derived background than mouse.

Protocol 11. PCR assays of hybrid DNA samples

Equipment and reagents

- Thermocycler and compatible reaction tubes or microtitre plates
- Agarose gel electrophoresis apparatus
- Agarose (SeaKem LE, FMC)
- 10 × TAE buffer (*Protocol 10*)
- Oligonucleotide PCR primers, desalted stocks at 10 µg/ml
- *Taq* DNA polymerase, 5 U/µl (Amplitaq, Perkin-Elmer)
- 10 × PCR buffer (*Protocol 10*)
- dNTP solution (*Protocol 10*)
- Loading buffer (*Protocol 10*)
- DNA size standards (100 bp ladder, Pharmacia)
- Ethidium bromide (stock solution, 10 mg/ml)
- Hybrid DNA (*Protocol 9*, 40 µg/ml)
- Light mineral oil (Sigma)

Method

1. Prepare a PCR master mix. The following are the ingredients for one reaction:
   - 2 µl 10 × PCR buffer

- 3.2 μl dNTP solution
- 1 μl of each PCR primer
- 0.24 μl (1.2 U) *Taq* DNA polymerase
- 7.6 μl water

2. Aliquot 5 μl of each template DNA into a suitable reaction tube or microtitre well. Each set of reactions should include a negative control with water in place of template DNA.

3. Add 15 μl of master mix (step 1) per reaction and overlay with one drop of mineral oil. Centrifuge briefly in a microcentrifuge to ensure that all of the contents are at the bottom of the tube.

4. Transfer to the thermocycler. Conditions may vary depending on the primers used, but the following conditions are generally suitable:

- 95°C × 7 min

followed by 30 cycles of:

- 95°C × 1 min
- Annealing temperature[a] × 1min
- 72°C × 1 min

followed by a final extension of:

- 72°C × 10 min

Incubate at 4°C until samples are removed from the cycler.

5. Add 3 μl of loading buffer, and analyse by electrophoresis on a 2% agarose gel in TAE buffer with size standards. Stain with ethidium bromide (20 μg/ml) and photograph on a UV transilluminator.

[a]The optimal annealing temperature may need to be determined empirically; it is normally 2°C below the T_m of the primer, which is estimated in degrees centigrade as $2 \times (A + T) + 4 \times (G + C)$.

4.6 Isozyme analysis

Although this method has been largely supplanted by PCR and other techniques, isozyme analysis has been used to characterize the human complement of somatic cell hybrids. It is an inexpensive method which can be performed on many samples.

The proteins present in the hybrid are resolved by electrophoresis, and a chosen enzyme is then detected by incubating the gel with appropriate reagents, such that the enzyme is revealed by (for example) the colour of its reaction products. The mobility of the host and donor proteins (if both are present) serves to distinguish them; enzymes composed of multiple subunits may form heterodimers with intermediate mobilities.

A description of isozyme analysis of hybrids is beyond the scope of this chapter, but a very thorough description is given by Siciliano and White (71).

5. Maintenance and stability of clones

Whole-cell hybrids lose the majority of human chromosomes in the first few cell divisions. By the time a hybrid is cloned, the rate of loss has slowed. However, each clone is a population of cells and, each time they are examined, a snapshot of the population is assayed. Consequently, our approach is to grow a large number of cells (four to eight T150 flasks or a roller bottle) and harvest them at the same time for all assays. With PCR assays, this will provide enough DNA for hundreds of experiments.

Any shock to the culture—such as removal of selective pressure, or freezing and thawing—can result in the selection of a new subpopulation of cells. Most often such shocks cause a loss of markers, but apparent gain of markers can be seen when a minor population of cells overtakes a culture.

Microcell hybrids have fewer chromosomes and tend to stabilize more quickly than whole-cell hybrids. However, because of the prolonged treatment with colcemid in the enucleation stage, there are often many chromosomal breaks (55). Consequently, each hybrid must be carefully analysed for the presence of broken chromosomes and for the integrity of a seemingly intact chromosome. Microcell hybrid clones appear to be extremely stable with prolonged culture and with freezing and thawing.

6. Mapping with somatic cell hybrids.

Until 1980, mapping with somatic cell hybrids depended on detecting expressed genes. The first human genes mapped were a result of complementation of rodent deficiencies (1), and species-specific proteins were detected by isozyme analysis or with species-specific antibodies (7, 72).

A revolution in gene mapping occurred with the implementation of the Southern blot (59) on somatic cell hybrid DNA, allowing genes to be mapped regardless of their pattern of expression (73). Just as isozymes from donor and host could be differentiated by electrophoretic properties, so even highly conserved genes from different species could be distinguished by restriction enzyme digestion. More recently, PCR analysis has further increased the ease with which any sequence can be mapped using suitable hybrids.

6.1 Southern analysis of somatic cell hybrids

DNA prepared for PCR screening using *Protocol 9* is not suitable for restriction digestion and Southern analysis, due to the presence of residual proteinase K and other contaminants. *Protocol 12* should be used to isolate hybrid DNA suitable for digestion and Southern analysis (such DNA may, of course, also be used for PCR assays). Probes derived from coding sequences (such as cDNA probes) will commonly hybridize to sequences in both the recipient and donor genomes (*Figure 4*). In such cases, a restriction enzyme

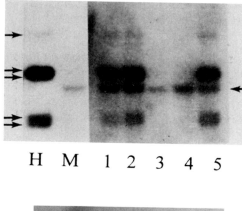

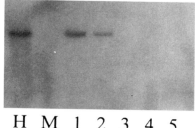

Figure 4. Southern hybridization of somatic cell hybrid DNA. Top: hybridization of a human cDNA probe to Southern blots of *Pvu*II digested DNA from human (H), mouse (M), and human–mouse hybrids (1–5). The probe detects a mouse-derived fragment (right-hand arrow) in all hybrids, as well as the human-derived fragments (left-hand arrows). Bottom: hybridization of a probe derived from a human non-coding sequence to Southern blots of *Eco*RI digested DNA from parental and hybrid cells (notation as above), showing no cross-reactivity with mouse-derived fragments.

must be found which effectively distinguishes between the donor and recipient genes. Samples of host and recipient DNAs (prepared according to *Protocol 12*) should be digested with a variety of restriction enzymes (typically 4 U per microgram of DNA for 3 h, under the conditions recommended by the enzyme suppler), and 5 μg samples resolved by agarose gel electrophoresis, blotted and hybridized with the probe. Once an enzyme has been found which gives distinct fragment sizes from the two DNAs, hybrid DNA samples may be analysed similarly. Probes directed against donor non-coding sequences usually produce little signal from the recipient background. The intensity of the hybridization signal is proportional to the average number of copies of a chromosome per hybrid cell in the population.

Any standard protocols for probe preparation, Southern blotting and hybridization may be used, such as those given in Chapter 12 (*Protocols 8 and 9*). If the probe hybridizes to repetitive sequences in the donor genome,

Susan L. Naylor

these can usually be blocked effectively with unlabelled Cot-1 DNA from the appropriate species (this chapter, *Protocol 13*).

Protocol 12. Isolation from tissue culture cells of purified DNA

Equipment and reagents
- Suspension buffer: 10 mM Tris pH 8.0, 2 mM EDTA, 10 mM NaCl
- Lysis buffer (5 ×): 1 × suspension buffer supplemented with 5% (w/v) SDS and 1 mg/ml Proteinase K (prepare fresh just prior to use)
- Chloroform/isoamyl alcohol (24:1)
- Pasteur pipette (not siliconized), the tip bent into a hook over a Bunsen flame
- Siliconized Pasteur pipette
- 3M sodium acetate pH 5.5
- Buffer-saturated phenol: melt purified phenol at 60°C and add an equal volume of 10 mM Tris–HCl pH 7.6, 1 mM EDTA. Shake and remove the aqueous layer. Repeat this a further two times, leaving some of the final aqueous layer. Check the pH of the aqueous layer, and adjust to pH 7.6 if necessary, before storing at 4°C.
- Isopropanol (100% and 70%)
- TE buffer: 10 mM Tris-HCl pH 7.6, 1 mM EDTA

Method

1. Isolate mitotic adherent cells by rapping the flask sharply and collecting the medium, or use suspension cells. Pellet at 1000 *g* for 10 min, resuspend in serum-free medium and pellet again under the same conditions. Decant the supernatant and chill the cell pellet to 4°C.[a]

2. Resuspend cells at ~ 10[7]/ml in chilled suspension buffer, and add 0.25 vol. of 5 × lysis buffer one drop at a time, mixing by gentle shaking.

3. Incubate at 55°C for 5 h to overnight, mixing occasionally.

4. Add an equal volume of buffer-saturated phenol and mix. Centrifuge at 15000 *g* (or, for smaller volumes, in a microcentrifuge at maximum speed) for 15 min at room temperature.

5. Remove aqueous phase using a siliconized Pasteur pipette which has been shortened to increase the bore size. If there is a lot of protein in the sample, it will collect at the interface.

6. Add an equal volume of chloroform/isoamyl alcohol, mix, and centrifuge at 15000 *g* (or in a microcentrifuge at maximum speed) for 15 min at room temperature. Remove the aqueous phase and place on ice.

7. Add 0.1 vol. of sodium acetate and 1 vol. of cold isopropanol, pH 5.5. Mix gently, and remove the precipitated DNA using the bent Pasteur pipette.[b]

8. Transfer the DNA to a clean tube containing 70% isopropanol and mix gently to wash.

9. Repeat step 8.

150

10. Remove the excess alcohol by gently blowing a stream of air over the pellet. Do not completely dry the pellet or it will not resuspend easily.

11. Redissolve the DNA in TE. Mammalian diploid cells typically yield 2–5 μg of DNA per 10^6 cells.

[a] If necessary, the pellet can be frozen at −80°C at this stage and thawed on ice for later use; however, DNA obtained from frozen cells is of lower quality.
[b] If precipitation is performed quickly, most of the RNA will be left in solution.

Protocol 13. Masking of repetitive probe sequences for Southern hybridization

Equipment and reagents

- Labelled probe DNA[a]
- Cot-1 DNA, 10 mg/ml[b]
- 20 × SSC: 3 M NaCl, 0.3 M sodium citrate
- Sodium dodecyl sulfate (SDS), 1% (w/v)

- 3 M sodium acetate pH 5.5
- Ethanol
- TE buffer: 10 mM Tris–HCl pH 7.6, 1 mM EDTA

Method

1. Adjust the probe volume to 100 μl with TE.

2. Add 5 μl of Cot-1 DNA[c] 50 μl of 20 × SSC, 25 μl distilled water, and 20 μl SDS.

3. Denature by placing in boiling water-bath for 5 min.

4. Incubate at 65°C for 20 min.

5. Add probe to filters without further denaturation.

[a] Any unincorporated nucleotides remaining from probe labelling should first be removed, for example by passage through a Sephadex G-50 column.
[b] This should be from the appropriate species. Human Cot-1 DNA can be obtained from BRL.
[c] Probes with extensive stretches of repetitive sequence may require higher amounts of Cot-1 DNA.

6.2 PCR typing of somatic cell hybrids

Because so little template is required for the PCR reaction, one large DNA isolation can be used to map hundreds of genetic loci. As hybrid panels are now available commercially, virtually all researchers can assign markers to chromosomes (or subchromosomal regions) by PCR typing of well-character-ized somatic cell hybrids.

The PCR protocols used for typing are identical to those discussed in Section 4.5, and the same cautionary notes concerning amplification from the host DNA apply. Primers designed against non-coding sequences are less

Table 6. Pooling strategy for human monochromosomal hybrids. The 24 hybrids, containing the indicated chromosomes, are combined pairwise into pools A-H (from ref. 74).

Chromosome	Pool A	B	C	D	E	F	G	H
1	◆	◆						
2		◆	◆					
3			◆	◆				
4				◆		◆		
5					◆	◆		
6						◆	◆	
7							◆	◆
8	◆		◆					
9		◆		◆				
10			◆		◆			
11			◆				◆	
12					◆		◆	
13						◆		◆
14	◆			◆				
15		◆			◆			
16			◆				◆	
17				◆				◆
18					◆			◆
19	◆				◆			
20		◆				◆		
21			◆					◆
22	◆						◆	
X		◆						◆
Y	◆			◆				

likely to amplify homologous sequences from the host genome. When designing primers for use on hybrids, the same criteria apply as for any PCR primers. Primer length should be ~ 18–20 bases, with G + C contents approximating 50% where possible. Complementary regions of > 4 bases (either within or between primers) should be avoided, as should consecutive runs of > 4 of any one base, and A or T bases at the 3' end should be avoided if possible. The annealing temperatures of the primers (calculated in degrees centigrade as $2 \times [A + T] + 4 \times [G + C]$) should be similar to each other, and should not exceed 68°C. Many computer programs exist to facilitate primer design; details of some of these appear in Appendix 2.

There have been a number of large-scale mapping projects based on somatic cell hybrids, and various strategies have been used to facilitate the typing of large hybrid panels. Sunden *et al.* (74) reported mapping 2900 STSs to specific chromosomes using monochromosomal hybrids. To expedite the mapping, they combined the 24 hybrids into eight pools (*Table 6*). In this way, the chromosomal location of each marker can be determined by only eight PCR typing reactions; for example, a marker detected in pools A and C must be present on chromosome 8. Note that any such pooling strategy must take account of the possibility of a multicopy sequence being present on more than one chromosome; in this example, such markers are recognized by being present in three or more pools (for example, pools A, C, and E for a multicopy marker on chromosomes 8 and 10), whilst markers confined to a single chromosome are found in only two pools.

7. Isolation of donor DNA from hybrid cells

Somatic cell hybrids can serve as sources of DNA from the donor chromosomes or chromosome fragments which they carry. Several different approaches have been taken for isolating donor sequences from the recipient cell background, as detailed below.

7.1 Screening of hybrid-derived clones for donor sequence motifs

Gusella and his co-workers made clone libraries from somatic cell hybrids containing human chromosome 11 (75). These libraries were then screened by hybridization with human-specific repeat sequences to identify those containing human-derived inserts. Such an approach can be applied to other donor species, provided that suitable donor-specific repeat motifs exist which can be used as hybridization targets.

However, such an approach has a number of disadvantages. Not only must many clones be screened to identify the minority which contain donor sequences, but only those clones containing the chosen sequence motif will be identified. Where these motifs are sparse (for example, around centromeres in the case of human *Alu* sequence motifs), the donor DNA will be poorly represented in the identified clones. Moreover, the necessary presence of repeated sequences in the isolated clones may hamper their subsequent analysis.

7.2 Repeat element-mediated PCR of hybrid DNA

The repeat element-mediated (REM) PCR approach has been particularly useful for isolating human- (or, more generally, donor-) specific sequences from hybrids (76). The hybrid's DNA is amplified with primers specific to donor repeat sequences such as (in the case of human) *Alu* elements. The orientation of the primers is chosen such that they amplify the sequences lying between the repeat elements, rather than the repeat elements themselves. The resulting donor PCR products are cloned, and can be sequenced (to allow the design of sequence-specific PCR primers) or used as probes to screen a library containing larger inserts.

Naturally, only those regions lying between two closely spaced repeat elements (typically less than a few kilobases apart) in the correct orientations will be amplified by this approach. As with repeat-based screening (Section 7.1), the distribution of the sequences isolated by this approach will reflect that of the repeat motif but, because amplification relies on the close proximity of *two* repeat elements, biases in distribution will be exaggerated. However, the high frequency of *Alu* sequences ensures that the distances between markers isolated by *Alu* PCR in humans is usually small enough to be bridged by YAC, P1, PAC, or BAC clones, allowing rapid contig assembly in the chosen region.

The optimal primers and PCR conditions for REM-PCR will, of course, depend on the donor species. In the case of human, relatively few sequences will be amplified by using LINE element (L1Hs) primers as described in *Protocol 9*, and the Alu2 primer (5'TCCAYTGCAATCCAGCCTG3'), or a combination of Alu2 and L1Hs primers, should be used under identical conditions to achieve better representation.

7.3 Isolation of donor hnRNA sequences

Somatic cell hybrids have also been used as sources human transcripts from a particular chromosome or region. Liu *et al.* (77) took advantage of the fact that unspliced RNA from the nucleus (hnRNA) often contains human-specific repeat elements in its introns. Starting with a hybrid containing 25 Mb of human DNA, they used hexamers complementary to the 5' intron splice site to prime cDNA synthesis from hnRNA. The cDNA was then cloned, and the library screened by hybridization to human DNA to identify those clones which contained human repeat elements, and hence represented unspliced human RNA. Corbo *et al.* (78) modified this approach by priming cDNA synthesis from hybrid hnRNA with human-specific *Alu* primers, thereby producing only cDNA of human origin.

These approaches parallel those described in Sections 7.1 and 7.2 and, like them, have drawbacks in that only those transcripts which contain repeat elements can be isolated. However, these approaches do provide an easy means to isolate human-specific coding regions from a hybrid cell.

8. Existing hybrid mapping panels

Many researchers will want to exploit the technology of somatic cell hybrids without making and characterizing their own hybrids. A number of panels of hybrids already exist, particularly for the human genome, for chromosomal or subchromosomal studies.

8.1 Whole-chromosome hybrid panels

Listed in *Table 7* are some of the available mapping panels whose members contain one or more whole chromosomes. There are many more panels in existence; those listed were chosen on the basis of frequency of use and of availability. The Coriell Cell Repositories NIGMS panel 1 originated from a

Table 7. Some available panels of human somatic cell hybrids

Source	Panel name[a]	Type	Evaluated by:[b]	Reference
Coriell Inst. NIGMS repository	1	Whole-cell hybrids	K, F, S, P	ccr@arginine.umndj.edu 79–81
Coriell Inst. NIGMS repository	2	Monochromosomal hybrids marked with a variety of selectable markers	K, F, S, P	ccr@arginine.umndj.edu 74, 81, 82
Dr R. Newbold, Brunel University, Uxbridge, UK[c]	Hytk	Monochromosomal hybrids marked with *hyg* and with herpes simplex thymidine kinase	K,F,C,R,P	17
Bios Laboratories	1	Whole-cell hybrids	K,S,P	www.bioslabs.com
Bios Laboratories	2	Monochromosomal hybrids marked with *gpt*	F,S	www.bioslabs.com
Dr N. Spurr, Human Genetic Resources, ICRF Clare Hall Labs., South Mimms, Herts, U.K.[c,d]	–	Monochromosomal hybrids	F,C	83
Uta Francke, Stanford Univ. Medical Centre, Palo Alto, C.A., USA[a]	–	Whole-cell hybrids	K, S, P	84
Dr. N. Spurr (see above)[c,d]	–	Whole-cell hybrids	K, S, P, I	85
D. T. Shows, Roswell Park Cancer Inst., Buffalo, NY, USA[c]	–	Whole-cell hybrids	K, F, S, I	86–88

[a]The members of these panels contain one or more intact human chromosomes.
[b]K: karyotyping; F: fluorescence *in situ* hybridization; S: Southern hybridization; P: PCR typing of specific markers/genes; C: chromosome painting; R: repeat-element-mediated PCR fingerprinting; I: isozyme analysis
[c]DNA from these sources is available on a collaborative basis.
[d]Currently undergoing reorganization.

panel of hybrids isolated by T. Mohandas (79, 80); a few additional hybrids have been added to this panel to ensure full coverage of the genome. These hybrids have been characterized by many techniques and are available both as DNA and as cell lines, and are supplied with extensive documentation. Bios also has a commercially available panel which originated from hybrids isolated by J. Wasmuth.

Those hybrid panels whose members each contain a single human chromosome are of particular use in genome mapping. Several such panels have been extensively characterized to ensure as far as possible that each member

Table 8. Details of known chromosomal contamination and deletions in the NIGMS monochromosomal panel 2, as assessed by fluorescence *in situ* hybridization (FISH), by screening for mapped genes, or by screening for mapped marker sequences

Principal chromosome[b]	Contamination/deletions determined by:[a]		
	FISH	Genes	Markers
1*	X, T	–	6
2			
3	T	–	–
4	–	7	9
5	–	–	–
6	T	5	5, 12
7	–	Del	–
8	–	Del	–
9	F, T	–	–
10	T	–	–
11	F, T	–	–
12	–	–	–
13	T	–	–
14	–	–	16
15	T	–	–
16	T	Del	
17	–	–	–
18	–	–	–
19	–	–	–
20*	F, T	4, 7, 8, 10, 22	4, 8, 10, 22
21	–	–	–
22	–	–	–
X	–	–	–
Y	F, T	–	4

[a] The human chromosomal origin of the contaminating material is indicated where known; F and T indicate contaminating human DNA of unknown chromosomal origin, either as independent fragments or translocated to the host rodent chromosomes, respectively. Del indicates that one or more sequences have been deleted from the human chromosome in the hybrid.
[b] Asterisks indicate cell lines which have been replaced in version 2 of this panel.

of the panel carries a single, intact human chromosome, or to characterize any minor deletions or fragments of other chromosomes. For example, the NIGMS panel 2 from the Coriell Cell Repositories been tested by karyotyping, by FISH with human DNA, by Southern analysis, and by PCR analysis of specific markers (*Tables 7* and *8*) (74, 81, 82). The Hytk panel of marked chromosomes developed by Cuthbert *et al.* (17) (*Table 7*) has also been well characterized by painting, *Alu* PCR, FISH, and by PCR for specific markers.

8.2 Regional hybrid mapping panels

Hybrids which contain portions of a chromosome have been isolated most often using human patient material containing a defined translocation or

Table 9. Some currently available regional mapping panels for human chromosomes

Chromosome	Hybrids	Intervals[a]	Source[b]	References
1	3	4	CCR (3)	90, 92
2	6	7	CCR (2)	90, 93, 94
3	8	13	CCR (11)	95
3	22	23	–	96
3	5	6	–	97
4	14	N.D.	CCR (6)	Wasmuth (pers. comm., GDB), 92, 90
5	13	14	CCR (10)	90, 98
6	8	5	–	99
6	2	N.D.	–	90
7	30	21	CCR (2)	100
8	N.D.	17	CCR*	101, 102
9	33	13	CCR (1); EHCB (17)	103
10	9	10	–	90
10	12	5	–	104
11	20	N.D.	CCR (5)	105
11	8	9	–	106, 107, 108
12	24	> 25	CCR (1)	109
13	27	18	CCR (2)	110
14	7	6	–	111
15	7	6	CCR (4)	112
16	95	72	CCR (5)	113
17	34	20	CCR (6)	114
18	20	16	CCR (6)	115
18	7	8	–	116
19	22	22	–	117, 118
20	7	4	–	103, 119
21	27	36	CCR*	120
22	27	N.D.	ATCC; EACC; CCR*	121
X	109	89	CCR*	91–93
Y	1		CCR (1)	

*Panel under development.
[a]The number of regions which can be distinguished using the hybrids. N.D.: not determined.
[b]Figures in brackets indicate the number of panel members available from the stated source. CCR: Coriell Cell Repositories; ATCC: American Type Culture Collection.

deletion (72, 89). Hybrids which have suffered spontaneous breaks have also been used in regional mapping, but often the breaks in these hybrids are complex and they should be used cautiously. Such hybrids should not be confused with radiation hybrids (Chapter 4), which typically contain a large proportion of the donor genome in the form of relatively small fragments.

Table 9 lists the regional mapping panels which have been reported for human chromosomes. Some of the references are to single chromosome workshops which have compiled a list of all the hybrids and their sources; others are to panels isolated in a particular laboratory. The majority of

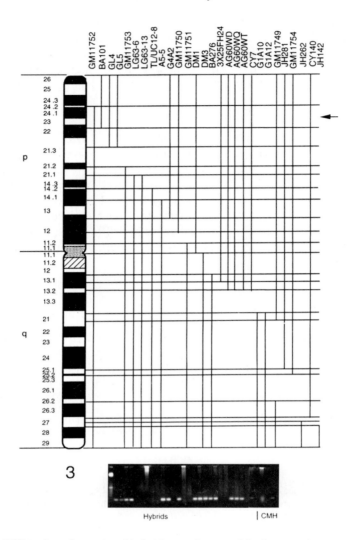

Figure 5. PCR typing of a regional hybrid mapping panel for human chromosome 3. The upper panel shows the chromosome, and the 28 regional human–mouse and human–Chinese hamster hybrids (named across the top). The human DNA present in each hybrid is indicated by vertical lines. The gel in the lower panel shows the amplification products of PCR with primers for the marker D3S3124 with template DNAs from the hybrids (in the same order as above), and from Chinese hamster (C), mouse (M), and human (H). The left-most lane of the gel contains DNA size standards. Comparison of these results with the diagram in the upper panel allows the marker to be localized to the interval arrowed.

chromosomes have at least a few hybrids which divide the chromosome; some, such as 16 (90) and X (91), have a large number of hybrids which effectively dissect the chromosome into small pieces. *Figure 5* shows the results of PCR typing of a marker on a panel of human chromosome 3 regional hybrids, allowing localization of the marker.

9. Summary

Somatic cell hybrids have become a key reagent in the localization of genes. Not only it is possible to quickly assign a gene to a chromosome with these reagents, but also there have been developed extensive panels to localize a gene to a specific region. Positional cloning efforts have also benefited from the use of hybrids as sources of DNA from a defined chromosome or region. The technology has been refined, and many hybrid panels are widely available, making the somatic cell genetic approach to mapping and cloning accessible to all researchers.

Acknowledgements

The following people are acknowledged for their help in compiling techniques and lists of resources: Ann Killary, Robin Leach, Val Sheffield, Nigel Spurr, Tom Shows, Uta Francke, David Nelson, Dawn Garcia, and Karen Kerbacher. The lists of hybrid resources are by no means complete and are meant only to be starting points for finding specific reagents. Many more excellent reagents are available.

References

1. Weiss, M. C. and Green, H. (1967). *Proc. Natl. Acad. Sci. USA*, **58**, 1104.
2. Ruddle, F. H. and Creagan, R. P. (1975). *Annu. Rev. Genet.*, **9**, 407.
3. Kohler, G. and Milstein, C. (1975). *Nature*, **256**, 495.
4. O'Brien, S. J., Peters, J., Searle, A. G., Womack, J. E., Johnson, P. A., and Graves, J. A. M. (1993). In *Human gene mapping*, (ed. A. J. Cuticchia and P. L. Pearson), p. 846. Johns Hopkins University Press, Baltimore and London.
5. Rao, P. N. and Johnson, R. T. (1970). *Nature*, **225**, 136.
6. Cox, R. D. and Lehrach, H. (1991). *Bioessays*, **13**, 193.
7. Creagan, R. P. and Ruddle, F. H. (1975). *Cytogenet. Cell Genet.*, **14**, 282.
8. McKusick, V. A. (1991). *FASEB J*, **5**, 12.
9. Fournier, R. E. K. and Ruddle, F. H. (1977). *Proc. Natl. Acad. Sci. USA*, **58**, 319.
10. Killary, A. M. and Fournier, R. E. K. (1995). In *Methods in enzymology* (ed. P. K. Vogt), p. 133. Academic Press, New York.
11. Stanbridge, E. J. (1992). *Cancer Surv.*, **12**, 5.
12. Anderson, M. J. and Stanbridge, E. J. (1993). *FASEB J.*, **7**, 826.

13. Minna, J. D. and Coon, H. G. (1974). *Nature,* **252**, 225.
14. Saxon, P. J., Srivatsan, E .S., and Stanbridge, E. J. (1986). *EMBO J.*, **5**, 3461.
15. Ning, Y., Lovell, M., Taylor, L., and Pereira-Smith, O. M. (1992). *Cytogenet. Cell Genet.*, **60**, 79.
16. Warburton, D., Gersen, S., Yu, M. T., Jackson, C., Handelin, B., and Housman, D. (1990). *Genomics,* **6**, 358.
17. Cuthbert, A. P., Trott, D. A. and Ekong, R. M. (1995). *Cytogenet. Cell Genet.,* **71**, 68.
18. Lugo, T. G., Handelin, B., Killary, A. M., Housman, D. E., and Fournier, R. E. (1987). *Mol. Cell. Biol.*, **7**, 2814.
19. Littlefield, J. W. (1964). *Science,* **145**, 709.
20. Littlefield, J. W. (1965). *Biochim. Biophys. Acta,* **95**, 14.
21. Klebe, R. J., Chen, T. R., and Ruddle, F. H. (1970). *J. Cell Biol.*, **45**, 74.
22. Kit, S., Dubbs, D. R., Piekarshi, L. J., and Hsu, T. C. (1963). *Exp. Cell Res.*, **31**, 297.
23. Beaudet, A. L., Roufa, D. J., and Caskey, C. T. (1973). *Proc. Natl. Acad. Sci. USA*, **70**, 320.
24. Douglas, G. R., McAlpine, P. J., and Hamerton, J. L. (1973). *Proc. Natl. Acad. Sci. USA*, **70**, 2737.
25. Killary, A. M., Lugo, T. G., and Fournier, R. E. (1984). *Biochem. Genet.,* **22**, 201.
26. Adair, G. M. and Siciliano, M. J. (1986). *Som. Cell Mol. Genet.*, **12**, 111.
27. Deaven, L. and Petersen, D. (1973). *Chromosoma,* **41**, 129.
28. Puck, T. T. and Kao, F. (1967). *Proc. Natl. Acad. Sci. USA*, **58**, 1227.
29. Siminovitch, L. (1976). *Cell,* **7**, 1.
30. Littlefield, J. W. (1966). *Exp. Cell Res.*, **41**, 190.
31. Kusano, T., Long, C., and Green, H. (1971). *Proc. Natl. Acad. Sci. USA*, **68**, 82.
32. Tischfield, J. A. and Ruddle, F. H. (1974). *Proc. Natl. Acad. Sci. USA*, **71**, 45.
33. Southern, P. J. and Berg, P. (1982). *J. Mol. Appl. Genet.*, **1**, 327.
34. Mulligan, R. C. and Berg, P. (1981). *Proc. Natl. Acad. Sci. USA*, **78**, 2072.
35. Lupton. S. D., Brunton, L. L., Kalberg, V. A., and Overell, R. W. (1991). *Mol. Cell. Biol.*, **11**, 3374.
36. Szybalski, W., Szybalska, E. H., and Ragni, G. (1962). *Natl. Cancer Inst. Monogr.*, **7**, 75.
37. Patterson, D., Jones, C., Morse, H., Rumsby, P., Miller, Y., and Davis, R. (1983). *Som. Cell Genet.*, **9**, 359.
38. Patterson, D. and Carnwright, D. V. (1977). *Som. Cell Genet.*, **3**, 483.
39. Law, M. L. and Kao, F. T. (1979). *Cytogenet. Cell Genet.*, **24**, 102.
40. Dana, S. and Wasmuth, J. J. (1982). *Mol. Cell. Biol.*, **2**, 1220.
41. Cirullo, R. E., Arredondo-Vega, F. X., Smith, M., and Wasmuth, J. J. (1983). *Som. Cell Genet.*, **9**, 215.
42. Jones, C., Patterson, D., and Kao, F. T. (1981). *Som. Cell Genet.*, **7**, 399.
43. Schild, D. (1990). *Proc. Natl. Acad. Sci. USA*, **87**, 2916.
44. Mayhew, E. (1972). *J. Cell Physiol,.* **79**, 441.
45. Kucherlapati, R. S., Baker, R. M., and Ruddle, F. H. (1975). *Cytogenet. Cell Genet.*, **14**, 362.
46. Barski, G., Sorieul, S., and Cornefert, F. (1961). *C. R. Acad. Sci. Paris*, **251**, 1825.
47. Klebe, R. J., Chen, T. R., and Ruddle, F. H. (1970). *J. Cell Biol.*, **45**, 74.
48. Okada, Y. and Tadokoro, J. (1962). *Exp. Cell Res*, **26**, 108.

49. Poole, A. R., Howell, J. I., and Lucy, J. A. (1970). *Nature*, **230**, 367.
50. Pontecorvo, G. (1976). *Som. Cell Genet.*, **1**, 397.
51. Davidson, R. L. and Gerald, P. S. (1977). *Meth. Cell Biol.*, **15**, 325.
52. Kao, K. N. and Michayluk, M. R. (1974). *Planta*, **115**, 355.
53. Ege, T. and Ringertz, N. R. (1974). *Exp. Cell Res.*, **87**, 378.
54. McNeill, C. A. and Brown, R. L. (1980). *Proc. Natl. Acad. Sci. USA*, **77**, 5394.
55. Leach, R. J., Thayer, M. J., Schafer, A. J., and Fournier, R. E. K. (1989). *Genomics*, **5**, 167.
56. Pinkel, D., Straume, T., and Gray, J. W. (1986). *Proc. Natl. Acad. Sci. USA*, **83**, 2934.
57. Bobrow, M. and Cross, J. (1972). *Nature New Biol.*, **238**, 122.
58. Ledbetter, S. A., Garcia-Heras, J., and Ledbetter, D. H. (1990). *Genomics*, **8**, 614.
59. Southern, E. M. (1975). *J. Mol. Biol.* **98**, 503.
60. Sakai, R. K., Scharf, S., and Faloona, F. (1985). *Science*, **230**, 1350.
61. Seabright, M. (1971). *Lancet*, **2**, 971.
62. Fournier, R. E. K. and Ruddle, F. H. (1977). *Proc. Natl. Acad. Sci. USA*, **74**, 319.
63. Archidiacono, N., Antonacci, R., Forabosco, A., and Rocchi, M. (1994). *Meth. Mol. Biol.*, **33**, 1.
64. Trask, B. and Pinkel, D. (1990). *Meth. Cell Biol.*, **33**, 383.
65. Ning, Y., Lovell, M., Cooley, L. D., and Pereira-Smith, O. M. (1993). *Genomics*, **16**, 758.
66. Liu, P., Siciliano, J., and Seong, D. (1993). *Cancer Genet. Cytogenet.*, **65**, 93.
67. Weissenbach, J., Gyapay, G., and Dib, C. (1992). *Nature*, **359**, 794.
68. Gyapay, G., Morissette, J., and Vignal, A. (1994). *Nature Genet.*, **7**, 246.
69. Buetow, K. H., Ludwigsen, S., and Scherpbier-Heddema, T. (1994). *Science*, **265**, 2055.
70. Theune, S., Fung, J., Todd, S., Sakaguchi, A. Y., and Naylor, S. L. (1991). *Genomics*, **9**, 511.
71. Siciliano, M. J. and White, B. F. (1987). In *Methods in enzymology*. (ed. M. M. Gottesman), p. 169. Academic Press, San Diego.
72. Ricciuti, F. C. and Ruddle, F. H. (1973). *Genetics*, **74**, 661.
73. Owerbach, D., Bell, G. I., Rutter, W. J., Brown, J. A., and Shows, T. B. (1981). *Diabetes*, **30**, 267.
74. Sunden, S. L. F., Businga, T., Beck, J., McClain, A., Gastier, J. M., Pulido, J. C. *et al.* (1996). *Genomics*, **32**,
75. Gusella, J. F., Keys, C. and VarsanyiBreiner, A. (1980). *Proc. Natl. Acad. Sci. USA*, **77**, 2829.
76. Ledbetter, S. A., Nelson, D. L., Warren, S. T., and Ledbetter, D. H. (1990). *Genomics*, **6**, 475.
77. Liu, P., Legerski, R., and Siciliano, M. J. (1989). *Science*, **246**, 813.
78. Corbo, L., Maley, J. A., Nelson, D. L., and Caskey, C. T. (1990). *Science*, **249**, 652.
79. Mohandas, T. K., Heinzmann, C., Sparkes, R. S., Wasmuth, J. J., Edwards, P., and Lusis, A. J. (1986). *Som. Cell Mol. Genet.*, **12**, 89.
80. Taggart, R. T., Mohandas, T. K., Shows, T. B., and Bell, G. I. (1985). *Proc. Natl. Acad. Sci. USA*, **82**, 6240.
81. Drwinga, H. L., Toji, L. H., Kim, C. H., Greene, A. E., and Mulivor, R. A. (1993). *Genomics*, **16**, 311.
82. Dubois, B. L. and Naylor, S. L. (1993). *Genomics*, **16**, 315.

83. Kelsell, D., Rooke, L., Warne, D., Povey, S., and Spurr, N. K. (1995). *Ann. Hum. Genet.*, **59**, 233.

84. Francke, U., Yang-Feng, T., Brissenden, J. E., and Ullrich, A. (1986). *Cold Spring Harbor Symp. Quant. Biol.*, **51**, 855.

85. Zhong, S., Wolf, C. R., and Spurr, N. K. (1992). *Hum. Genet.*, **90**, 435.

86. Shows, T. B., Sakaguchi, A. Y., and Naylor, S .L. (1982). In *Advances in human genetics* (ed. H. Harris and K. Hirschhorn), p. 341. Plenum Press, New York.

87. Shows, T. B. (1983). In *Isozymes: current topics in biological and medical research* (ed. M. C. Rattazzi, J. G. Scandalios, and G. S. Whitt), p. 323. Alan R. Liss, New York.

88. Shows, T. B., Eddy, R., and Haley, L. (1984). *Som. Cell Mol. Genet.*, **10**, 315.

89. Grzeschik, K.-H., Allderdice, P. W., Grzeschik, A., Opitz, J .M., Miller, O. J., and Sinicalco, M. (1972). *Proc. Natl. Acad. Sci. USA*, **69**, 69.

90. Callen, D. F., Doggett, N. A., and Stallings, R. L. (1992). *Genomics,* **13**, 1178.

91. Schlessinger, D., Mandel, J., Monaco, A. P., Nelson, D. L., and Willard, H. F. (1993). *Cytogenet. Cell Genet.*, **54**, 147.

92. Schaefer, L., Ferrero, G. B. and Grillo, A. (1993). *Nature Genet.*, **4**, 272.

93. Philippe, C., Cremers, F. P. M. and Chery, M. (1993). *Genomics*, **17**, 147.

94. Tsukamoto, K., Tohma, T., and Ohta, T. (1992). *Hum. Mol. Genet.*, **1**, 315.

95. Drabkin, H. A., Wright, M., and Jonsen, M. (1990). *Genomics*, **8**, 435.

96. Leach, R. J., Chinn, R., and Reus, B. E. (1994). *Genomics*, **24**, 549.

97. Sieburth, D., Jabs, E. W., and Warrington, J. A. (1992). *Genomics*, **14**, 59.

98. McPherson, J. D., Morton, R. A., and Ewing, C. M. (1994). *Genomics*, **19**, 188.

99. Boyle, J. M., Hey, Y., and Myers, H. (1992). *Genomics*, **12**, 693.

100. Tsui, L. C., Donis-Keller, H., and Grzeschik, K. (1995). *Cytogenet. Cell Genet.*, **71**, 1.

101. Wagner, M. J., Ge, Y., Siciliano, M., and Wells, D. E. (1991). *Genomics*, **10**, 114.

102. Parrish, J. E., Wang, Y., Wagner, M. J., and Wells, D. E. (1994). *Som. Cell Mol. Genet.*, **20**, 143.

103. Zhou, C., Goudie, D. and Carter, N. (1992). *Ann. Hum. Genet.*, **56**, 215.

104. Mole, S. E., Jackson, M. S., Tokino, T., Nakamura, Y., and Ponder, B. A. (1993). *Genomics*, **15**, 457.

105. Couillin, P., Le Guern, E., and Vignal, A. (1994). *Genomics*, **21**, 379.

106. Evans, K. L., van Heyningen, V., and Porteous, D. J. (1995). *Eur. J. Hum. Genet.*, **3**, 42.

107. Hunt, D., van Heyningen, V., Jones, C., McConville, C., and Benham, F. J. (1994). *Ann. Hum. Genet.*, **58**, 81.

108. Junien, C., van Heyningen, V., Evans, G., Little, P., and Mannens, M. (1992). *Genomics*, **12**, 620.

109. Kucherlapati, R. S., Craig, I., and Marynen, P. (1994). *Cytogenet. Cell Genet.*, **67**, 245.

110. Washington, S. S., Warburton, D., and Chakravarti, A. (1995). *Cytogenet. Cell Genet.*, **70**, 1.

111. Cox, D. W. (1994). *Cytogenet. Cell Genet.*, **66**, 2.

112. Malcolm, S. and Donlon, T. A. (1994). *Cytogenet. Cell Genet.*, **67**, 1.

113. Doggett, N. A. and Callen, D. F. (1995). *Cytogenet. Cell Genet.*, **68**, 165.

114. Fain, P. R. (1992). *Cytogenet. Cell Genet.*, **60**, 177.

115. Rojas, K., Silverman, G. A., and Hudson, J. R. J. (1995). *Genomics*, **25**, 329.

116. Markie, D., Jones, T. A., Sheer, D. A., and Bodmer, W. F. (1992). *Genomics*, **14**, 431.
117. Bachinski, L. L., Krahe, R., and White, B. F. (1993). *Am. J. Hum. Genet.*, **52**, 375.
118. Brook, J. D., Knight, S. J. L. and Roberts, S. H. (1991). *Hum. Genet.*, **87**, 65.
119. Rao, P. N., Hayworth, R., Akots, G., Pettenati, M. J., and Bowden, D. W. (1992). *Genomics*, **14**, 532.
120. Graw, S. L., Gardiner, K., Hall-Johnson, K., Hart, I., Joethem, A., Walton, K. *et al.* (1995). *Som. Cell Mol. Genet.,*
121. Scambler, P. J. (1994). *Cytogenet. Cell Genet.*, **67**, 277.

┌─────────┐
│ **7** │
└─────────┘

The use of flow-sorted
chromosomes in genome mapping

MARK T. ROSS and CORDELIA F. LANGFORD

1. General introduction

Flow-sorted chromosomes are an invaluable resource for genome mapping. Sorted material can be cloned efficiently in *E. coli* to produce chromosome-specific libraries with moderate insert sizes which are suitable for contig building, gene identification, and genomic sequencing. Alternatively, small-insert libraries, which provide a source of chromosome-specific landmarks, can be generated either directly or following amplification of the sorted DNA by the polymerase chain reaction. This chapter gives a brief introduction to flow-sorting itself, but the principal emphasis is on the use of flow-sorted material for the construction of libraries and as a PCR template for the chromosomal assignment of markers.

2. Flow-sorting of chromosomes

Flow-sorting has been dealt with extensively elsewhere, and a full treatment cannot be given here. However, refs 1–3 give details of the theory and practice of flow-sorting.

2.1 Establishing the flow karyotype

Highly pure samples of individual chromosomes of human and other organisms can be obtained by flow-sorting. Chromosome preparations are stained with fluorescent dyes and quantitative differences in the fluorescence character-istics of chromosomes are analysed using a flow cytometer, to develop a flow karyotype. Greatest resolution is achieved using a combination of two fluoro-chromes: the most commonly used are Hoechst 33258 and Chromomycin A3, which bind preferentially to AT-rich and GC-rich regions, respectively. When the intensity of Hoechst 33258 fluorescence is plotted against that of Chromomycin A3 for each chromosome analysed, a bivariate flow karyotype results (4) in which the chromosomes are resolved by DNA content and base

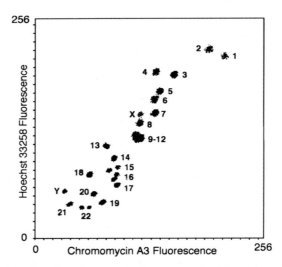

Figure 1. Human bivariate flow karyotype. Chromomycin A3 and Hoechst 33258 fluorescence intensities are plotted in arbitrary units. Each cluster of points corresponds to one chromosome type, with the exception of chromosomes 9–12 which appear as a single cluster.

pair composition. Any discrete chromosome peak on the flow karyotype can be selected using the cytometer workstation software and sorted to a high degree of purity (> 95%). The sorting process uses electrostatic deflection to direct charged droplets of the sheath fluid containing the chromosome of choice into a collection tube. Since droplets can be charged either positively or negatively (and hence deflected to one side or the other), it is possible to sort two chromosomes simultaneously into separate collection tubes.

Human chromosomes lend themselves well to flow-cytometric analysis and sorting because of their large range of sizes and base pair compositions. All but chromosomes 9–12 of man can be resolved on the bivariate flow karyotype (*Figure 1*). There has also been increasing interest in flow-sorting of chromosomes from a wide range of mammals including mouse (5), rat (6, 7), and dog (8), which are valuable as human medical genetic models, and also from farm animals (9). In the pig, for example, all 18 autosomes and the sex chromosomes can be resolved on the bivariate karyotype (10, 11). In the mouse, it is not possible to separate all chromosomes using a single strain. However, by exploiting the chromosome size heteromorphisms and homozygous translocations which occur in different strains, it becomes possible to purify each chromosome (5).

2.2 Choice of cell line and chromosome preparation

Chromosomes may be isolated from a range of cell types. In the past, because of the limited resolution of early flow karyotypes, human chromosome-specific libraries have been constructed by sorting from rodent–human hybrid cell lines (12). Improvements in chromosome sorting procedures and instrumentation have now made this unnecessary for all chromosomes other than 9–12. When chromosomes are to be used for direct cloning, the chances of chromosomal rearrangements should be minimized in the choice of cell type. The most suitable cells for human chromosome preparation, therefore, are peripheral blood lymphocytes. If these are difficult to obtain in sufficient quantity, an acceptable alternative is to use minimally-passaged transformed lymphoblastoid cell lines (see *Protocol 1*) or fibroblasts.

For good quality chromosome preparations it is essential that cells are healthy and growing optimally. Established cell lines are arrested in metaphase before chromosome isolation, by the addition of colcemid (4). Peripheral blood lymphocytes, on the other hand, must be stimulated to divide before colcemid treatment, by the addition of one or more mitogens (e.g. phytohaemagglutinin (13), lipopolysaccharide (5), or a combination of phytohaemagglutinin and pokeweed mitogen (8)) to the culture medium. The optimal moment for adding colcemid to stimulated lymphocytes (i.e. that which results in the highest proportion of metaphase cells) varies between organisms and should be determined empirically.

For the chromosome preparation, cells which have been arrested in metaphase are swollen in a hypotonic buffer, then the cell membrane is removed using a combination of detergent and mechanical disruption to release the chromosomes into a stabilizing buffer (14). This procedure is described in *Protocol 1*.

Protocol 1. Lymphoblastoid cell culture and chromosome isolation

Equipment and reagents

- Tissue culture medium: 500 ml of RPMI, 100 ml of fetal bovine serum, supplemented with 2 mM L-glutamine, 100 U/ml penicillin, and 100 µg/ml streptomycin
- Tissue culture incubator (37°C; 5% CO_2)
- 50 ml centrifuge tubes
- Colcemid solution (10 µg/ml; Gibco–BRL)
- Hypotonic swelling solution: 75 mM KCl, 0.2 mM spermine, 0.5 mM spermidine, 0.2 µm filtered
- Ethidium bromide (1 mg/ml)
- Turck's stain: 1% (v/v) acetic acid, 0.1 mg/ml gentian violet
- Hoechst 33258 (1 mg/ml; Sigma)
- Polyamine isolation buffer: 80 mM KCl, 20 mM NaCl, 2 mM EDTA, 0.5 mM EGTA, 15 mM Tris, 0.2 mM spermine, 0.5 mM spermidine, 3 mM dithiothreitol, 0.25% (v/v) Triton X-100, adjusted to pH 7.2 with HCl, 0.2 µm filtered
- 1 M $MgSO_4$
- Chromomycin A3 (2 mg/ml in ethanol; Sigma)
- 0.1 M trisodium citrate
- 0.25 M sodium sulphite
- Microscope with phase-contrast and fluorescence capability

Method

1. Grow the lymphoblastoid cell line to confluence in 50 ml of tissue culture medium in a 5% CO_2, 37°C incubator.[a]

Protocol 1. *Continued*

2. Add a further 25 ml of tissue culture medium to the flask. Gently break up any cell clumps and distribute the 75 ml between three 250 ml flasks. Add 25 ml of fresh medium to each flask so that the cell concentration is one-third of that of the original culture. Incubate for 24 h in a 5% CO_2, 37 °C incubator.

3. Add 0.5 ml of colcemid solution to each flask and mix gently. Incubate for 6 h at 37 °C.

4. Transfer the cells to 50 ml centrifuge tubes and pellet at 200 *g* for 10 min. Decant the supernatants and blot the edges of the tubes on an absorbent paper towel to remove residual medium.

5. Gently resuspend each cell pellet in 10 ml of hypotonic swelling solution and incubate at room temperature for 15 min.

6. Assess the proportion of cells arrested in metaphase by staining 10 µl of the cell suspension with 10 µl of Turck's stain and viewing in a haemocytometer with a phase-contrast microscope.[b]

7. Transfer the cells to 25 ml centrifuge tubes and pellet at 400 *g* for 10 min. Decant the supernatants and blot the tubes as before.

8. Gently resuspend each pellet in 1.5 ml of ice-cold polyamine isolation buffer and incubate on ice for 10 min.

9. Vortex for 15 sec at a speed which causes the suspension to swirl around the wall of the tube.

10. Remove 10 µl of the suspension onto a microscope slide and stain it with 1 µl of 1 mg/ml ethidium bromide. Using a fluorescence microscope, check that the chromosomes are free in solution. If clumps are present, vortex for further periods of 3 sec until few clumps are apparent.[c]

11. Centrifuge the chromosome suspension at 100 *g* for 1 min to pellet any remaining clumps.

12. Transfer 1.4 ml of the supernatant to a tube suitable for use on the flow sorter. Add 70 µl of Chromomycin solution and mix immediately.

13. Add 3.5 µl of 1 M $MgSO_4$ and 7 µl of Hoechst stain to the chromosome suspension. Mix well and incubate the preparation for at least 1 h on ice.

14. Add 175 µl of sodium citrate and 175 µl of sodium sulphite solution 15 min prior to flow-sorting. Mix the solution well and incubate on ice.

[a] Each 50 ml flask will yield 4.5 ml of chromosome suspension, which is sufficient to sort approximately 1.5×10^6 copies of two autosomes.
[b] Metaphase chromosomes, stained purple, are visible filling the cell.
[c] Care should be taken to avoid over-vortexing which increases the number of broken chromosomes in the preparation.

2.3 Chromosome sorting

Various dual-laser flow sorters which can be modified for chromosome analysis are available commercially (Coulter Electronics, Becton-Dickinson, Cytomation). Ideally, lasers should have a power of 300 mW and be aligned such that the chromosomes pass first through the UV beam (exciting Hoechst 33258 stain) then through the 457.9 nm beam (exciting Chromomycin A3). Spatial separation and optical alignment of the laser beams should be carried out according to the manufacturer's recommendations. A laser beam focusing lens should be selected which avoids lateral beam expansion and, hence, maximizes the light incident at the chromosome detection point. Light collection, dual beam delay, and sorting settings should be adjusted according to the manufacturer's recommendations. Fine optical adjustments using fluorescent microspheres (Coulter Electronics) should be carried out daily in preparation for chromosome analysis, such that the coefficient of variation (c.v.) of signals at the Hoechst and Chromomycin detectors is 1.6 or less.

3. The uses of flow-sorted material in genome mapping

3.1 Direct cloning of flow-sorted material using cosmid vectors with two *cos* sites

The major application of chromosome sorting in genome mapping is as a source of material for making chromosome-specific libraries by direct cloning. Currently, the combination of a cosmid vector and an *E. coli* host is the best system for achieving a high coverage of a sorted chromosome (15–17). Large numbers of cosmid clones can be generated from the small amounts of DNA that result from sorting. Cloned insert sizes lie in the range from 32–46 kb, and the libraries produced are easily stored and manipulated in microtitre plate format.

3.1.1 Cosmid vector preparation

Many cosmid vectors have been described (see ref. 18). The protocols described here are based upon vectors with two cohesive terminus (*cos*) sites (19). Vector arms, each carrying a single *cos* site, are generated by first linearizing the vector by restriction digestion (and dephosphorylating the ends), then by digesting at the (*Bam*HI) cloning site. A productive construct is created by the ligation of both arms to a genomic DNA fragment. In contrast to the situation with vectors having a single *cos* site, the products of vector-only ligation events are unproductive, that is, they are too small to be packaged into a λ bacteriophage head. For this reason, genomic DNA fragments can be dephosphorylated, and an excess of vector molecules can be used in the ligation, allowing the efficient cloning of very small quantities of genomic DNA.

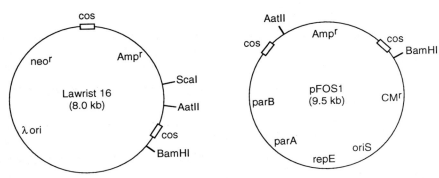

Figure 2. Schematic illustration of Lawrist 16 (P. de Jong, unpublished data) and pFOS1 (20) vectors. The restriction sites shown are those which are used to prepare vector arms. The *Bam*HI cloning site is flanked by T7 and T3 promoters and *Sfi*I sites (not shown) for Lawrist 16; and by T7 and Sp6 promoters and *Not*I sites (not shown) for pFOS1. λ ori is the λ phage origin of replication of Lawrist 16. The oriS and repE genes control the replication of the pFOS1 vector, while parA and parB regulate the copy number. Amp[r] = ampicillin resistance, CM[r] = chloramphenicol resistance, neo[r] = kanamycin resistance.

One of the vectors described (Lawrist 16; Pieter de Jong, unpublished data) replicates to moderately high copy number because of its λ bacteriophage origin of replication (*Figure 2*). This may account for the high proportion of deletions seen in some cosmid libraries (20). The second ('fosmid') vector, pFOS1 (20), has an *E. coli* F-factor origin of replication, and so is propagated at only one to two copies per cell (*Figure 2*). This is thought to result in more stable inserts. *Protocol 2* gives details of the preparation of vector arms in readiness for cloning.

Protocol 2. Preparation and testing of cosmid vector arms

Equipment and reagents

- Restriction endonucleases *Sca*I, *Aat*II, and *Bam*HI (New England Biolabs)
- Low salt restriction endonuclease buffer (LSRE): 50 mM Tris–HCl pH 7.5, 10 mM MgCl$_2$
- Calf intestinal alkaline phosphatase (CIAP; 1 U/μl, Boehringer Mannheim)
- Nitrilotriacetic acid (150 mM, trisodium salt; Sigma)
- Phenol/chloroform/isoamyl alcohol (25:24:1)
- Chloroform/isoamyl alcohol (24:1)
- TE buffer: 10 mM Tris–HCl, 1 mM EDTA pH 8.0
- 3 M sodium acetate pH 5.2
- Ethanol (100% and 70%)
- T4 DNA ligase (New England Biolabs; diluted to 50 U/μl in manufacturer's 1 × ligase buffer)
- Lawrist 16 or pFOS1 vector DNA (prepared by caesium chloride gradient centrifugation according to ref. 18)[a]
- Size standards (bacteriophage λ DNA digested with *Hind*III)
- Agarose gel electrophoresis apparatus
- Agarose (electrophoresis grade)
- TAE buffer: 0.04 M Tris–acetate, 1 mM EDTA
- UV transilluminator and gel photography system
- Ethidium bromide (1 mg/ml stock)

A. *Preparation of vector arms*

1. Digest 20 μg of vector DNA for 2 h at 37 °C in a total volume of 100 μl with 90 U of *Sca*I (Lawrist 16) or *Aat*II (pFOS1), using the enzyme manufacturer's recommended buffer conditions.

2. Remove a 5 μl aliquot, and process this control sample according to *Protocol 2B*.

3. To the remainder of the digestion, add 275 μl of LSRE buffer and 35 μl of CIAP. Incubate for 45 min at 37 °C.

4. Inactivate the CIAP by adding 45 μl of nitrilotriacetic acid and incubating for 25 min at 68 °C.

5. Extract the DNA with an equal volume of phenol/chloroform/isoamyl alcohol, then with chloroform/isoamyl alcohol, each time re-extracting the organic phase with an equal volume of TE buffer. Finally, extract the aqueous phase with ether until no whiteness appears at the interface of the two phases. Place the tube at 68 °C for a few minutes with the cap open to evaporate the ether, then precipitate the DNA overnight at −20 °C with 0.1 vol. of 3 M sodium acetate and 2 vol. of absolute ethanol.

6. Pellet the DNA in a microcentrifuge at 14 000 r.p.m. for 15 min at room temperature and remove the supernatant. Wash the DNA pellet with 1 ml of 70% ethanol, then centrifuge for 2 min, and remove supernatant as before. Air dry, then redissolve the DNA in 177.5 μl of TE. Remove a 5 μl aliquot and store at −20 °C for use in *Protocol 2B*.

7. To the remaining DNA, add 20 μl of the manufacturer's recommended buffer (10 ×) and 7.5 μl of *Bam*HI (20 U/μl). Incubate for 90 min at 37 °C. Carry out organic extractions and precipitate the DNA as described in steps 5 and 6. Dissolve the vector arms in 100 μl of TE. Remove a 5 μl aliquot for use as a control in *Protocol 2B*. Store the prepared arms at −70 °C in small aliquots.

B. *Testing of vector arms*

1. To the 5 μl sample from step A2, add 45 μl of TE. Extract the DNA with an equal volume of phenol/chloroform/isoamyl alcohol, then with chloroform/isoamyl alcohol, and finally with ether. Precipitate the DNA as described in steps A5 and A6, then redissolve the DNA in 5 μl of TE.

2. To this sample and those from steps A6 and A7, add 11 μl of TE and 2 μl of the manufacturer's recommended 10 × T4 DNA ligase buffer.

3. Divide each sample into two 9 μl aliquots. To one add 1 μl of T4 DNA ligase. To the other, add 1 μl of 1 × ligase buffer. Incubate for 16 h at 14 °C.

Protocol 2. *Continued*

4. In order to test the quality of the vector arms, electrophorese the samples on a 0.7% agarose gel[b] in TAE buffer. Include *Hin*dIII digested λ DNA markers and undigested vector DNA. Stain in ethidium bromide solution (0.6 μg/ml) and photograph with UV transillumination.

---

[a] Owing to the two *cos* sites, both vectors are unstable unless grown in the *E. coli* strain pop2136 (21), a temperature-sensitive mutant which expresses the λ bacteriophage repressor (cI) gene at the permissive temperature (30°C). Bacterial growth should, therefore, be carried out at 30°C in the presence of appropriate antibiotics (100 μg/ml of ampicillin and 20 μg/ml of kanamycin for Lawrist 16; 25 μg/ml of chloramphenicol for pFOS1).

[b] A single band should be observed for the *Sca*I or *Aat*II digested vector DNA (8.0 kb for Lawrist 16, 9.5 kb for pFOS1). Following CIAP treatment, no concatamers of this fragment should be observed in the presence of DNA ligase. The doubly digested samples should display two bands which can be ligated in all combinations.

3.1.2 Flow-sorting for cosmid library construction

The amount of DNA required for a cosmid library of a given coverage is proportional to chromosome size. However, in practice, a certain minimum quantity of DNA is needed to allow monitoring of the partial digestion. The experimental design requires a total of approximately 1 μg of DNA (e.g. 3×10^6 human X chromosomes or 9×10^6 chromosome 22) (22). Only approximately 60 ng of this DNA will be used in the final library construction (which in our hands has produced cosmid libraries of 11 equivalents of the X chromosome and 23 equivalents of chromosome 22). With practice, and if the quality of the sorted material is uniform, the amount of starting DNA can be reduced (see below).

In our laboratory, using the Coulter Epics Elite ESP sorter, chromosomes are sorted at a rate of approximately 50/sec (total event rate—4000/sec) in sheath buffer into tubes coated with yeast tRNA, which both acts as a carrier for DNA precipitation and also prevents the sorted chromosomes from adhering to the tube.

With current instrumentation it is feasible to sort human chromosomes to > 95% purity. For library construction, a suitable method for quality control should be used (*Protocol 3*).

Protocol 3. Purity testing of sorted chromosomes

Equipment and reagents

- Sheath fluid: 100 mM NaCl, 10 mM Tris base, 1 mM EDTA, 0.5 mM sodium azide, pass through a 0.2 μm filter
- 4% (v/v) formaldehyde solution (BDH)
- 3:1 methanol/acetic acid
- 70% formamide/2 × SSC (20 × SSC is 3 M NaCl, 0.3 M sodium citrate)
- Microscope slides
- Diamond pen
- Yeast tRNA, 1 μg/μl
- Fluorescence microscope with FITC and DAPI filter blocks
- Fluorescein labelled chromosome-specific DNA probe (Boehringer Mannheim)

- Hybridization buffer: 50% (v/v) deionized formamide, 10% (w/v) dextran sulfate, 2 × SSC, 1 × Denhardt's solution, 40 mM sodium phosphate buffer pH 7.0
- 50% formamide/2 × SSC
- DAPI (4,6-diamidino-2-phenylindole; Sigma) solution (80 ng/ml in 2 × SSC, protected from the light)
- Ethanol dehydration series (70%, 70%, 90%, 90%, and 100% ethanol)

Method

1. Coat sterile 1.5 ml microcentrifuge tubes with yeast tRNA by adding 12.5 µl of tRNA solution and flicking the tube to distribute the liquid over the inside surface.

2. From a cooled chromosome preparation, sort batches of 250 000 chromosomes in sterile sheath buffer into cooled sterile 1.5 ml Eppendorf tubes coated with 12.5 µg of yeast tRNA.

3. From each tube, transfer 10 µl (containing approximately 5000 chromosomes) to a 0.5 ml sterile Eppendorf tube. Add 0.5 µl of 4% formaldehyde solution, gently mix, and incubate at room temperature for 5 min.

4. Transfer the fixed chromosomes to a clean microscope slide and allow the spot to air dry. Mark the dried spots with a diamond pen.

5. Dehydrate the slides by incubating for 1 min each in a series of 70%, 70%, 90%, 90% and 100% ethanol at room temperature. Allow the slides to air dry.

6. Fix the slides by incubating for 30 min in methanol/acetic acid at room temperature. Air dry the slides.

7. Denature the chromosomes by incubating in 70% formamide/2 × SSC for 2 min at 65°C.

8. Hybridize 20 ng of the chromosome-specific DNA probe in 15 µl of hybridization buffer to the denatured chromosomes for at least 2 h according to Chapter 9, *Protocol 6*.

9. Wash the slides according to Chapter 9, *Protocol 7*, steps 1–6.

10. Counterstain the chromosomes by incubating the slides in DAPI solution for 5 min at room temperature.

11. Rinse, dehydrate, air dry, and mount the slides as in Chapter 9, *Protocol 7*, steps 16–18.

12. Analyse the slides using a standard fluorescence microscope with FITC and DAPI filter blocks, and score at least 100 chromosomes for the presence or absence of a FITC signal (*Figure 3*). The proportion of chromosomes with a FITC signal corresponds to the purity of the preparation.

3.1.3 Partial digestion of flow-sorted DNA

Partial digestion of genomic DNA is achieved using a mixture of *Mbo*I restriction enzyme and *dam* methylase. The competing reactions of restric-

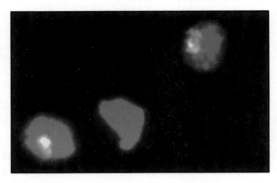

Figure 3. Assessment of sorted chromosome purity by FISH using a chromosome 22-specific probe (*Protocol 3*). The chromosomes are pseudocoloured red, and the hybridization signal yellow. Of the three chromosomes seen, two display hybridization of the probe. The morphology of the chromosomes is a consequence of the processes of flow-sorting and purity assessment.

tion and methylation are allowed to proceed to completion using an excess of the enzyme mixture. This allows control of the average fragment size even when the total amount of DNA is variable (23). A range of enzyme mixtures with different ratios of the *Mbo*I to *dam* methylase is used, and the optimal digestion is then used for library construction.

Protocol 4. Partial digestion of flow-sorted DNA

Equipment and reagents

- 0.25 M EDTA, 10% sodium lauroyl sarcosine
- Proteinase K (20 mg/ml in water; Gibco–BRL)
- Phenylmethylsulfonyl fluoride (PMSF, 4 mg/ml in ethanol; Sigma)
- 5 M NaCl
- TE buffer (*Protocol 2*)
- Enzyme diluent: 20 mM Tris–HCl pH 7.5, 100 mM KCl, 50 μg/ml gelatine, 0.1% (v/v) β-mercaptoethanol, 50% (v/v) glycerol
- Restriction endonuclease *Mbo*I (diluted to 0.1 U/μl with enzyme diluent; New England Biolabs.)
- *dam* methylase (8 units/μl; New England Biolabs)
- Enzyme mixtures[a]
  A 4 μl *Mbo*I + 36 μl *dam* methylase
  B 4 μl A + 8 μl *dam* methylase
  C 4 μl A + 12 μl *dam* methylase
  D 1 μl A + 9 μl *dam* methylase
  E 2 μl C + 8 μl *dam* methylase

- 10 × TAK buffer: 300 mM Tris–HCl pH 7.9, 600 mM potassium acetate, 90 mM magnesium acetate, 5 mM dithiothreitol, 3 mg/ml BSA, 800 μM *S*-adenosyl methionine
- Calf intestinal alkaline phosphatase (CIAP); freshly diluted from 1 U/μl to 0.02 U/μl in sterile distilled water
- Ethanol (70% and 100%)
- Bacteriophage λ DNA size standards (undigested, and digested with *Hind*III)
- Agarose gel electrophoresis apparatus
- Agarose (electrophoresis grade)
- UV transilluminator and gel photography system
- Nitrilotriacetic acid (150 mM, trisodium salt; Sigma)
- TAE buffer: 0.04 M Tris–acetate, 1 mM EDTA
- Ethidium bromide (1 mg/ml stock)
- Flow-sorted chromosomes (250 000 per tube in 450 μl of sheath buffer; the total number of chromosomes required corresponds to 1 μg of DNA)

174

Method

1. To each tube of 250 000 chromosomes,[b] add 50 μl of EDTA/sodium lauroyl sarcosine and 5 μl of Proteinase K. Incubate overnight at 42°C.[c]

2. Inactivate the Proteinase K by adding PMSF to 0.04 mg/ml, and incubating for 40 min at room temperature.

3. Precipitate the DNA using 20 μl of 5 M NaCl and 1 ml of absolute ethanol. Mix very gently by inversion to avoid shearing the DNA, and place the samples overnight at −20°C.

4. Pellet the DNA in a microcentrifuge at 14 000 r.p.m. for 15 min at room temperature. Locate the pellet of nucleic acid,[d] then remove the supernatant carefully using a pipette. Wash the pellet with 1 ml of 70% ethanol, then spin again for 7 min. Remove the supernatant as before, and allow the pellet to air dry.

5. Add TE buffer directly to the pellet to give a DNA concentration of approximately 8 ng/μl. Mix very gently, and allow the DNA to dissolve for at least 2 h at room temperature.

6. Pool DNA samples to give six tubes each containing 18 μl (approximately 150 ng).

7. To each of six separate 150 ng samples,[e] add 2 μl of 10 × TAK buffer. Remove and set aside 6 μl from each tube (45 ng) as an undigested control.

8. To five of the samples, add 1 μl of one of the enzyme mixtures A–E respectively. As a control, add 1 μl of *dam* methylase alone to the sixth tube. Mix gently, then incubate these six samples and the six 6 μl control samples for 3 h at 37°C.

9. Add 1 μl of CIAP to the six digested samples, and continue the incubation of all 12 samples for a further 30 min at 37°C.

10. Remove 7 μl (45 ng) of DNA from each of the six digested tubes. Store these and the six undigested controls on ice until step 13.

11. Add 11 μl of TE and 2.2 μl nitrilotriacetic acid to the digested samples, and incubate them at 68°C for 20 min to inactivate the CIAP.

12. In order to precipitate the DNA, add 1.2 μl of 5 M NaCl and 60 μl of absolute ethanol, and place the tubes at −20°C overnight.[f]

13. Electrophorese the undigested and digested samples from steps 7 and 10, at 1.5 V/cm overnight at room temperature on a 0.3% agarose gel in TAE buffer. Include undigested and *Hind*III-digested λ DNA as size markers.

14. Stain the DNA for 45 min in 0.5 μg/ml of ethidium bromide, then destain the gel for 1 h in distilled water. Analyse the gel (see *Figure 4*).

Protocol 4. *Continued*

15. Pellet the optimal digestion (from step 12)^g as in step 4, and dissolve the DNA in 4 μl of TE^h before proceeding to *Protocol 5*.

^{a} Enzyme mixtures should be stored at −20°C and used within two weeks.
^{b} Efficient cloning of inserts in the 40 kb size range requires starting DNA with an average size of at least 200 kb. Therefore, the DNA samples should be handled gently until digestion is completed.
^{c} Tubes can be then stored at 4°C for several months.
^{d} The pellet is most easily located by positioning the tube with its hinge facing outwards in the microcentrifuge
^{e} If the quality of the sorted chromosomes is consistent, then it is possible with experience to omit certain of the partial digestions and so reduce the quantity of starting material.
^{f} Digestions can be stored at this stage if necessary.
^{g} The optimal digestion should show clear digestion, but no evidence of DNA running ahead of the 23.1 kb λ *Hind*III fragment.
^{h} At this stage the DNA can be seen as specks on the side of the tube when viewed in a bright light. Work the bead of TE over the appropriate part of the side of the tube.

3.1.4 Ligation and packaging

Protocol 5 describes the ligation of the digested DNA to the vector arms, and packaging into infective particles ready for plating on *E. coli*.

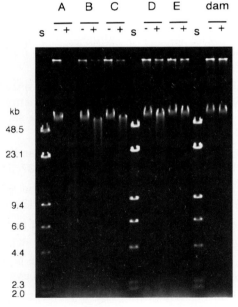

Figure 4. Analysis of partial *Mbo*I digestion of flow-sorted chromosomal DNA. The size markers (s) are a mixture of uncut (25 ng) and *Hind*III-digested (100 ng) λ phage DNA. For each chromosomal DNA, a sample is shown following incubation without (−) or with (+) the enzyme mixture indicated (A–E; see *Protocol 4*). The final sample (dam) was incubated without (−) or with (+) *dam* methylase only. In this case, partial digestion E was selected for cosmid library.

Protocol 5. Ligation of vector arms to digested DNA and packaging

Reagents

- 10 × ligase buffer: 400 mM Tris–HCl pH 7.6, 100 mM MgCl₂, 10 mM DTT
- 6 mM ATP
- TE (*Protocol 2*)
- T4 DNA ligase (400 U/μl; New England Biolabs)
- Gigapack gold II packaging extract (Stratagene)

- λ diluent: 10 mM Tris–HCl pH 7.5, 10 mM MgSO₄
- 5 × SM: 500 mM NaCl, 50 mM MgSO₄, 250 mM Tris–HCl pH 7.5, 0.05% (w/v) gelatine, 50% (v/v) glycerol
- Digested DNA (from *Protocol 4*)
- Vector arms (from *Protocol 2*)

Method

1. Ligate the 4 μl of digested DNA from *Protocol 4* to 400 ng (2 μl) of vector arms from *Protocol 2* in an 8 μl reaction containing 0.8 μl of 10 × ligation buffer, 0.5 μl of 6 mM ATP, and 0.7 μl of T4 DNA ligase. Set-up also a test of self-ligation of the vector arms in which the flow-sorted DNA is replaced with 4 μl of TE*ᵃ*. Incubate for 16 h at 14 °C.

2. Package the entire ligation for 2 h according to the manufacturer's instructions.

3. Stop the reaction by adding 500 μl of λ diluent and 132 μl of 5 × SM and mixing gently by inversion. Freeze small aliquots of the packaged bacteriophage on dry ice, and store them at −70 °C.

*ᵃ*This control allows an estimate for the frequency of non-recombinant clones, which can arise from a failure to modify completely the sites for *Sca*I or *Aat*II.

3.1.5 Plating libraries on *E. coli*

The *E. coli* plating strain used for cosmid libraries should have a *rec*A genotype and additionally should be deficient in certain restriction systems (*mcr*A *mcr*BC *mrr hsd*RMS). The host strain traditionally used is DH5αMCR (Gibco–BRL). However, recent sequencing data (Sanger Centre sequencing teams, unpublished data) have shown that this strain harbours a transposon (Tn1000; 24) which can integrate into the cosmid DNA. An alternative strain with the appropriate genotype but lacking this transposon is DL795 (Northumbria Biologicals). Since this strain is not widely characterized, our current policy is to plate half of the library on each strain. The procedure for plating is described in *Protocol 6*.

3.2 Direct cloning using other systems

Libraries of clones with either smaller or larger inserts than those of cosmids can be produced from flow-sorted material. Van Dilla and Deaven (12) reported the generation of both bacteriophage and cosmid libraries from the same starting material, with a more highly digested sample being used for the smaller-insert phage libraries. However, phage libraries offer no real advantages over cosmids for genome mapping since their inserts are smaller and their storage in an ordered format is less straightforward.

Protocol 6.   Plating cosmid and fosmid libraries on *E. coli*

Equipment and reagents
- LB broth: 1% (w/v) Bacto tryptone, 0.5% (w/v) Bacto yeast extract, 1% (w/v) NaCl
- 10 mM MgSO$_4$
- λ diluent (*Protocol 5*)
- Sterile toothpicks

- LB agar plates (85 mm and 22 × 22 cm) containing kanamycin (30 μg/ml) for Lawrist 16 or chloramphenicol (25 μg/ml) for pFOS1
- LB freezing broth: LB broth containing antibiotics as above and 7.5% (v/v) glycerol

A. *Preparation of plating cells*

1. Inoculate 40 ml of LB broth in a 100 ml conical flask with a fresh single colony of the *E. coli* plating strain. Incubate overnight with shaking at 37°C.

2. Pellet the cells at 4500 *g* for 15 min. Resuspend the cell pellet in 20 ml of sterile 10 mM MgSO$_4$, and store the plating cells at 4°C for up to three weeks.

B. *Testing the titre of the library*

1. Set-up the following plating reactions and negative controls:
   (a) 1 μl chromosome-specific library phage + 99 μl λ diluent.
   (b) 50 μl vector arms self-ligation phage + 50 μl λ diluent.
   (c) 100 μl λ diluent alone.
   (d) 10 μl 5 × SM + 90 μl λ diluent.

2. Add 100 μl of plating cells (step A2) to each sample, and incubate at room temperature for 20 min. In addition, prepare a fifth sample of 100 μl of plating cells alone.

3. Add 1 ml of LB broth to each of the five samples, then incubate for 45 min at 37°C.

4. Pellet the cells in a microcentrifuge for 2 min at 6500 r.p.m. Pour off the LB broth to leave approximately 30 μl. Resuspend the cells in the residual broth, and plate them on LB agar containing kanamycin (for cosmids) or chloramphenicol (for fosmids). Incubate the plates overnight at 37°C, and count the colonies.[a]

C. *Plating the library for picking*

1. Mix a volume of packaged phage sufficient for 2–5000 colonies (as determined in step B4) with λ diluent to a total volume of 100 μl.

2. Follow steps B2 and B3.

3. Plate the entire sample on a 22 × 22 cm plate containing LB agar and the appropriate antibiotic.

4. Incubate the plates overnight at 37°C, then pick the colonies using sterile toothpicks into microtitre dishes containing LB freezing broth.

5. Grow the cultures overnight, then store frozen at −70°C.

[a]The result obtained with the vector self-ligation should indicate that less than 1% of clones are non-recombinant.

Small-insert plasmid libraries may be generated from the partial digests found unsuitable for cosmid library construction (*Protocol 4*; M. T. R. unpublished observation). The DNA is simply digested to completion with a second restriction enzyme, then is purified by organic extraction, precipitated with ethanol, and ligated to an appropriately digested and dephosphorylated plasmid vector. The sequencing of clones from such libraries can be used to produce chromosome-specific sequence-tagged sites (STSs) (25) which can provide landmarks for both physical and genetic mapping.

The production of libraries with larger inserts than cosmids would be highly desirable for large scale genome mapping and sequencing, but is currently problematic. Yeast artificial chromosome libraries have been reported for flow-sorted chromosomes 9, 16, and 21 (26, 27), but the low yield of DNA from chromosome sorting and the inefficiency of yeast transformation combine to make flow-sorted YAC library construction a daunting task. Much greater promise is held by the P1 artificial chromosome (PAC) (28) and bacterial artificial chromosome (BAC) (29) systems, which are capable of propagating fragments in excess of 100 kb, and produce clones which can be manipulated and stored in the same way as cosmids.

3.3 PCR amplification of flow-sorted DNA

3.3.1 Amplification of specific sequences

An early application of the flow-sorting of chromosomes was in the assignment of cloned genes to specific chromosomes by 'spot blot' analysis, in which at least 10 000 copies of each human chromosome are flow-sorted into separate spots onto hybridization membranes then probed with labelled DNA fragments (30–33). Such analyses can now be carried out by PCR, in which a few hundred sorted chromosomes are tested as templates for the amplification of specific STSs (34) (*Protocol 7*). The major application of this technique is currently in the analysis of translocation and deletion breakpoints (35, 36): the derivative chromosomes appear as novel peaks on the flow karyotype and can be sorted and then tested with a variety of markers to define the interval containing the translocation breakpoint or deletion.

Protocol 7. PCR amplification of STSs from sorted chromosomes

Equipment and reagents

- 10 × TAPS2 buffer: 250 mM TAPS pH 9.3, 166 mM $(NH_4)_2SO_4$, 25 mM $MgCl_2$, 0.17% (w/v) BSA, 100 mM β-mercaptoethanol
- dNTP mixture: 2.5 mM each dATP, dCTP, dGTP, dTTP (Pharmacia)
- PCR primer mixture (20 μM each primer)
- *Taq* polymerase (5 U/μl; Amplitaq, Perkin-Elmer)
- Thermocycler and compatible reaction tubes
- Light mineral oil (Sigma; not necessary if using an 'oil-free' thermocycler)

Method

1. Sort 500 chromosomes into a tube containing 35.5 μl of sterile distilled water.

2. Add 5 μl of TAPS2 buffer, 5 μl of primer mixture, 4 μl of dNTP mixture, and 0.5 μl *Taq* polymerase. Mix the reagents. Prepare also positive and negative controls containing, respectively, 20 ng of genomic DNA or distilled water in place of the sorted chromosomes. Overlay each reaction with 30 μl of mineral oil, unless using an 'oil-free' thermocycler.

3. Thermocycle as follows:
   - 94°C × 10 min

   followed by 40 cycles of:
   - 93°C × 15 sec
   - annealing temperature (as appropriate to primers) × 15 sec
   - 72°C × 30 sec

   followed by:
   - 72°C × 5 min

4. Analyse the PCR products by agarose gel electrophoresis[a].

[a]As flow-sorting never achieves 100% purity of the chosen chromosome, STSs from other chromosomes may give faint signals.

3.3.2 Non-specific amplification of flow-sorted chromosomes for construction of small-insert libraries

Small-insert libraries, suitable sources of chromosome-specific landmarks, can be produced rapidly by amplification of flow-sorted chromosomes followed by cloning of the PCR products. There are three methods by which the amplification of multiple sequences from the flow-sorted chromosomes can be achieved:

(a) IRS-(interspersed repeat sequence) PCR (37, 38). This uses PCR primers complementary to repetitive elements scattered throughout the genome,

8

The use of microdissected chromosomes in genome mapping

UWE CLAUSSEN, GABRIELE SENGER, and ILSE CHUBODA

1. Introduction

The mapping and cloning of defined regions of the genome requires a high density of DNA markers for the region of interest. Flow-sorting, as discussed in the previous chapter, provides a source of intact chromosomes. The physical dissection of banded metaphase chromosomes complements flow-sorting, making it possible to isolate small, defined parts of chromosomes in pure form.

The first scientific paper to describe microdissection and microcloning (1) described the dissection and cloning DNA from *Drosophila* polytene chromosomes. Later, the technique was applied to mammalian genomes (2, 3), using unbanded chromosomes and a cloning strategy that was too coarse and inefficient to saturate a specific chromosome region with DNA markers. A breakthrough was achieved by an improved dissection technique which allows the excision of single chromosomal bands (or, with sufficiently fine needles, of a prophase sub-band containing < 5 fg of DNA), and by the development of PCR-mediated cloning (4–6). Such an approach is the fastest way to isolate DNA markers from defined subchromosomal regions.

Here, we describe the microdissection procedure and DNA amplification of fragments using a degenerate oligonucleotide primer (DOP) (7). Material generated in this way can be used to construct chromosome region-specific DNA libraries or, as also described here, as a probe for fluorescence *in situ* hybridization (FISH) experiments. Most of the protocols described here have been developed for use with human chromosomes, but can equally be applied to other mammalian species with little or no modification. Microdissection of plant chromosomes can also be successfully performed, though the preparation of metaphase spreads for microdissection requires modifications (8).

Uwe Claussen, Gabriele Senger, and Ilse Chudoba

2. Instruments and procedures for microdissection

2.1 Instrumentation

Microdissection is performed on an inverted microscope (e.g. IM 135, Zeiss) under bright-field illumination. A Petri dish acts as a 'work-table', supporting coverslips carrying metaphase spreads for dissection, and a moist chamber containing a collection droplet into which the fragments are transferred (*Figure 1*). The moist chamber is essential to prevent evaporation of the collection drop. A rectangular hole in the Petri dish gives access for the microscope objective. Two objectives are necessary: ×10 (for manipulation of the microdrops, for the search of metaphase spreads, and for monitoring the transfer of excised chromosome fragments) and ×100 (oil immersion; for microdissection). The electronically controlled micromanipulator (MR Mot; Zeiss) is attached to the inverted microscope.

In addition, a pipette puller (e.g. Narishige) is essential for the preparation of micropipettes and microdissection needles. A small volume pump (Microgen T4; Orbis Labortechnik) is useful, though not essential, for handling the minute liquid drop in which the chromosome fragments are collected.

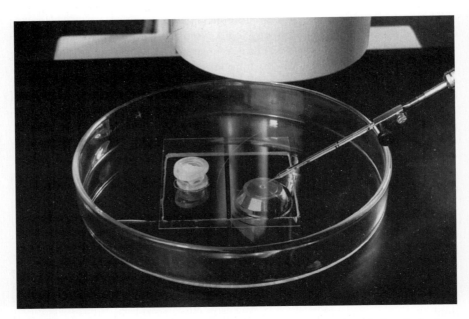

Figure 1. Microdissection set-up. The Petri dish has a rectangular window in its base, over which are placed the coverslips carrying the metaphase spreads (right, with microscope objective visible below and glass needle above) and the moist chamber housing the collection drop (left).

Most of the instruments for microdissection must be prepared by the user. The preparation of these instruments is described in *Protocol 1*.

Protocol 1. Preparation of instruments for microdissection

Equipment and reagents
- 2 mm diameter solid glass rods
- 230 mm long glass Pasteur pipettes
- Pipette puller (e.g. Narishige)
- 150 mm Petri dishes (Nunc)
- Glass microscope coverslips
- Siliconizing solution: 1% dichlorodimethyl silane (Merck, No. 803452) in CCl_4 (Merck, no.2222)
- 1 mM EDTA
- 90% ethanol
- 25% HCl
- Silicone rubber gaskets–these are taken from Cryotubes (Nunc, 3-66656)
- Embedding material (Eukitt, Riedel de Haen)
- Cellulose swabs

A. Preparation of micropipettes

1. Using the pipette puller, extend 230 mm Pasteur pipettes according to the manufacturer's instructions. Break the extended tips to a length of about 15 mm, discarding those which show sharp edges. The diameter of the opening should be between 60 μm and 100 μm.

2. Rinse the micropipettes with siliconizing solution. After evaporation of the solvent, rinse with 1 mM EDTA, heat at 100 °C for 1 h, and sterilize by exposure to short wave ultraviolet light for at least 30 min. About three siliconized pipettes are necessary for one experiment, and should be discarded after a single use.

B. Preparation of microneedles

1. Using the pipette puller, extend 2 mm diameter solid glass rods, following the manufacturer's instructions. If the extended tips are too thin for dissection of chromosomes, they must be carefully broken off under the microscope. Avoid needles with long extended tips as they are not rigid enough. One needle can be used for several experiments, but it must be thoroughly cleaned with collection drop solution (*Protocol 4*) and sterilized by exposure to short wave UV light in a laminar flow-hood for at least 30 min before each use.

C. Preparation of Petri dish

1. A 'window' in the bottom of the Petri dish (which acts as a 'workbench') allows the microscope objective to approach as close as possible to the underside of the coverslip carrying the metaphase spreads. Cut a rectangular hole (5.5 × 5.5 cm) in the middle of the bottom of the Petri dish with a hot scalpel.[a]

Protocol 1. *Continued*

D. *Preparation of coverslips*

1. Clean all coverslips by incubation in 25% HCl at room temperature for several days, then rinse with distilled water, and store in 90% ethanol.

2. Coverslips to be used for the collection drop and the moist chamber (below) are dried and siliconized as in step A-2.

E. *Construction of moist chamber*

1. Using embedding material, affix a silicone rubber gasket to the centre of a siliconized coverslip (step D2); this acts as the bottom of the moist chamber.

2. The lid of the moist chamber is constructed from a lid of a cryotube with a small piece of a cellulose swab fixed inside with embedding material. Before use, the moist chamber is sterilized by exposure to short wave UV light for at least 30 min.

---

[a] Alternatively, a similar hole can be cut in the bottom of a glass Petri dish. Glass dishes have the advantage that they do not tend to acquire an electrostatic charge, which can interfere with the handling of the microdissected fragments.

2.2 Preparation and banding of metaphase spreads on coverslips

The metaphase spreads from which fragments are microdissected are prepared in essentially the same way as for fluorescence *in situ* hybridization (FISH) (Chapter 9). Cultures of amniotic fluid cells, PHA-stimulated lymphocyte cultures, or any cell line may be used.

One special precaution, however, is essential when preparing spreads for microdissection. All solutions must be sterile and free of foreign DNA and nucleases, as these will either contaminate or degrade the minute quantities of DNA involved. Contamination with exogenous DNA is particularly a problem if it is intended to make clone libraries from the microdissected fragments. The solutions used in preparing the spreads (and those which subsequently come into contact with the microdissected fragments) should be stored in single-use aliquots, and all steps should be performed under sterile conditions.

Protocol 2 describes the modifications to basic metaphase preparation techniques which are required.

Prior to microdissection, the metaphase spreads can be GTG-banded, as described in *Protocol 3*, to facilitate chromosome identification.

Protocol 2. Preparation of metaphase spreads for microdissection

Equipment and reagents

- As in Chapter 9, *Protocols 1* and *2*, with the following exceptions:
- BrdU and ethidium bromide solutions (Chapter 9, *Protocol 1*) and the ethanol series (Chapter 9, *Protocol 2*) are not required

- In place of pre-cleaned microscope slides (Chapter 9, *Protocol 2*), substitute coverslips as prepared in *Protocol 1* of this chapter, and rinsed in sterile distilled water at 4°C prior to use
- PBS (Merck No. 7294) is required

Method

As in Chapter 9, *Protocols 1* and *2*, with the following modifications:

1. Omit the incubations with BrdU and with ethidium bromide (Chapter 9, *Protocol 1*, steps 1 and 2).
2. Substitute coverslips for microscope slides in Chapter 9, *Protocol 2*.
3. Immediately after fixation and air drying (Chapter 9, *Protocol 2*, step 9), wash the spreads in PBS or 70% ethanol.
4. Store the spreads in 70% ethanol at −20°C (do *not* proceed to steps 10–12 of Chapter 9, *Protocol 2*).

Protocol 3. GTG-banding of metaphase spreads

Equipment and reagents

- Four sterile 50 ml plastic centrifuge tubes (Falcon; Nunc)
- Phosphate buffer pH 6.88 (Merck, No. 7294)
- Lyophilized Bacto trypsin (Difco): prepare a 5% (w/v) solution in distilled water, filter sterilize, and store in 100 µl aliquots at −20°C

- Giemsa (Merck, No. 9204)
- Sterile distilled water
- Metaphase spreads on coverslips (*Protocol 2*)

Method

1. Prepare the following solutions in the 50 ml centrifuge tubes:
   (a) Phosphate buffer at room temperature.
   (b) One aliquot of frozen trypsin in 35 ml phosphate buffer (final concentration 80 µg/ml) at 37°C.
   (c) 3 ml Giemsa mixed with 35 ml phosphate buffer, filter sterilized, at room temperature.
   (d) Sterile distilled water.
2. Remove the coverslip(s) from storage in 70% ethanol and allow to air dry.
3. Incubate in solution (a) for 60 sec.

Protocol 3. *Continued*

4. Incubate in solution (b) for 10-30 sec.

5. Incubate in solution (c) for 2-3 min.

6. Dip the coverslips briefly into solution (d) and air dry immediately.

2.3 Microdissection

Protocol 4 describes the microdissection procedure. Once again, to reduce the risk of degrading or contaminating the fragments, it is essential that all solutions be sterile and free of nucleases or DNA. They should be stored in small aliquots which are discarded after a single use.

Protocol 4. Microdissection

Equipment and reagents

- Collection drop solution: 10 mM Tris–HCl pH 7.5, 10 mM NaCl, 0.1% SDS
- Proteinase K (Boehringer Mannheim, 1413 783)
- Light mineral oil (Sigma M-5904): incubate at 100°C for 2 h to inactivate nucleases
- Instruments prepared as in *Protocol 1*
- GTG-banded metaphase spreads on coverslips (*Protocol 3*)
- Stainless steel box with lid (large enough to hold the moist chamber—*Protocol 1E*)
- Microgen T4 pump (Orbis Labortechnik)

Method

1. See *Figure 1*. Place the Petri dish with rectangular window on the inverted microscope. Place on it a coverslip of banded metaphase spreads, overlying the window, with the chromosomes uppermost. Next to this place the base of the moist chamber (coverslip carrying silicone gasket).

2. Add Proteinase K to a final concentration of 0.5 mg/ml to one aliquot of collection solution. Place a 10–100 nl droplet of this solution into the centre of the moist chamber using a micropipette connected to the Microgen T4 pump.[a]

3. Moisten the cellulose swab in the lid of the chamber with sterile distilled water.

4. For precise dissection, place the chromosome of interest perpendicularly to the needle and position the tip in front of the chromosome band to be excised.

5. Lower the tip of the needle carefully onto the coverslip. This leads to a forward movement of the tip and excision of the chromosome material (*Figure 2*).

6. Touch the excised fragment carefully with the tip of the needle. Usually, the excised piece sticks to the tip after a few trials.

---

7. Elevate the needle, remove the lid of the moist chamber, and move the Petri dish to position the collection drop beside the centre of the visual field.

8. Place the collection drop in the centre at a magnification of ×250 (×10 objective; phase-contrast; optovar ×2.5).

9. Lower the needle carefully until the tip with the excised chromosome fragment is visible above the collection drop.

10. Wash the tip of the needle with the adhering chromosome fragment in the collection drop. This step should be visually controlled to be sure that the chromosome fragment is inside the collection drop.

11. Repeat steps 4–10 until the required number of fragments has been collected.

12. Overlay the collection drop with mineral oil.

13. Prepare a fresh solution of 0.5 mg/ml Proteinase K in collection drop solution. Fuse the collection drop with an equal volume of this solution.

14. Place the moist chamber in a stainless steel box with lid and incubate at 60°C for 2 h in a water-bath.

15. For DOP-PCR amplification of the fragments, proceed to *Protocol 5*.

[a]The size of the collection drop can be controlled visually and regulated by the heat from a desk lamp bulb.

3. DNA amplification of microdissected fragments using DOP-PCR

The amplification of microdissected material is mainly hampered by the small amount of DNA (typically a few tens of femtograms) obtained from one experiment. For this reason, it is essential that the amplification system used be as efficient as possible. In this respect, T7 DNA polymerase (Sequenase) and *Taq* polymerase Stoffel fragment have been found to be the most effective enzyme combination for amplification, and superior to the normal *Taq* polymerase.

Amplification is achieved by means of degenerate oligo primed- (DOP-) PCR (7), which was first applied to the amplification of DNA from microdissected fragments by Meltzer *et al.* (9). Initial amplification is performed using a degenerate primer and a low annealing temperature, with T7 polymerase. Subsequent amplification is performed with a higher annealing temperature and *Taq* Stoffel fragment polymerase.

The amplification protocol described below (*Protocol 5*) gives good

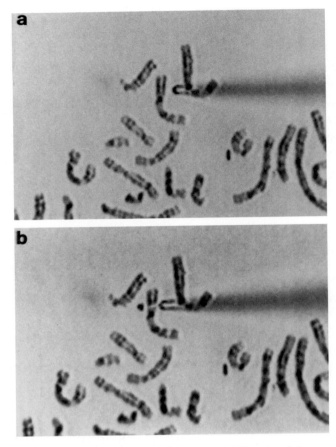

Figure 2. Excision of a defined chromosome region. (a) The tip of the needle is placed perpendicularly to the GTG-banded human chromosome 11 (centre of the three upper-most chromosomes). The tip of the needle was then lowered, causing the tip to move forward and excise the region of interest. (b) The needle has been moved back, leaving the fragment to the right of the chromosome, ready for transfer to the collection drop.

amplification from three to five microdissected fragments. In special cases, even a single fragment can be amplified effectively, but the sequence representation of the PCR products is less good than when using larger numbers of fragments. This is apparent from its weaker signal when used as a probe for FISH. It should go without saying that all stages of the amplification procedure must be performed under rigorously clean conditions. Not only must nucleases be excluded, but also any trace of contaminating DNA, which would easily overwhelm the minuscule amount of DNA in the fragments.

Protocol 5. DOP-PCR amplification of microdissected fragments

Equipment and reagents

- Sequenase, diluted 1:7 in Sequenase dilution buffer (Version 2.0, US70775, Amersham): 2 μl required per reaction
- PCR mixture I: 0.6 × Sequenase buffer, 5 μM DOP (5′ CCG ACT CGA GNN NNN NAT GTG G 3′), 200 μM each dATP, dCTP, dGTP, dTTP (Perkin-Elmer, N808-0007); 5 μl required per reaction
- PCR mixture II: 1 × AmpliTaq Stoffel fragment buffer (Perkin-Elmer), 220 μM each dNTP (Perkin-Elmer), 1.1 μM DOP, 2.5 mM MgCl₂; 45 μl required per reaction

- PCR mixture III: 1 × AmpliTaq Stoffel fragment buffer, 2.5 mM MgCl₂, 1 U/μl AmpliTaq DNA polymerase Stoffel fragment (Perkin-Elmer N808-0107); 5 μl required per reaction
- Thermocycler with heated lid, and compatible reaction tubes
- Light mineral oil (Sigma)
- Microdissected chromosome fragments (*Protocol 4*)
- EDTA (0.5 M) pH 8.0

Method

1. Place 5 μl of PCR mixture I in a reaction tube.

2. Draw a large volume of oil into the micropipette to overcome the capillaric pressure, then draw up the collection drop containing the microdissected fragments and transfer it to the reaction tube.

3. Transfer the tube to the thermocycler and cycle as follows:
   - 96 °C × 5 min

   followed by eight cycles of:
   - 30 °C × 2 min, during which add 0.25 μl (0.3-0.4 U) of the diluted Sequenase
   - 37 °C × 2 min
   - 95 °C × 1 min

4. Cool to 30 °C and add 45 μl of PCR mixture II. Perform 32 cycles as follows, adding 5 μl of PCR mixture III during the first 56 °C step:
   - 94 °C × 1 min
   - 56 °C × 1 min
   - 72 °C × 2 min

   followed by a final extension of:
   - 72 °C × 8 min

5. Add EDTA to a final concentration of 5 mM. The PCR products can now be stored at −20 °C.

---

4. Fluorescence *in situ* hybridization

The amplification products of microdissected fragments can be used as probe for FISH studies, as illustrated in *Figure 3*. This process, first described by Trautmann *et al.* (10) on the one hand allows the quality and coverage of the

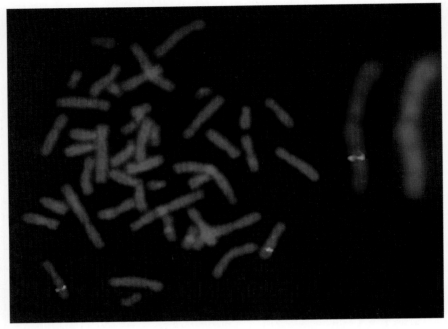

Figure 3. Fluorescence *in situ* hybridization of DOP-PCR amplified material from microdissected human 7q31.1 to a normal human metaphase. Inset at right are two images of a hybridized chromosome 7 at higher magnification, showing the hybridization signal (green) and DAPI staining (blue).

fragment-derived material to be assessed or, on the other, allows unknown chromosome fragments to be identified by a reverse painting approach.

The DOP-PCR products generated in *Protocol 5* may be labelled for use as FISH probes by performing a further re-amplification to incorporate biotinylated dUTP, as described in *Protocol 6*.

Protocol 6.   Biotin labelling of DOP-PCR products

Equipment and reagents

- Labelling mixture: 1 × Replitherm buffer (Biozyme, No. 100001), 100 μM dTTP, 200 μM each dATP, dCTP, dGTP, 100 μM biotin-16-dUTP[a] (Boehringer), 2 μM DOP (*Protocol 5*), 2.5 mM MgCl$_2$, 0.03 U/μl AmpliTaq DNA polymerase
- Thermocycler and compatible reaction tubes
- Light mineral oil (Sigma; unless using 'oil-free' thermocycler)
- DOP-PCR amplification products (*Protocol 5*)
- QIAquick purification kit (Qiagen)

Method

1. Place 20 μl of labelling mixture and 2 μl of DOP-PCR products in a reaction tube and (unless using an oil-free thermocycler), overlay with 30 μl of light mineral oil.

---

2. In the thermocycler, perform 20 cycles of:
   - 94 °C × 1 min
   - 56 °C × 1 min
   - 72 °C × 2 min

3. Purify the PCR products using the QIAquick purification kit, isolating the PCR products in a final volume of 50 μl.

---

[a] Digoxigenin-16-dUTP may be substituted for biotin-16-dUTP if desired. In this way, two probes can be labelled with different colours for simultaneous hybridization. Further details of two-colour FISH can be found in Chapter 9.

The biotin labelled products may now be used as a probe for hybridization to metaphase spreads. Further detailed information on FISH will be found in Chapter 9, but the brief protocol given below (*Protocol 7*) has been found to be effective.

Protocol 7. FISH using biotinylated DOP-PCR fragments

Equipment and reagents

- Metaphase spreads (*Protocol 2*, or see Chapter 9)
- Denaturation solution: 70% formamide, 2 × SSC pH 7.0 (20 × SSC is 3 M NaCl, 3 M sodium citrate)
- Hybridization mixture: 50% formamide (ICN Biochemicals, No. 800685), 10% dextran sulfate, 2 × SSC, 1% Tween 20 pH 7.0
- Wash solution I: 50% formamide, 2 × SSC pH 7.0
- Wash solution II: 2 × SSC pH 7.0
- Wash solution III (SSCT): 4 × SSC, 0.05% Tween 20 pH 7.0
- SSCTM: wash solution III plus 5% low-fat dried milk (Marvel)
- Purified biotinylated DOP-PCR products (*Protocol 6*)
- Human Cot-1 DNA (BRL)
- Ethanol series in Coplin jars (70% at −20 °C; 95% and 100% at room temperature)
- 24 × 24 mm glass coverslips
- Rubber cement (Cow Gum, or similar)
- FITC-conjugated avidin (Vector laboratories, A-2011)
- Vectashield antifade solution (Vector laboratories) supplemented with 0.1 μg/ml DAPI
- Biotinylated anti-avidin (optional; Vector laboratories, BA-0300)
- Lyophilizer

Method

1. Mix 10 μl of the purified, biotinylated DOP-PCR products (*Protocol 6*) with 30 μg of Cot-1 DNA and lyophilize.

2. Redissolve the mixture in 12 μl of hybridization mixture, denature at 75 °C for 5 min, chill on ice, then incubate at 37 °C for 30 min.

3. Incubate a slide carrying metaphase spreads at 70–75 °C in denaturation solution for 3 min.

4. Dehydrate the slide by immersion for 1–2 min in the 70%, 95%, and 100% ethanol solutions and allow to air dry.

5. Apply the probe to one half of the slide, cover with a coverslip and seal the edges with rubber cement.

Protocol 7. *Continued*

6. Incubate at 37°C overnight in a moist chamber.

7. Remove the coverslip and wash the slide three times for 5 min each at 42°C in wash solution I.

8. Wash a further three times for 5 min each at 42°C in wash solution II.

9. Wash briefly in wash solution III at room temperature.

10. Incubate the slide in SSCTM for 15 min at 37°C.

11. Rinse briefly in SSCT.

12. Prepare a 1:500 dilution of FITC-conjugated avidin in SSCTM. Place 100 μl of this solution on the slide, overlay with a coverslip and incubate at 37°C for 30 min in a moist chamber.

13. Wash the slide three times in SSCT.

14. Dehydrate and air dry as in step 4.

15. Mount a coverslip using antifade solution supplemented with 0.1 μg/ ml DAPI. The slide is now ready for microscopic examination (see Chapter 9).

16. If the signal is too weak, it may be enhanced as follows. Carefully remove the coverslip and wash the slide in 100% ethanol until free of antifade solution. Air dry.

17. Repeat step 10.

18. Wash briefly in SSCT.

19. Place 100 μl of a 1:100 dilution of biotinylated anti-avidin in SSCTM on the slide, overlay with a coverslip and incubate at 37°C for 30 min.

20. Wash the slide three times in SSCT.

21. Repeat steps 12–15.

---

5. Cloning of microdissected fragments

Direct cloning of microdissected material is possible (11), but is inadvisable. Not only are large numbers (> 100) of fragments required for even a small yield of recombinant clones, but the purification, digestion, and ligation of the fragments in nanolitre volumes is extremely laborious.

We recommend, therefore, that clone libraries be made from the DOP-PCR amplification products of microdissected fragments (*Protocol 5*). The DOP-PCR primer contains a *Xho*I cleavage site, which can be used for cloning into suitably prepared vectors, as described in Lüdecke *et al.* (6). This requires only a small number of fragments (three to five), and is less difficult than the PCR-mediated cloning procedure described in an earlier paper (4). It is, however, more prone to contamination as any DNA present in the

8. Add 2–3 ml fixative. Mix thoroughly by gentle swirling.

9. Centrifuge at 250 *g* for 5 min.

10. Remove supernatant and loosen cell pellet thoroughly as in step 5.

11. Add 5 ml fixative and mix.

12. Centrifuge at 250 *g* for 5 min.

13. Repeat steps 10–12 twice more.

14. Resuspend cells finally in a sufficient volume of fixative to give the desired cell density (typically 2–5 ml for 20 ml of lymphoblastoid culture).

15. Transfer fixed cells to smaller 10 ml screw-capped centrifuge tubes.

16. Store cells at −20 °C until required.[c]

Note. Lymphoblastoid cell lines should be handled in a class II biological safety cabinet until step 10. Gloves should be worn.

[a] As with any long-term cell culture, lymphoblastoid cells should be monitored regularly for acquired chromosomal changes, and prolonged culture should be avoided. It is preferable to maintain frozen stocks of cells from an early passage and establish fresh cultures whenever new chromosome preparations are required.

[b] This protocol can also be applied to PHA-stimulated lymphocytes 48–72 h after initiation of the short-term PHA culture.

[c] The suspension can be stored at −20 °C for one to two years at least, provided that there is a reasonable volume (> 1 ml) and the tubes are well sealed to limit evaporation. Care should be taken to avoid water contamination of the fixative solution. Samples may absorb water vapour if repeatedly exposed to the air for extended periods and esterification will eventually result in significant water contamination of stored material. Slides should be prepared and the tubes returned to the freezer without delay.

2.1.2 Preparation of slides

Microscope slides must be clean and free from oily residues before use. Packaged pre-cleaned slides can be kept in ethanol, and briefly polished with a lint-free tissue before use. Uncleaned slides should be soaked for several hours in a freshly prepared detergent solution (e.g. 5% Decon 90), rinsed very, very thoroughly in running tap-water, then rinsed with distilled water, and two changes of ethanol before storing in fresh ethanol.

Fixed chromosomal material is spotted onto clean, dry slides and allowed to dry, as described in *Protocol 2*. It is useful to check the quality of the metaphase spreads under a phase-contrast microscope at this point, and define the area of material for hybridization by marking the limits with a diamond pencil. Avoid scoring the slide too vigorously as the resultant unevenness may interfere with coverslipping later.

Slides prepared in this way are best used after they have aged slightly, because fresh slides (as with many cytogenetic banding techniques) do not maintain good morphology through the hybridization procedure. The optimum ageing time depends upon the conditions in the laboratory, but is typically

24 hours. If the slides are required on the same day it is necessary to employ a rapid ageing strategy. The easiest method is to bake the slides at 60–65 °C for at least 15 minutes.

Protocol 2. Slide preparation

Equipment and reagents
- Pre-cleaned microscope slides, stored in ethanol
- Fixative (3:1 methanol/glacial acetic acid), freshly prepared
- Acetone
- Ethanol series: jars containing 70%, 70%, 90%, 90%, 100% ethanol
- Sealed storage box containing desiccant (e.g. silica gel)

Method
1. Prepare a Coplin jar of fresh 3:1 fixative.
2. Place a tube of fixed metaphase suspension (*Protocol 1*) on ice. Allow cell suspensions stored at −20 °C time to equilibrate.
3. Remove clean slides from ethanol and lightly polish them dry with lint-free tissue. Place the slides on a horizontal surface.
4. Mix the cell suspension by gently flicking the tube or, if necessary, using a Pasteur pipette. Take a small volume of suspension in the pipette and drop a single drop onto a slide.
5. Immediately follow with a drop of fixative while the first drop is still spreading, taking care that this drop is not applied too violently. Label the slide and air dry.
6. Examine the slide under a phase-contrast microscope.
7. If the chromosomes are overspread, make a second slide allowing more time between the first and second drops, or position the pipette nearer the slide. Conversely, if the chromosomes are not well spread, drop the cells from a slightly greater height, or add two drops of fixative. Underspreading may result if the cell density is too high, in which case metaphase quality may be improved by diluting the sample with fixative.
8. When the required number of slides have been made, mark the limits of the cell spots with a diamond pencil.
9. Fix the slides in a Coplin jar of fixative at room temperature for 30–60 min, then air dry.
10. Dehydrate the slides through a fresh ethanol series, 1–2 min in each of 70%, 70%, 90%, 90%, and 100% ethanol. Air dry.
11. Fix in acetone at room temperature for 10 min. Air dry.
12. Store slides in a sealed box containing desiccant at room temperature for two to three weeks, or at −20 °C for much longer periods.

2.2 Probe preparation

2.2.1 Choice of probe

The size of the probe is critical to the success of the procedure. The greater the linear extent of the genomic target, the greater the potential size of the probe hybridization complex, and the greater the chance of producing sufficient recognizable signal at the target site. When using cloned probes, vector sequences probably also contribute to the signal by promoting network formation over the site of hybridization.

Cloned DNA can be prepared for use as a FISH probe by any standard method which produces sufficient quantities (several micrograms) of pure DNA. For plasmids, cosmids, PACs, and BACs, we favour the standard alkaline lysis procedure for preparing DNA from 10 ml overnight cultures, followed by phenol/chloroform extraction. An RNase step should be included, but there is no need to remove the RNase. We find that most commercial kits give low yields of DNA (particularly of large-insert clones), making them uneconomical for large scale use.

Single-copy genomic clones over ~ 5 kb in length can be readily detected by FISH. Smaller inserts (such as typical cDNA sequences) are more problematic: hybridization efficiency diminishes with probe length and the size of the hybridization signals are difficult to distinguish from non-specific background. For these small probes, FISH localization may require spot counting procedures similar to those used for isotopic mapping. Despite widespread efforts, only about a quarter of cDNAs with inserts smaller than 1.5 kb can be localized by FISH (7). Again, the genomic extent of the target gene (or genes) appears to be the determining factor. A small cDNA from a gene with numerous introns is more likely to generate visible probe complex than one from a relatively uninterrupted gene. The successful mapping of cDNAs therefore depends on the genomic organization of the gene of interest. At present, single PCR products (e.g. STSs) cannot be used reproducibly for FISH.

At the other end of the scale, FISH of YAC clones presents certain complexities. The most straightforward approach is to use total yeast DNA as probe. As the cloned insert is only a small part of the total DNA, increased amounts of probe may be required for YAC mapping compared with bacterial clones. After hybridization, the short arms of the human acrocentric chromosomes are often decorated by probe derived from yeast ribosomal sequences. This varies from clone to clone and can be difficult to suppress.

An alternative approach is to selectively amplify the human sequences in the YAC from total yeast DNA using PCR primers for human-specific interspersed repetitive elements such as *Alu* repeats. Many protocols exist for this (8) (see also Chapters 6 and 7). Success is dependent on an adequate distribution of *Alu* elements in the clone being amplified. However, although the average distribution of *Alu* repeats is approximately one every 4 kb, these are not evenly distributed throughout the genome, being relatively rare in regions corresponding to cytogenetic G-dark bands. Thus not all clones may be amplified sufficiently for successful FISH mapping, and it is possible for small co-ligation events to be overlooked (9).

2.2.2 Probe labelling for FISH   *My probe is labelled with DIG*

Successful FISH depends on generating sufficient adequately labelled probe. For maximum hybridization efficiency, the probe fragments should be about 200–500 base pairs long after labelling. Longer fragments tend to form aggregates which will bind non-specifically giving greatly increased, coarse background signal.

Most labelling procedures are based on the enzymatic incorporation of hapten-conjugated nucleotides by nick translation, random primer extension, or the polymerase chain reaction. Hapten-dUTP is usually used instead of, or in addition to, dTTP in the labelling mixture. Biotin-11-dUTP or biotin-16-dUTP, and digoxigenin-11-dUTP are the most frequently used. Fluoro-chrome-conjugated nucleotides are also available, but these are used less frequently for gene mapping because of their lower signal intensity.

Nick translation is the method most widely used. The double-stranded target DNA is randomly nicked by DNase I, then DNA polymerase I (Kornberg holoenzyme) binds to the nicked DNA and replaces bases progressively from 5′ to 3′ to generate a newly synthesized strand. Labelled nucleotides in the reaction mixture are incorporated into the new DNA. Since the nicks are random, the distribution of nicks (and therefore the length of DNA fragments) is related both to the relative concentrations of DNase I and template DNA and to the duration of the reaction. It is necessary to titrate a stock solution of enzyme (*Protocol 3*) to determine the best reaction conditions—particularly as different DNase I preparations vary in activity.

Protocol 3.   Titration of DNase I for nick translation

Equipment and reagents

- Deoxyribonuclease I (D4527, Sigma)
- 10 × nick translation buffer: 0.5 M Tris–HCl pH 7.5, 0.1 M MgSO₄, 1 mM dithiothreitol, 500 µg/ml bovine serum albumin
- Enzyme diluent: 50% glycerol (v/v) in 1 × nick translation buffer

- Control DNA samples[a]
- Water-bath or thermocycler set at 14°C
- 1 kb size marker (Gibco–BRL)
- 0.5 M EDTA, pH 8.0

Method

1. Prepare a 1 mg/ml stock solution of DNase I in enzyme diluent. This can be stored at −20°C for several years.

2. Prepare a 1 µg/ml working solution of DNase I in enzyme diluent from the stock solution. Mix thoroughly by vortexing. This stock will remain

---

stable for at least a year of regular use, if kept on ice when in use and stored at −20°C.

3. Prepare several DNase I digestion reactions in microcentrifuge tubes on ice, containing 2 μg control DNA, 5 μl 10 × nick translation buffer, 1–2 μl DNase I working solution, made up to 50 μl final volume with sterile distilled water. Compare different amounts of DNase I working solution and different DNA samples.

4. Incubate the reaction tubes at 14°C.

5. After 20 min incubation, transfer 10 μl aliquots from each reaction to labelled 0.5 ml microcentrifuge tubes containing 1 μl of 0.5 M EDTA pH 8.0 (to inactivate the DNase I). Store the aliquots on ice.

6. Remove additional 10 μl aliquots at 10 min intervals, giving samples with a total time range of 20–60 min.

7. Run all samples against a 1 kb size marker on a standard 1% agarose gel.

8. Choose the DNase I concentration and incubation time which give fragment smears with a size range of 200–700 bp.

[a] Use any DNA samples, prepared by laboratory method of choice, which are known to cut cleanly with restriction enzymes.

If the titrated DNase I repeatedly fails to cut adequately during titration, further purification of the DNA may be necessary, or the reaction conditions may need to be modified to include additional Ca^{2+} to maintain active enzyme conformation (10). The exact amount of $CaCl_2$ will need to be titrated also but final concentrations in the range of 50–200 μM should restore efficient DNA nicking. Once titrated, the DNase I is used in the labelling reaction described in *Protocol 4*.

Protocol 4.   Nick translation for labelling FISH probes

Equipment and reagents

- 10 × nick translation buffer (see *Protocol 3*)
- 0.5 mM dNTPs: 2 μl each of Pharmacia 100 mM dATP, dCTP, and dGTP, and 1194 μl sterile distilled water (store aliquots at −20°C)
- 1 mM biotin-16-dUTP or digoxigenin-11-dUTP (Boehringer)
- DNase I, titrated 1 μg/ml working solution (*Protocol 3*)

- DNA polymerase I (10 U/μl; D9380, Sigma)
- DNA to be labelled (0.1–1 μg/μl)
- 0.5 M EDTA pH 8.0
- 3 M sodium acetate pH 7.0
- Absolute ethanol, stored at −20°C
- 80% ethanol, stored at −20°C
- Waterbath or thermocycler set at 14°C

Method

1. Add the following components in order in a 1.5 ml microcentrifuge tube, scaling up as necessary if larger quantities of probe are required:

Protocol 4. *Continued*

- 2.5 μl 10 × nick translation buffer
- sterile distilled water to make a final volume (including all ingredients) of 25 μl
- 1.9 μl dNTPs
- 0.7 μl dUTP
- 1 μl DNase I working solution (or other volume determined by titration)
- 0.5 μl DNA polymerase I (5 U)
- ~ 1 μg DNA

2. Mix well by lightly flicking the tube. Spin briefly in a microcentrifuge to bring down the solution.

3. Incubate at 14°C for the desired time (usually 40–45 min, as determined in *Protocol 3*).

4. Add 0.1 vol. 0.5 M EDTA to inactivate the enzymes.

5. Add 0.1 vol. 3 M sodium acetate pH 7 and 1 ml ice-cold absolute ethanol to precipitate the DNA. Mix thoroughly by inversion.

6. Incubate at −70°C for 30 min (or −20°C overnight).

7. Microcentrifuge at maximum speed for 10 min. A white pellet should be clearly visible. Discard the supernatant.

8. Add 1 ml ice-cold 80% ethanol, without disturbing the pellet, and microcentrifuge immediately for 10 min. Discard the supernatant immediately, leaving the pellet as dry as possible. The pellet will now be transparent and difficult to see. (The pellet tends to loosen: take care that labelled probe is not accidentally discarded.)

9. Air dry the pellet, but do not over-dry as this can make it difficult to redissolve.

10. Add 10 μl TE buffer and stand the tube on ice for 10 min. Flick the tube to resuspend the probe. Store at −20°C until required.

For each labelling reaction, fragment size is easily checked by running samples on an agarose gel. If FISH requirements are modest, commercially available nick translation kits (such as those from BRL or Boehringer Mannheim) may be more convenient.

2.3 Optional slide pre-treatment

In general, no slide pre-treatment should be necessary for mapping larger clones such as cosmids, PACs, BACs, or YACs. These produce strong hybridization signals which are easily distinguished above any background. If the cytological preparations have been prepared adequately there will be no traces of cell membrane or cytoplasm to impede access of the probe to the chromosomal target.

The mapping of smaller-insert clones, or clones which are difficult to map cleanly, can be assisted by treating the slides with RNase A to eliminate any RNA-related background signal, and Proteinase K digestion to remove contaminating cellular protein and to partially dissociate chromosomal proteins. *Protocol 5* gives a suitable pre-treatment.

Protocol 5.   Additional slide pre-treatment for mapping smaller insert clones

Reagents

- 2 × SSC (prepare as 20 × SSC stock: 3 M NaCl, 0.3 M sodium citrate)
- RNase A, 100 μg/ml in 2 × SSC, warmed to 37 °C

- Proteinase K, 0.06 μg/ml in 2 mM CaCl₂, 20 mM Tris pH 7.4, warmed to 37 °C

Method

1. Incubate the prepared slides in a Coplin jar of RNase A for 1 h at 37 °C.
2. Rinse the slides twice in 2 × SSC at room temperature.
3. Rinse briefly in distilled water.
4. Incubate in a Coplin jar of Proteinase K at 37 °C for 7 min.
5. Rinse briefly in two changes of 2 × SSC.
6. (a) Proceed immediately to denaturation in 70% formamide (*Protocol 6*).

   (b) Alternatively, rinse briefly in distilled water and dehydrate through an ethanol series as in *Protocol 2*, step 10. Store with desiccant until required.

2.4 Probe hybridization

Protocol 6 gives details of the probe hybridization procedure. Unlabelled Cot-1 or placental DNA is added to complex probes such as cosmids or YACs to suppress hybridization to repetitive elements such as *Alu* sequences which are present in all chromosomes (11, 12). Interspersed repeats in the labelled probe can produce an '*Alu*-banding' pattern of hybridization, reflecting the preferential location of these sequences in the G-light (R) bands of the chromosomes. Incomplete suppression of these repeats is often seen as 'breakthrough' hybridization in regions such as the distal short arms of chromosome 1, and on chromosome 19, in particular. Even small probes (such as cDNAs) may require suppression in this way, and it is best to try probe mixes with and without the unlabelled competitor DNA. Such suppression should, of course, be omitted when using probes intended to detect simple sequence repeats.

The slides are prepared for hybridization by denaturing in formamide, quenching in ice-cold ethanol, and then dehydrating, to ensure that single-stranded targets for probe hybridization are exposed.

Protocol 6. Hybridization

Equipment and reagents

- Hybridization buffer: 50% deionized formamide,[a] 2 × SSC (*Protocol 5*), 10% dextran sulfate, 0.1% SDS, 1 × Denhardt's solution (0.02% Ficoll, 0.02% polyvinylpyrrolidone, 0.02% BSA),[b] 40 mM sodium phosphate pH 7. Store 10 ml aliquots of autoclaved 50% dextran sulfate in 50 ml Falcon tubes at −20°C. 50 ml batches of hybridization buffer can then be made by adding the other reagents to an aliquot of dextran sulfate. Mix the hybridization buffer very thoroughly, as the dextran sulfate is extremely viscous, and store 1 ml aliquots at −20°C.

- Cot-1 DNA (Gibco–BRL) or sonicated human placental DNA (Sigma)[c]
- 70% formamide: 70% formamide, 30% (v/v) 2 × SSC (*Protocol 5*), stored at 4–8°C and reused for a week
- Ice-cold 70% ethanol (in a Hellendahl jar, stored at −20°C, replaced weekly)
- 32 × 22 mm coverslips, stored in ethanol
- Rubber cement (Cow Gum, or similar)
- 65°C and 37°C water-baths
- Ethanol series (as for *Protocol 2*)

Method

1. Pre-warm a Coplin jar of 70% formamide to 65°C in preparation for step 6.

2. In a 0.5 ml microcentrifuge tube add:
   - 0.5 μl labelled DNA (30–50 ng)[d]
   - 1 μl Cot-1 DNA (1 μg) or sonicated placental DNA (10 μg)[c, d]
   - 14 μl hybridization buffer

3. Mix thoroughly and spin briefly in a microcentrifuge.

4. Denature the probe mix at 65°C for 10 min.

5. Transfer the probe mix to 37°C to pre-anneal for 15 min–3 h.[e] Meanwhile, proceed with steps 6–8.

6. Denature the slides in 70% formamide at 65°C for 2 min.[f]

7. Quench the slides in 70% ice-cold ethanol.

8. Dehydrate the slides through the ethanol series, allowing 30–60 sec in each jar. Air dry.

9. Pipette the probe mix onto the slides, and cover with polished 22 × 32 mm coverslips. Seal the coverslip edges with rubber cement.

10. Incubate overnight at 37–42°C.

[a] Use mixed bed resin (Bio-Rad AG 501-X8).
[b] Denhardt's solution improves hybridization efficiency, but is not essential for standard FISH.
[c] This DNA may be omitted for some probes; see Section 2.4.
[d] There is no need to precipitate the probe and competitor DNA before dissolving in hybridization buffer, as the buffer can be diluted (by probe) by as much as 20% without significant effect on hybridization efficiency.
[e] If Cot-1/placental DNA is omitted in step 2, this step should also be omitted. Instead, the probe should be snap-chilled on ice or kept at 65°C until it is used in step 9.

[f]The addition of extra slides reduces the temperature of the formamide approximately 0.5°C per slide, but sufficient denaturation still occurs at temperatures down to at least 58°C in 2 min. Stronger denaturation is not usually compatible with good chromosome morphology, particularly when banding is required. Older slides stored at room temperature become more difficult to denature and it may be necessary to extend the denaturation times when working through a batch of slides, for example by adding 15 sec per week after slide preparation. However, each laboratory will need to experiment with conditions to allow for inevitable local variations. Note that steam from the water-bath can negate the preparative slide-ageing steps.

2.5 Detection of hybridization signals
2.5.1 Single colour detection of biotinylated probes

For single colour detection of a biotinylated probe there are several different protocols available. The original 'three layer' method uses successive incubations in avidin–FITC, biotinylated anti-avidin, and then avidin–FITC again (13). A modified 'two layer' method uses avidin–FITC followed by FITC-conjugated anti-avidin. In the still simpler 'single layer' method, these same two components are added to the slide simultaneously in a single incubation (14). Choice of method will depend on the sensitivity required, which increases with the number of layers. Typically, however, the two or three layer method is most effective.

Protocol 7 describes the one, two, and three layer techniques. Incubations can be extended if necessary but prolonged incubation may produce increased non-specific background signal. Slides must not be allowed to dry as this also results in increased background.

Protocol 7. Single colour detection for biotinylated probes

Exposure of dye-containing solutions and slides to light should be minimized during this procedure, and incubations should not be performed in direct light. Under no circumstances must the slides be allowed to dry out during this procedure.

Equipment and reagents

- Wash solution (4 × TNFM): 4 × SSC, 0.05% Tween 20, 5% non-fat milk powder,[a] filtered through several layers of Whatman No. 4 filter paper (prepare SSC as 20 × stock: 3 M NaCl, 0.3 M sodium citrate)
- Three Coplin jars containing 2 × SSC, warmed to 42°C
- Two Coplin jars of 50% formamide (50% formamide, 50% 2 × SSC, v/v), warmed to 42°C
- One Coplin jar of 4 × TNFM, warmed to 37°C

- 0.08 μg/ml DAPI in 2 × SSC in foil-covered (light-proof) Coplin jar
- Antifade mountant (Citifluor AF1 or Vecta-shield)
- Clear nail varnish
- Laboratory sealing film (e.g. Nescofilm or Parafilm)
- 4 × SSC, 0.05% Tween 20
- Coverslips, 22 × 32 mm

Protocol 7. *Continued*

- Immunochemical detection solutions, diluted in 4 × TNFM. Prepare sets (a), (b), or (c) according to chosen detection strategy. Prepare 100 μl of each solution (i.e. 'one volume') per slide plus a minimum of 50 μl excess, unless stated otherwise below.

  (a) One layer detection: 8 μg/ml avidin–FITC DCS (0.5 vol.), 8 μg/ml fluorescein-conjugated anti-avidin D (0.5 vol.)

  (b) Two layer detection: 4 μg/ml avidin–FITC DCS, 4 μg/ml FITC-conjugated anti-avidin D (Vector)

  (c) Three layer detection: 4 μg/ml avidin–FITC DCS (Vector) (2 vol.), 4 μg/ml biotinylated anti-avidin D (Vector)

Stand each solution for 10 min at room temperature, then microcentrifuge 10 min to pellet any protein complexes. When using these solutions, take the supernatant, avoiding the protein pellet.

Method

1. Using fine forceps, carefully remove the dried rubber cement from the slides. Soak off the coverslips in the first jar of warmed 2 × SSC (approximately 5 min).

2. Transfer the slides to the first jar of 50% formamide and incubate at 42°C for 5 min.

3. Transfer the slides to the second jar of 50% formamide and again incubate at 42°C for 5 min.

4. Wash for 5 min each in the two jars of 2 × SSC at 42°C.[b]

5. Transfer the slides to the jar of 4 × TNFM and incubate at 37°C for 15–30 min.

6. Replace the 2 × SSC in the three jars at 42°C with 4 × TNFM.

7. Drain each slide briefly.

   (a) For one layer detection: mix equal volumes of each detection mix and immediately pipette 100 μl onto each slide.

   (b) For two- or three-layer detection: apply 100 μl of avidin–FITC solution.

8. Cover with a 25 × 50 mm strip of Nescofilm. Incubate the slides in a humidified box at 37°C for 20–60 min.

9. Wash slides in each jar of 4 × TNFM at 42°C for 5 min.

   (a) For one layer detection: proceed to step 14.

   (b) For two or three layer detection: continue with step 10.

10. Drain each slide and apply 100 μl of:

    (a) For two layer detection: FITC anti-avidin.

    (b) For three layer detection: biotinylated anti-avidin.

    In either case, cover with a strip of Nescofilm and incubate as before.

11. Replace the wash solutions in the Coplin jars with fresh 4 × TNFM and warm to 42°C.

12. Wash slides in each jar of 4 × TNFM at 42°C for 5 min.

    (a) For two layer detection: proceed to step 14.

    (b) For three layer detection: continue with step 13.

13. Repeat steps 7, 8, and 9.

14. Wash twice in 4 × SSC, 0.05% Tween 20 at room temperature.

15. Stain in 0.08 μg/ml DAPI for 2–3 min.

16. Rinse in 2 × SSC and dehydrate through an ethanol series. Air dry.

17. Apply 20 μl aliquots of antifade solution to clean 22 × 32 mm coverslips.

18. Overlay with slides, blot and seal with nail varnish.[c]

[a] The use of non-fat milk does not appear to be necessary for most purposes but it is possible that it gives slightly greater signal intensity and is therefore useful for FISH mapping of smaller clones.

[b] The stringency of these washes can be increased by diluting the salt down to 0.1 × SSC and/or raising the temperature of the washes to 45°C. Chromosome-specific centromeric repeat probes are usually washed at high stringency as cross-hybridization with related sequences on other chromosomes occurs at lower stringencies.

[c] These slides can be stored in the dark at 4°C for several months without substantial deterioration.

2.5.2 Multiple colour detection

Multiple colour detection is achieved by using successive incubation mixes containing the relevant affinity molecules. Care must be taken to choose a detection strategy avoiding cross-reacting antibodies. For example, sheep anti-digoxigenin cannot be used in conjunction with biotinylated anti-avidin raised in goat, as the subsequent anti-sheep antibody will bind to both species. In this case cross-reactivity can be avoided by separating the various components into a multilayer protocol and using the anti-sheep antibody before the goat anti-biotin. If possible, choose antibodies from unrelated species. *Protocol 8* describes a suitable method for two-colour detection.

Protocol 8. Two-colour detection for combinations of biotinylated and digoxigenin-labelled probes

Reagents

- As for *Protocol 7*, substituting the following immunochemical detection solutions:
  (a) 4 μg/ml avidin Texas Red DCS (Vector)
  (b) 4 μg/ml biotinylated anti-avidin D plus 1 μg/ml mouse anti-digoxigenin (Boehringer) or 1:500–1000 dilution of mouse anti-digoxin FITC (Sigma)

  (c) 4 μg/ml avidin Texas Red DCS plus 10 μg/ml goat anti-mouse FITC conjugate (Sigma, F0257) or 7.5 μg/ml rabbit anti-mouse FITC conjugate (Sigma, F7506)

Method

1. As for *Protocol 7*, using the three layer detection procedure. Solutions a, b and c above should be substituted for the first avidin–FITC, the biotinylated anti-avidin, and the second avidin–FITC solutions respectively.

2.6 Chromosome banding

Molecular cytogenetics aims to relate a given labelled probe to a specific chromosome region or regions. Therefore chromosomal identification is essential in many applications. FISH is quite compatible with simultaneous detection of hybridization signal and chromosome banding. There are a number of different approaches to achieve this end, including *Alu* banding (15, 16), and replication banding with BrdU substituted into chromosomal DNA (17–19).

Many of the fluorochromes used for staining chromosomes have a preference for AT- or GC-rich tracts of DNA. Moreover, the hybridization procedures modify chromatin to some degree to alter the distribution of non-specific staining. Consequently, since chromosome bands are known to have clear differences in base pair distribution, with R-bands being GC-rich and G-bands being AT-rich, it is possible to produce clear chromosome banding after FISH by simply counterstaining the chromosomes with a suitable fluorochrome.

The most widely used fluorochromes are propidium iodide (PI), which binds to DNA non-specifically, and DAPI (4',6-diamidino-2-phenylindole) which has a strong AT preference. Many of the published protocols recommend using these stains at much higher concentrations than necessary, resulting in saturated staining and complete obliteration of the inherent banding pattern. Simply reducing the concentration of PI to 0.4 µg/ml in antifade reveals traces of chromosome banding. Similarly, staining in 0.08 µg/ml DAPI (see *Protocol 7*) prior to mounting produces an adequate QFH-type banding pattern. This banding is not seen to the same extent when the DAPI is mixed with the alkaline antifade mountant. Differences in staining of the heterochromatic blocks, such as 9qh, may be observed after mounting in different antifade solutions. The presence or absence of glycerol and the pH of the mounting medium are critical to many fluorescent banding techniques (20).

The addition of 0.8 µg/ml DAPI enhances the R-banding pattern faintly seen with 0.4 µg/ml PI. PI fluorescence is suppressed in the heterochromatic regions of 1qh, 9qh, 16qh, and Yqh where the DAPI binds most strongly. Even stronger band definition is seen when BrdU is incorporated into the chromosomes for the last stage of replication prior to harvest (21).

The choice of fluorochrome or combination of fluorochromes is influenced by the type of imaging system being used. Propidium iodide cannot be used for monochrome digital imaging with multiple band-pass filters as it is excited over a wide range of wavelengths. Consideration must be given also to the properties of the fluorochromes used for detecting the hybridization signal. For example, propidium iodide is unsuitable for use with antibodies conjugated to Texas Red or Rhodamine.

3. Interpretation of metaphase results

3.1 Chromosome band assignments

Having successfully visualized the probe hybridized to its chromosomal target, this target must be identified unequivocally. Clear signal should be visible at specific chromosomal locations. Where the ratio of probe signal to background 'noise' is large and the hybridization efficiency good, as is usually the case with cosmid or YAC probes, specific signal appears as paired spots for each chromatid in most metaphase spreads (see *Figure 1a*). Examination of a small sample of metaphases will quickly confirm the localization, without the laborious spot counting required for isotopic *in situ* hybridization. However, it is always possible that weaker secondary signals, indicative of other related sequences or perhaps clone chimerism, may be present. A minimum of ten metaphases should be examined thoroughly, despite the presence of a strong specific signal.

When the non-specific background is relatively high, it is quite possible to overlook secondary signals and results cannot be regarded with the same degree of confidence. As clone size decreases, the signal-to-noise ratio drops, and a simple inspection of a metaphase will not yield clear results. In these cases spot counting will frequently be useful. All chromosomal signals from 15–25 metaphases are marked on a chromosome ideogram worksheet of about 550 band resolution, noting any doublets. Specific localization will be revealed by the accumulating score of signals above the background level. Even when an initial inspection of the slides is not promising, a result may still be possible with a little extra effort. Nevertheless, spot counting is appropriate only for mapping small-insert clones such as cDNAs and should not be used to overcome the defects of poor technique.

Although the simultaneous visualization of chromosome banding and hybridization signal enables clones to be placed with considerable accuracy, the signal is not necessarily seen at its precise origin. The chromosomal DNA has been partially released by denaturation and hybridization has been detected by a immunochemical complex lying over the chromosome. Moreover, the size of hybridization signal is frequently larger than many of the bands in prometaphase chromosomes. Therefore, chromosome band assignment is not necessarily clear-cut. High-resolution localizations should be interpreted as the most likely position within a certain range.

3.2 Clone ordering on metaphase chromosomes

Pairwise ordering on prometaphase chromosomes is possible for clones separated by more than 1–1.5 Mb (22, 23). When probes are close, statistical analysis of a large number of signals is required. The position of red and green signals along the chromosome axis is scored relative to the centromere for each chromatid, thereby generating a possible maximum of four scores from

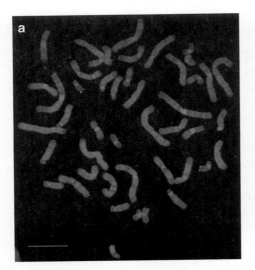
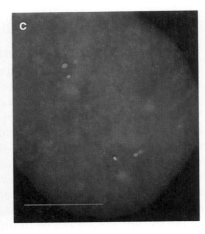
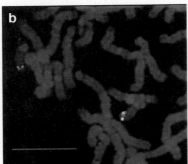
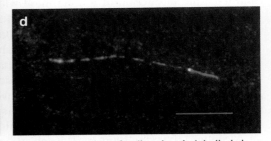

Figure 1. FISH mapping with cosmid clones. (a) Localization of a digoxigenin labelled clone (green) to chromosome 22q13 and a biotin labelled clone (red) to Xq25 on a DAPI-banded normal male metaphase. (b) Metaphase ordering of two clones in 22q13. The digoxigenin labelled clone (green) was proximal to the biotinylated clone (red) in 47 of 62 informative chromatids, in a total of 20 metaphases analysed. (c) Interphase FISH on a DAPI stained fibroblast nucleus, using three clones from chromosome 22, one labelled with biotin (red), one with digoxigenin (green), ordered relative to a mixture of biotin and digoxigenin labelled probe for the third clone (orange). The orange and red signals are separated by approximately 2.5 Mb. (d) DNA fibre FISH with two clones in 22q12.3, from a minimal overlapping set of cosmids prepared for sequencing. Scale bars = 10 μm. Images were obtained using a Zeiss Axioskop microscope fitted with a Photometrics cooled CCD camera and motorized filterwheel controlled by SmartCapture imaging software (Vysis, UK).

each metaphase (see *Figure 1b*). At least 15 metaphases should be scored, but larger samples will be required as the distance between clones decreases. It is probable that the actual molecular distances permitting metaphase resolution will vary between chromosome regions.

Additional ordering of clones which map to a single chromosome region can be achieved using chromosomal rearrangements with breakpoints involving that region (24, 25). Large numbers of clones can be placed readily into intervals defined by a panel of rearrangements (see Chapter 6). However, even apparently simple balanced rearrangements may be more complex at a molecular level.

4. Interphase FISH mapping

The interphase chromosome is significantly less condensed than its metaphase counterpart. Nevertheless, hybridization of specific whole-chromosome paints reveals that individual chromosomes occupy relatively compact domains within the nucleus (12, 26). This discrete chromosomal organization means that it is possible to study the spatial distribution of individual chromosomes, and chromosomal segments, in interphase.

Trask *et al.* (27) reported that the measured distance between clones hybridized to interphase nuclei is related to their molecular separation. This relationship has been shown to be linear over distances up to approximately 1 Mb in specific regions of the X chromosome (22, 28). Senger *et al.* (29) have shown a similar relationship in the HLA class II region in 6p21. More recently it has been shown that measured interphase distances have a linear relationship with molecular distances up to 2 Mb (30). This relationship is based on a statistical model, the Gaussian chain, developed to describe the spatial distribution of large polymers (31). In essence, it can be shown that the square of the mean measured interphase distance between two clones is proportional to their molecular separation. Further studies suggest that a similar relationship may extend over greater molecular distances (32).

This relationship between molecular and measured distances implies that it is possible to determine the probable order of a set of clones from the orientation of interphase hybridization signals. Orders are obtained by scoring the relative positions of signals from three clones labelled with different haptens (see Section 6). The three clones can be uniquely identified by using a mixture of biotin and digoxigenin labelled probe for one clone (producing an 'orange' signal), in addition to the red (biotin) and green (digoxigenin) signals for the other two probes (see *Figure 1c*).

4.1 Isolation of nuclei for interphase FISH

Cells for interphase analysis must be in the G_1–G_0 phase of the cell cycle. Once replication has begun, nuclear volume changes and the appearance of replicated chromosome signals makes analysis both confusing and potentially misleading. Nuclei in the G_0 phase of the cell cycle can be prepared from uncultured lymphocytes isolated from peripheral blood or from contact-inhibited cultured fibroblasts. The latter are preferable, both because of their larger nuclear volume (giving better resolution), and because contact inhibition ensures a high proportion of G_0 nuclei.

Protocol 9. Preparation of fibroblast interphase nuclei

Equipment and reagents
- Normal male fibroblasts (obtained from ECCAC or ATCC)
- 75 mM KCl, pre-warmed to 37°C
- Trypsin–EDTA solution (Sigma, T3924), warmed to 37°C
- See also *Protocol 1*

Method

1. Allow fibroblast cultures to reach confluency using standard culture conditions.
2. Leave undisturbed for a further four to seven days, to ensure that mitotic activity has reached a minimum.
3. Trypsinize cells according to standard subculturing procedures to produce a suspension of single cells.
4. Transfer cells to a 20 ml centrifuge tube, neutralizing the trypsin–EDTA with reserved culture medium.
5. Rinse the tissue culture flask with several aliquots of reserved culture medium to collect any remaining cells. Pool in the 20 ml tube.
6. Centrifuge at 250 *g* for 5 min.
7. Remove the supernatant and loosen the cell pellet by lightly flicking the base of the tube.
8. Add 10 ml pre-warmed 75 mM KCl and resuspend the cells.
9. Centrifuge at 250 *g* for 5 min.
10. Repeat steps 7 and 8.
11. Incubate the cells at 37°C for 10 min.
12. Proceed as for *Protocol 1*, step 8 onwards.

4.2 Preparations for interphase FISH

Slides are prepared as described previously but there is no need to preserve chromosome morphology by ageing the slides. Hybridization efficiency is highest with fresh slides so they can be prepared on the day of hybridization. However, older slides can also be used as for metaphase mapping.

Interphase analysis is not recommended with probes with hybridization efficiencies of less than 80% (i.e. hybridizing to less than 80% of metaphase targets), since less than 50% of nuclei will be successfully hybridized with three probes, and the number of informative nuclei will be even fewer due to unfavourable orientation of some nuclei. Therefore the larger-insert clones, particularly cosmids, are more useful for interphase mapping. Very large clones like YACs may cause some difficulties with interphase measurements because of the multilobular nature of the hybridization signal (reflecting the organization of the DNA in looped chromosomal domains), and the presence of repetitive sequences may make analysis unreliable.

The use of probe master mixes simplifies the making of slide mixes (since a number of probe combinations may be needed for any experiment), while preserving the original stock of labelled probe from repeated freeze/thawing. It is essential that each probe gives strong, clean signal which is of comparable intensity to the others used. Background signal from a poor probe will quickly render analysis meaningless, and any attempt to amplify weak signals will certainly increase the observed level of background.

For three-colour analysis, the reference clone is best allocated the 'orange' combination to avoid any difficulty in interpretation. In instances where a pair of clones is very close, a two-colour scheme is preferable, since it may not be possible to distinguish between overlapping red and green signals and the 'orange' of the reference clone (particularly as this 'orange' is often a mixture of red and green signals rather than a uniform orange spot).

Protocol 10. Interphase hybridization

Equipment and reagents

- Probe master mixes as required: 4 μl labelled probe (300–400 ng), 6 μl Cot-1 DNA or sonicated placental DNA, 40 μl hybridization buffer (*Protocol 6*)—store at −20°C (stable for at least one year)
- Slides of interphase nuclei prepared according to *Protocol 9*

Method

1. Pool 5 μl aliquots of probe master mixes in the desired combinations, i.e.:

    (a) Two biotin labelled probes, or one biotin and one digoxigenin labelled probe, for pairwise distance analysis.

    (b) Two biotin labelled probes plus a digoxigenin labelled probe for clone ordering.

    (c) One biotinylated probe, one digoxigenin labelled probe, and equal parts of biotinylated and digoxigenin labelled forms of the third clone (20 μl total volume), to enable three-colour analysis for both distance and order.

    Make up the volume to 15 μl with hybridization buffer if required.

2. Proceed as for *Protocol 6*, from step 3.

3. Detect hybridization using *Protocol 8*.

4.3 Assembling interphase mapping data

Examination of interphase FISH signals reveals that there is substantial variability in the relative positions of any pair of linked clones (22, 28). Much of the observed variation results directly from viewing the three-dimensional interphase nuclei in a single image plane, flattened onto microscope slides. The resultant signal distribution then reflects the different possible orientations of individual interphase chromosomes as the drop of fixed nuclei strikes the slide.

In order to allow for variability, several slides should be analysed. Each probe combination should be hybridized in hapten-reversed duplicate (i.e. with the colours of the two probes reversed), and at least 30 cells should be scored from each slide. All slide analysis should be performed 'blind', and biased selection of nuclei minimized. Some selection of nuclei for analysis is unavoidable since no clear data can be obtained from overlapping homologous chromosomes. Scan areas of the slide containing nuclei of regular shape, uniform size and DAPI staining intensity. Screening is most conveniently performed at high magnification using the filter combination for the biotin labelled probe(s). Nuclei showing distinct signals are then imaged, and the relative positions of the other probe(s) can then be recorded.

4.3.1 Interphase order analysis

Interphase signals are rarely aligned, so orders are determined by relative distances from a third orienting clone. The results are distributed into the different classes possible (red–red–green [RRG] or RGR for two-colour experiments; orange–red–green [ORG], OGR, ROG for three-colour experiments). The number of uninformative signal patterns and absent signals are also noted as these may be useful in assessing experimental reliability. Inspection of the data in the informative classes will usually indicate the probable clone order, which can be confirmed by applying a statistical test such as the standard χ^2 for goodness of fit, assuming an expected random frequency of 0.5 for two informative classes. The probability of the data arising by chance can then be obtained from χ^2 distribution tables (one degree of freedom).

Statistically significant interphase orders are obtained when the distances between the adjacent probes are approximately equal, and is similarly clearcut and highly significant when the ratio of distances increases from 1:1 to 2:1, and probably to 3:1. As the ratio increases, more nuclei have to be analysed to achieve statistical significance, reflecting increasing flexibility in the physical location of each locus (see *Table 1*). Therefore, interphase orders must be viewed with caution when reference clones are a substantial distance from those being ordered. Orders over distances greater than 4–5 Mb may be incorrect in some cases due to the flexibility of the target, and should be supported by as much corroborative data as possible. The use of a range of reference clones, both proximal and distal to the region of interest, greatly increases the chance of detecting potential conformational anomalies in the interphase nucleus.

Table 1. Interphase order analysis for the locus pair AK1 and SPTAN1 in 9q34[a].

Clones A–B–C	Orders ABC	ACB	BAC	N.I.[b]	N.H.[b]	Total	χ^2	P
D9S60–AK1–SPTAN1	26	13	8	5	0	52	4.33	< 0.05
AK1–SPTAN1–D9S61	35	2	9	6	8	60	15.36	< 0.001
AK1–SPTAN1–D9S115	35	2	7	3	5	52	18.67	< 0.001
AK1–SPTAN1–ASS	49	7	32	6	8	102	3.57	n.s.[b]
AK1–SPTAN1–ABL	22	9	19	0	14	64	0.2	n.s.
AK1–SPTAN1–DBH	16	4	22	2	10	54	0.95	n.s.
AK1–SPTAN1–D9S67	22	10	46	4	18	100	8.47	< 0.005

[a]The total distance from D9S60 to D9S67 is approximately 12–14 Mb, based on cumulative interphase distance estimates. AK1 and SPTAN1 are separated by approximately 1.5 Mb. Note that the order with the most distal clone, D9S67, appears inverted. D9S60 is 3–4 Mb proximal to AK1. Distal to SPTAN1 the approximate distances are 0.5 Mb to D9S61, 1.5 Mb to D9S115, 3 Mb to ASS, 6 Mb to DBH, and 7–8 Mb to D9S67.
[b]N.I. not informative. N.H. not hybridized (no signal). n.s. not significant.

4.3.2 Measurement of interphase distance

The most convenient description of the cytological relationship between two linked DNA clones is contained in the mean and variance of the total set of interphase distance measurements, even though these measurements are not normally distributed (as the variance increases with the mean). To minimize variation, ideally, all analyses should be performed on the same cell preparation using slides prepared in a single batch. However, the amount of variation

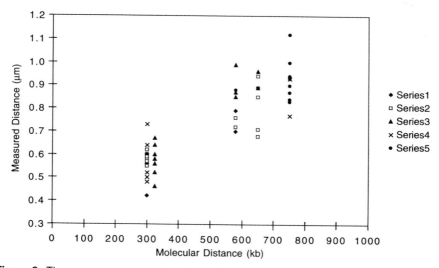

Figure 2. The mean measured interphase distance varies from slide to slide. Measurements from separate slides, from five different series of slides made from the same sample of fixed fibroblast nuclei, are shown for five different probe pairs. The variation between slide series appears no greater than the variation within a single series.

Table 2. Comparison of mean measured interphase distances (in μm) for loci in 9q32-33 indicates a probable locus order

	ALAD	ORM	D9S59	HXB	C5	GSN	D9S60
D9S58	1.63	1.77	1.86				
ALAD		0.91	0.96	1.47	2.02		
ORM			0.26	1.05	1.79	1.79	
D9S59				1.01	1.74	1.89	
HXB					1.30	1.34	
C5						0.59	1.78
GSN							1.66

among different slide batches, using the same cell preparation, appears to be no greater than that occurring between slides made on the same day (see *Figure 2*). The analysis of several slides for each probe combination should increase the level of confidence in the data obtained.

In general, the broad confidence limits for interphase distance measurements mean that it is often not possible to order small numbers of clones confidently by comparing average distances, and simple order analysis is usually more successful. Relative ordering of larger numbers of clones can be achieved by measuring as many combinations in the region as possible and arraying the average distances in increasing size in a two-dimensional matrix (see *Table 2*) (30).

4.3.3 Estimation of molecular distance

Since the variance of interphase distance measurements is large, a sufficient number of measurements must be made to obtain a reliable estimate. Unfortunately, the mean measured distance has no direct physical meaning in the absence of specific information about relative packing of different regions of interphase chromosomes. Therefore, calibration curves must be generated to relate the measured interphase distances between clones to true molecular distances. Measurements are made of interphase distances for a set of reference clones of known molecular separation, covering a range of at least 100–500 kb. Each combination should be replicated at least once. Where possible, the reference clones should come from a chromosomal region similar to the one studied, i.e. similar R- or G-band localization, GC content, etc. Linear regression analysis is used to generate a best-fit line relating measured distance to molecular separation (see *Figure 3*). Subsequent interphase distance measurements can be plotted onto the graph or calculated from the regression equation to derive estimates of molecular distance and confidence intervals. It is useful to include at least one slide with a subset of reference clones in every hybridization to monitor any potential slide-to-slide variation. A new calibration curve should be produced for every new cell preparation.

These calibration curves do not pass through the origin. This may be due

25. Kievits, T., Dauwerse, J. G., Wiegant, J., Devilee, P., Breuning, M. H., Cornelisse, C. J., *et al.* (1990). *Cytogenet. Cell Genet.*, **53**, 134.
26. Pinkel, D., Landegent, J., Collins, C., Fuscoe, J., Seagraves, R., Lucas, J., *et al.* (1988). *Proc. Natl. Acad. Sci. USA*, **85**, 9138.
27. Trask, B., Pinkel, D., and van den Engh, G. (1989). *Genomics*, **5**, 710.
28. Trask, B. J., Massa, H., Kenwrick, S., and Gitschier, J. (1991). *Am. J. Hum. Genet.*, **48**, 1.
29. Senger, G., Ragoussis, J., Trowsdale, J., and Sheer, D. (1993). *Cytogenet. Cell Genet.*, **64**, 49.
30. van den Engh, G., Sachs, R., and Trask, B. J. (1992). *Science*, **257**, 1410.
31. Doi, M. and Edwards, S. F. (1988). *The theory of polymer dynamics.* Oxford University Press, Oxford.
32. Yokota, H., van den Engh, G., Hearst, J. E., Sachs, R. K., and Trask, B. J. (1995). *J. Cell Biol.*, **130**, 1239.
33. Wiegant, J., Kalle, W., Mullenders, L., Brookes, S., Hoovers, J. M. N., Dauwerse, J. G., *et al.* (1992). *Hum. Mol. Genet.*, **1**, 587.
34. Parra, I. and Windle, B. (1993). *Nature Genet.*, **5**, 17.
35. Fidlerová, H., Senger, G., Kost, M., Sanseau, P., and Sheer, D. (1994). *Cytogenet. Cell Genet.*, **65**, 203.
36. Haaf, T. and Ward, D. C. (1994). *Hum. Mol. Genet.*, **3**, 697.
37. Heiskanen, M., Karhu, R., Hellsten, E., Peltonen, L., Kallioniemi, O. P., and Palotie, A. (1994). *BioTechniques*, **17**, 928.
38. Vogelstein, B., Pardoll, D. M., and Coffey, D. S. (1980). *Cell*, **22**, 79.
39. Haaf, T. and Ward, D. C. (1994). *Hum. Mol. Genet.*, **3**, 629.
40. Gerdes, M. G., Carter, K. C., Moen, P. T., and Lawrence, J. B. (1994). *J. Cell Biol.*, **126**, 289.
41. Huberman, J. A. and Riggs, A. D. (1966). *Proc. Natl. Acad. Sci. USA*, **55**, 599.
42. Lark, K. G., Consigli, R., and Toliver, A. (1971). *J. Mol. Biol.*, **58**, 873.
43. Heng, H. H. Q., Squire, J., and Tsui, L.-C. (1992). *Proc. Natl. Acad. Sci. USA*, **89**, 9509.
44. Houseal, T. W., Dackowski, W. R., Landes, G. M., and Klinger, K. W. (1994). *Cytometry*, **15**, 193.
45. Bensimon, A., Simon, A., Chiffaudel, A., and Croquette, F. (1994). *Science*, **265**, 2096.
46. Weier, H.-U. G., Wang, M., Mullikin, J. C., Zhu, Y., Cheng, J.-F., Greulich, K. M., *et al.* (1995). *Hum. Mol. Genet.*, **4**, 1903.

10

Contig assembly by fingerprinting

SIMON G. GREGORY, CAROL A. SODERLUND,
and ALAN COULSON

1. Introduction

A previous chapter in this series (1) described, primarily, the physical mapping of the 100 Mb *Caenorhabditis elegans* genome by fingerprinting of cosmid clones, and the linking of the contigs thus derived by YAC hybridization.

At that time, the primary function of the map was to enhance the molecular genetics of the organism by facilitating the cloning of known genes, and to serve as an archive for genomic information. However, a clonal physical map—even with good alignment to the genetic map—carries only a tiny proportion of the information present in the genome. Consequently, the current objective of the *C. elegans* genome project (2) is to establish of the entire genomic sequence. The bacterial clone map, although incomplete by virtue of the uncloneability of regions of the genome in cosmid vectors (a factor which we shall discuss later in this chapter), has proved a sound basis for the systematic sequence analysis. The sevenfold cosmid coverage has a resolution sufficient to enable the selection of a subset of cosmids for sequencing such that, on average, each clone contributes 30 kb of unique sequence to the whole. Sequencing projects based on bacterial clone maps (3–5) of a number of other genomes of a range of sizes are also well advanced, in particular *Saccharomyces cerevisiae* (15 Mb; complete), *Schizosaccharomyces pombe* (15 Mb), and *Drosophila melanogaster* (150 Mb). Although it has recently been demonstrated that small bacterial genomes can be sequenced by direct shotgun sequence analysis of the entire genome with no prior mapping (6), the ability to interrelate and map clone sets, whether derived by random selection or in a directed manner, is still the most convenient route to the sequence analysis of larger genomes.

The methods we describe here are currently being applied to projects that aim to produce large amounts of contiguous high quality sequence from human chromosomes. The bacterial clone restriction fingerprinting procedure is capable of integrating clones from a number of sources in different vectors. For example, randomly selected cosmids derived from a flow-sorted chromosome-specific library can be matched to fosmids and to larger-insert bacterial clones such as PACs (P1 artificial chromosomes) (7) and BACs (bacterial artificial chromosomes) (8) derived by probing from whole genome libraries. (The construction of adequate PAC or BAC libraries from flow-sorted chromosomes has not yet been realized.) These larger-insert clones are coming to the fore as the raw material for sequencing for reasons of efficiency. Moreover, it has been predicted (though not yet proven) that they may represent regions of genome not found in cosmid libraries.

This chapter will also mention the approaches which can be used to finger-print yeast artificial chromosomes (YACs). We refer to protocols for *Alu* PCR fingerprinting of YACs as applied to a pilot project on chromosome 22 and a combined *Alu* PCR fingerprinting/hybridization approach on the X chromosome (9).

Finally, we will describe the recent developments in image analysis, data processing and analytical computing as applied to contig assembly.

2. Principle of restriction fingerprinting

Restriction fingerprinting is a procedure for clone comparison based on the matching of characteristic restriction fragment sets ('fingerprints') between clones (*Figure 1*). If the fingerprints of two clones are found to share a signifi-cant number of fragments then it may be assumed that these two clones over-lap, the shared fragments representing the region of the overlap. In a typical 'shotgun' project, clones are picked at random from a library representing the genome or region of interest, their fingerprints are determined, and over-laps are established. Initially, of course, few overlaps will be detected but, as the project proceeds, small clusters of overlapping clones will be identified. As further clones are added, these small isolated contigs merge into a decreasing number of larger contigs, ultimately leading (in the ideal case) toward a single contig spanning the region of interest.

In the simplest case, fingerprints account for all of the DNA contained within a clone, as with those produced from a single or multiple enzyme digestion followed by agarose gel electrophoresis and detection by staining with ethidium bromide or SyBr Green (FMC). In some cases, however, it is preferable to analyse and compare only a subset of the restriction fragments from each clone. This can be achieved by digesting the clone with two or more restriction enzymes and labelling (fluorescently or radioactively) only those fragment ends produced by one of the enzymes. Then, only the subset of fragments which have one or both ends generated by the chosen enzyme will be represented in the 'fingerprint' of the clone.

In some situations, it is preferable to create and compare two or more independent fingerprints per clone, each generated by a different restriction enzyme (or combination of enzymes). This is particularly true when analysing

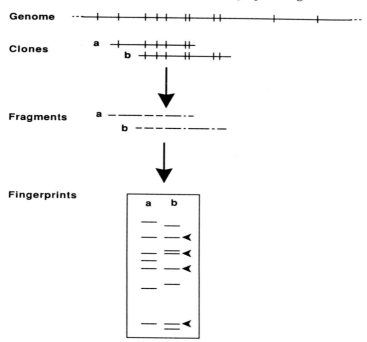

Figure 1. Principle of restriction fingerprinting. Restriction enzyme sites in the genomic DNA and in two overlapping clones (a, b) are indicated as short vertical lines. Digestion of the clones yields two populations of fragments which, when compared by gel electrophoresis, have some members in common (arrowheads). The common fragments indicate that the two clones overlap. For simplicity, the cloning vector is not shown; it will give rise to one or more fragments which (apart from those spanning the vector–insert junctions) will be identical for all clones.

fragments by agarose gel electrophoresis: the lack of resolution and (particularly for smaller fragments) of sensitivity of such gels means that the alignment of two clones on the basis of a single fingerprint from each may be unreliable.

The number of fragments produced by digestion of a clone depends on the base composition of the DNA and on the restriction enzyme used. For example, a *Hin*dIII / *Sau*3AI digestion in which only the *Hin*dIII sites are labelled (*Protocol 2*) will give on average 23 bands from a 40 kb cosmid derived from the genome of *C. elegans* (36% GC).The construction of fingerprint contigs on the human Y chromosome (10) is an example of using a more frequent cutter, *Hin*fI, to produce 30–40 bands per cosmid. The optimal enzyme (or enzyme combination) is that which will produce enough bands to allow the rapid identification of overlaps between smaller insert clones (such as cosmids or fosmids) without producing an unmanageably large number of fragments from larger clones such as BACs or PACs. Our fingerprinting studies on a variety of human chromosomes (including 13, 6, X, and 22) indicate that one should aim for a digestion protocol which will give an average of 15–23 fragments per 40 kb of cloned insert. The enzyme combination which best achieves this aim should be verified empirically. However, an approximate estimate of the frequency of cleavage by a restriction enzyme may be made on the basis of its cleavage sequence and on the GC content (refined, if possible, by reference to the flow karyotype) of the genome or chromosome to be mapped.

The electrophoresis system by which fingerprints are obtained is also an important consideration: the gel matrix will have direct effects on the separation and resolution of the fragments. We prescribe using 4% denaturing polyacrylamide gels run in wide sequencing tanks so as to include a maximum number of clones on each gel.

The power of our fingerprinting technique lies in its ability to incorporate any type of bacterial clone; to produce fragment sets that are sufficiently resolved and informative to identify overlaps between clones; and, by the use of specialized analytical programs, to identify overlaps between large numbers of clones of different types.

3. Expected coverage and rate of completion of contigs

Based on experimental experience (1–6) and on theoretical predictions, we estimate that maximum coverage of a region or chromosome with bacterial clones can be achieved by 10- to 20-fold representation. For example, a bacterial clone library with an average insert size of 40 kb and representing a 40 Mb region of the genome would give maximum coverage after 10 000 to 20 000 clones had been fingerprinted. Any gaps which remain at this point are likely to be due to lack of representation of certain sequences in the clone library, in which case fingerprinting of further clones will be of no help.

It is obviously desirable to monitor the progress of any large-scale contig assembly project. The increase in coverage contributed by the addition of new clones, and the number of independent contigs into which the clones fall are important factors in deciding when to stop fingerprinting and to begin directed work to close the remaining gaps between unlinked contigs. Typically, the number of independent contigs will rise rapidly during the early stage of the project, when most new clones do not overlap with any others. This number will reach a peak as more of the newly-added clones overlap with those already in the database, and will then fall as the addition of further clones links previously independent contigs. Finally, the curve will tend toward a plateau indicating that no greater coverage is being achieved by the addition of more clones. *Figure 2* indicates the progress of our chromosome

Chromosome 22

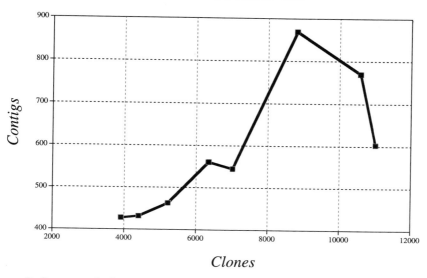

Clones

Figure 2. Progress of a fingerprinting project on human chromosome 22. The number of independent contigs is plotted as against the total number of clones fingerprinted. It rises initially, then peaks, and declines as the addition of further clones bridges the gaps between contigs. If continued, the number of contigs would be expected to reach a minimum value, reflecting the presence of genomic regions not represented in the bacterial clone library.

22 project; the current number of clones (approximately 11 000) corresponds to an average of 7.3-fold coverage of the chromosome, and the number of contigs has not yet declined to a plateau. In our experience, the closing of gaps in the final stages of a project is more efficiently achieved by using large-insert clones to bridge specific gaps rather than by performing further finger-printing of clones from a different library.

4. Fingerprinting strategies

There are several different fingerprinting strategies and types of bacterial clone which can be used when attempting to build a contig of either a whole chromosome or genome, or of a defined region. The choice of strategy is dependent upon the size of the region to be covered, the availability of a pre-existing large scale physical map (such as a well-characterized YAC contig), and the presence and density of markers such as sequence-tagged sites (STSs) which can aid in the identification or localization of the bacterial clones.

We have adopted three different strategies successfully, depending upon the circumstances. Where an entire chromosome is to be mapped, we use either a strategy of fingerprinting randomly-chosen cosmid clones from a chromosome-specific cosmid library (Section 4.1); or that of using previously mapped markers such as STSs as a 'scaffold' on which to assemble a contig of larger-insert bacterial clones (Section 4.2). The third approach is applicable when the region of interest is already covered by one or more YACs; the YACs can be used as a substrate to identify corresponding bacterial clones from a whole-genome library and assemble them into an overlying contig, an approach which we term 'regional contig assembly' (Section 4.3).

We feel that large-insert bacterial clones (BACs and PACs) represent the future of bacterial clone contig assembly due to their ease of manipulation and the smaller number of clones needed to cover a given region. As yet, however, chromosome-specific libraries of such clones are not available, and this limits the ways in which they can be used.

The following sections describe these strategies in more detail, along with the protocols necessary for their implementation. Later sections deal with the analysis of data generated by these approaches.

4.1 Fingerprinting of whole-chromosome cosmid libraries

It is now possible to produce chromosome-specific cosmid libraries for all chromosomes which can be purified by flow-sorting (see Chapter 7). Randomly selected clones from such a library can then be fingerprinted in order to assemble a bacterial clone contig. The speed at which a contig can be constructed by this 'whole-chromosome fingerprinting' approach is potentially greater than the mapping of clones by the regional hybridization approach (Section 4.3). Whole-chromosome fingerprinting allows bacterial clone mapping to proceed in parallel with the construction of a large-scale YAC map, and avoids the problems encountered in identifying bacterial clones with YAC probes. We estimate that fingerprinting an eight- to tenfold excess of clones would achieve 80–90% coverage of a chromosome prior to any directed gap closing efforts (Section 5).

We are currently undertaking a pilot project on chromosome 22 which incorporates the whole-chromosome fingerprinting strategy. The project involves fingerprinting cosmid clones from a well-characterized flow-sorted library and incorporating PACs isolated from a whole-genome library identified by hybridization of single copy probes. Contigs constructed from the fingerprinted PACs act as a seed for the immediate supply of clones for sequencing as well as providing positional information for a database containing randomly arrayed cosmids. Naturally, similar approaches can be applied to the construction of contig maps for entire small genomes, in the same way as for individual chromosomes from larger genomes.

It is assumed that a suitable library of bacterial clones already exists as a source of the raw material for fingerprinting. *Protocol 1* describes the preparation of clone DNA from glycerol stocks of the bacterial colonies. The method we describe is a derivation of the 'microprep' first described by Gibson and Sulston (11) incorporating the alkaline lysis procedure of Birnboim and Doly (12). The method is applicable to the extraction of PAC, BAC, fosmid, and cosmid DNA for fingerprinting and typically yields 100 ng to 200 ng, depending on the copy number of the vector and growth of the host.

Protocol 1. Microprep of bacterial insert clones

Equipment and reagents

- 1 ml deep-well microtitre plates (Beckman)
- Deep-well microtitre plate caps (Beckman)
- Plate sealers (Dynatech; Cat. No. 001-010-5701)
- U-bottom microtitre plates, and lids with condensation rings (Greiner)
- Orbital shaker (New Brunswick, model G24)
- Centrifuge with rotor holding microtitre trays (e.g. Sorval–RT 6000D)
- Multichannel pipette (5-50 µl, Flow Laboratories)
- Sterile cocktail sticks or 96-pin replicating tool (Denley)

- 2 × TY supplemented with 30 µg/ml kanamycin, 30 µg/ml ampicillin, or 12.5 µg/ml chloramphenicol as appropriate for the host/vector system used
- Solution I: 50 mM glucose, 10 mM EDTA, 25 mM Tris-HCl pH 8.0
- Solution II: 2 M NaOH, 1% (w/v) SDS in sterile, double distilled water
- Solution III: 3 M NaOAc pH 5.2
- Lithium chloride, 4.4 M
- Sterile double distilled water
- Ethanol or isopropanol (96%)
- Ethanol (70% in water)
- TE: 10 mM Tris–HCl pH 7.4, 1 mM EDTA

Method

1. Dispense 500 µl of 2 × TY (with appropriate antibiotic) into each well of a deep-well microtitre plate.

2. Inoculate each well from one of the clone glycerol stocks, either with sterile cocktail sticks or (if the clone library is stored in 96-well format) with the 96-pin inoculating tool. Seal the wells and place the plate in an incubator to shake overnight at 37°C for 18 h.

3. Remove 250 µl from the overnight culture to a sterile microtitre plate[a] and centrifuge at 1000 g for 4 min (PAC and BAC clones) or 2 min (cosmids, fosmids).

4. Discard the supernatant and disrupt the pellet using a vortexer set on intermediate speed or with a toothpick if the pellet remains adherent to the bottom of the well.

5. Add 25 µl of solution I to each well and tap the plate gently to resuspend the pellet.

6. Add 25 µl of solution II to each well and tap the plate gently to mix. Leave the plate at room temperature for 5 min. The solution in the wells should clear as the bacterial cells in culture lyse.

7. Add 25 µl of solution III to each well, and again tap the plate gently to mix, and leave at room temperature for 5 min. Cover the plate with a plate sealer and vortex vigorously for ten sec.[b]

8. Centrifuge the plate at 1800 g for 10 min at 4°C.

Protocol 1. *Continued*

9. Transfer 75 µl of the supernatant from each well to a fresh microtitre plate containing 150 µl of 96% ethanol or 100 µl of isopropanol. Ensure that none of the pellets are carried over. Seal the plate and leave at −20°C for 30–60 min.

10. Centrifuge the plate at 1800 *g* for 10 min at 4°C. Discard the supernatant and drain the plate by placing it face-down on a clean tissue, ensuring that the pellets are not dislodged from the bottom of the well.

11. Add 25 µl of ddH$_2$O to each well and resuspend the pellets by tapping the side of the plate or by using a sterile toothpick.

12. Add 25 µl of 4.4 M lithium chloride per well, mix by tapping, and leave at 4°C for 1 h.

13. Spin the plate at 1800 *g* for 10 min at 4°C.

14. Transfer the supernatants to a fresh plate containing 100 µl of 96% ethanol per well, ensuring that none of the pellets is carried over. Leave for 1 h or overnight.

15. Spin the plate at 1800 *g* for 10 min and discard the supernatant. Add 200 µl of 70% ethanol to each well and spin again at 1800 *g* for 10 min. Discard supernatant and ensure that the pellet has air dried to transparency before proceeding to the next step.

16. Resuspend pellet in 10 µl of TE[c] by slowly pipetting the TE up and down at the base of the well. The dissolved DNA can now be used for fingerprinting or stored the microtitre plate at −20°C.[d]

[a] The remaining 250 µl can be diluted with an equal volume of glycerol and stored at −20 or −70°C, or can be kept as a reserve in case the DNA prep fails.
[b] We routinely vortex the samples to break up the pellet prior to transfer of the supernatant using a Biomek 2000 robot; if the procedure is being carried out by hand no vortexing is required.
[c] This volume can be reduced to 6–8 µl if the yield of DNA is low.
[d] We store our micropreps for up to two weeks at −20°C before fingerprinting. We have found that storage of samples for a longer period of time reduces the quality of the fingerprints and also leads to evaporation of the sample aliquot.

In *Protocol 2* we describe our procedure for fingerprinting bacterial clones using digestion with two enzymes, in which the fragment ends created by one of the enzymes are labelled. As described above, this protocol gives the optimum fingerprint pattern from human-derived bacterial clones, and should prove applicable to the majority of mammalian chromosomes with minor modifications. The combination of enzymes and radioactive nucleotide will be determined by base composition, as mentioned in Section 2. Where possible, incorporate the appropriate radioactive nucleotide at the first position of the digest fragment (i.e. complementary to the innermost base of the 5′ overhang of the fragment) and a dideoxynucleotide at the second base to ensure irreversible completion of the labelling reaction. The fingerprinting method we employ uses the *Hind*III / *Sau*3AI digestion system first described by Coulson *et al.* (13), with a modification to the labelling procedure. In a one-step reaction (G. Hong, personal communication) we digest simultaneously with both enzymes and label in the first position of only the *Hind*III ends with [α-^{32}P]dATP before closure with the ddGTP terminator. This has allowed us to reduce the time required from approximately 3.5 hours (as required by the previous two-step protocol) to one hour.

Protocol 2. Restriction digest fingerprinting

Equipment and reagents

- U-bottom microtitre plates (Greiner)
- Microtitre plate lids with condensation rings (Greiner)
- Dideoxy GTP (ddGTP), 10 mM (Pharmacia)
- Double distilled water
- 10 × NEB2 buffer (New England Biolabs)
- *Hind*III (20 U/μl, Boehringer Mannheim)
- *Sau*3AI (50 U/μl, New England Biolabs)

- AMV reverse transcriptase (10 U/μl, N.B.L.)
- [α-^{32}P]dATP (800 Ci/mmol; Amersham)
- Formamide dyes: 98% (v/v) deionized formamide, 2% (v/v) 0.5 M EDTA, 0.1% (w/v) bromophenol blue, 0.1% xylene cyanol[a]
- 2 μl repeat dispenser (Hamilton Company)
- Multichannel pipette (2–10 μl, Flow Laboratories)

Method

1. Using a multichannel pipette, transfer 2 μl of the microprepped samples (*Protocol 1*) to a sterile microtitre plate.

2. On ice in a 1.5 ml Eppendorf tube prepare the labelling pre-mix. The following recipe is sufficient for 96 samples:

   - 155 μl double-distilled H$_2$O
   - 40 μl 10 × NEB2 buffer
   - 5 μl ddGTP
   - 4 μl *Hind*III
   - 5 μl *Sau*3AI
   - 4 μl AMV RT
   - 8 μl [α-^{32}P]dATP[b]

3. Within a radiation safety cabinet add 2 μl of the labelling pre-mix to the side of each of the wells of the microtitre plate. Cover the plate with cling film and spin the premix into the DNA aliquot before placing the microtitre plate into a radiation safety box for incubation at 37°C for 1 h.

4. Terminate the reaction by adding 4 μl of formamide dyes per well using the repeat dispenser. The samples can analysed immediately (*Protocol 4*) or stored at −20°C in a radiation safety box.[c]

[a] Formamide is a suspected carcinogen, teratogen, and mutagen, and is known to be toxic, harmful, and an irritant.
[b] Standard radiation safety procedures should be followed for this and subsequent steps.
[c] Samples run and labelled on the same day produce the best quality fingerprints. Samples can be stored over a weekend if required.

We always include radioactively labelled DNA size standards on the gels used for analysing fingerprints. These are most easily prepared by *Sau*3AI digestion of bacteriophage λ DNA, as described in *Protocol 3*.

Protocol 3. Preparation of labelled λ/Sau3AI size standards

Equipment and reagents
- TE: 10 mM Tris–Cl pH 7.4, 1 mM EDTA
- 10 × NEB2 buffer (New England Biolabs)
- Bacteriophage λDNA (New England Biolabs—500 μg/ml)
- *Sau*3AI (Amersham—50U/μl)
- 10 mM dGTP (Pharmacia)
- 10 mM ddTTP (Pharmacia)
- [α-³⁵S] dATP (400 Ci/mmol; Amersham)[a]
- AMV reverse transcriptase (10 U/μl, N.B.L.)
- Formamide dyes (*Protocol 2*)
- Standard sample dyes: 10% glycerol, 0.1% (w/v) bromophenol blue
- Minigel apparatus

Method

1. To a 1.5 ml microcentrifuge tube add 171 μl TE, 25 μl 10 × NEB2 buffer, 16.5 μl λ DNA, and 5 μl *Sau*3AI. Incubate at 37°C for 1 h.

2. To verify that digestion is complete, remove a 5 μl aliquot, add an equal volume of standard sample dyes, and resolve on a 2% agarose minigel (90 V, 40 min) before staining with ethidium bromide, and visualizing with UV illumination.

3. Prepare the labelling reaction:
   - 43.5 μl digested λ DNA (step 1)
   - 2.0 μl 10 mM dGTP
   - 2.5 μl 10 mM ddTTP
   - 4 μl [α-³⁵S]dATP
   - 1 μl AMV reverse transcriptase

   Incubate at 37°C for 30 min, then add an equal volume (53 μl) of formamide dyes.

4. To verify that labelling has been successful, analyse 1 μl, 2 μl, and 3 μl aliquots on a 4% polyacrylamide gel (*Protocol 4*).

Analysis of samples is performed on conventional polyacrylamide sequencing gels. The procedure which we have found most effective is described in *Protocol 4*.

Protocol 4. Polyacrylamide gel preparation and electrophoresis

Equipment and reagents
- AR urea (BDH)
- 10 × TBE: 108 g Tris, 55 g boric acid, 9.3 g Na₂EDTA, made to 1 litre
- Double distilled water
- Ammonium persulfate solution (APS), 10% (w/v) (Sigma)

- *N,N,N',N'*-tetramethylethylenediamine (TEMED)
- Acetic acid (Fisher Scientific)
- Ethanol
- Methacryloxypropyl-trimethyloxysilane (bonding solution; Fluka)
- 'Repelcote' (dimethyldichlorosilane solution; BDH)[a]
- Acrylamide solution, 19:1 acrylamide/bisacrylamide ('Easigel'; Scotlab)[a]

- Detergent solution (Decon Laboratories Ltd.)
- Vertical sequencing sank (Gibco–BRL— model S2)
- Glass sequencing plates, 0.4 mm spacers, Gel Tape (Gibco–BRL)
- 60-well fingerprinting comb, 0.4 mm (Scientific Imaging Systems)
- Bulldog clips

A. *Polyacrylamide gel preparation*

1. Dissolve 42 g of urea in 45 ml of double distilled water and 10 ml of 10 × TBE in a 250 ml beaker

2. Clean both the large and small glass plates with detergent and rinse thoroughly with water. In a fume-hood, siliconize one side of the larger plate by covering liberally with 'Repelcote' solution and leave to dry. Apply a small amount of the bonding solution (3 ml of ethanol, 50 μl of 10% acetic acid solution, and 3 μl of bonding solution) to one side of the smaller glass plate.[b] When both plates are dry, liberally wipe the coated surfaces with ethanol, using a separate tissue for each plate, and assemble the plates with the 0.4 mm spacers. Tape the sides and bottom of the plates to seal.

3. Add 10 ml of the 40% acrylamide stock to the gel mix and leave to stir. Add 80 μl of TEMED and 800 μl of 10% APS to the gel solution and mix well. Draw approximately 60 ml of the gel mix into a syringe and inject into the top corner of the gel. When pouring, the gel plates should be raised to a 60° angle and tilted toward one corner. As the space between the plates fills with the gel mix, lower the plates so that the top end comes to rest approximately 5 cm from the level of the bench.

4. Push in the comb so that the tops of each well are below the end of the plate, and clamp across the width of the comb with bulldog clips to ensure good contact between the plates and the comb.

5. Leave the gel to set for at least an hour or, for overnight storage, wrap the top of the gel with cling film.

B. *Gel loading and running*

1. Remove the tape from the bottom of the plates and any excess polymerized acrylamide adjacent to the comb. Carefully remove the bulldog clips and comb from the gel, taking great care not to damage the wells. Ensure that the rubber seals are stuck to the spacers and flush with the top of the small plate.

2. Clamp the cast gel into the sequencing tank and add 500 ml of 1 × TBE to each buffer chamber.

Protocol 4. *Continued*

3. Denature the microtitre plate of samples in an 80 °C oven for 10 min. We use a metal block to ensure oven temperature fluctuations do not affect denaturation.

4. In a 1.5 ml microcentrifuge tube with a pierced lid, place 1–1.5 µl of labelled λ markers for every six samples to be fingerprinted, allowing an extra 2 µl for evaporation. Place the tube in a boiling water-bath for 5 min.

5. Positions of the λ marker and samples (groups of six samples, separated by markers) should be marked on the gel plates below the level of the wells to aid when loading.

6. Flush a group of six sample wells thoroughly to remove urea which has diffused out of the gel (failure to do so will lead to fuzzy bands), then load the six samples, completely filling each well. Leave the microtitre plate of samples behind a protective radiation screen whilst loading.

7. When all samples are loaded, flush the marker wells and load 1–1.5 µl of boiled markers (step B4) in each.

8. Gels should be run under uniform conditions to attain continuity. Electrophorese gels at a constant 74 W (~ 1400 V and ~ 55 mA) until the bromophenol blue dye is 2.5 cm from the bottom of the gel.

9. After the run is complete, dismantle the gel. The gel should adhere to the small plate.

10. Immerse the gel (on the glass plate) in a tray of 10% acetic acid for 10 min to fix the DNA.

11. Immerse in a tray of water for 25 min.

12. Dry the gel onto the plate at 80 °C for 25 min.

13. Expose to X-ray film for three days, develop the autoradiograph and proceed to analysis (Section 7).

[a] Acrylamide and 'Repelcote' are suspected carcinogens. Non-permeable nitrile gloves should be worn at all times when handling these chemicals. Use of 'Repelcote' should be restricted to the confines of a fume-cupboard. Acrylamide is less harmful once polymerized.
[b] On no account should the plates be accidentally exchanged. If a gel is poured between two plates which are (or have previously been) both coated with bonding agent, they will be inseparable.

4.2 Assembly of large-insert bacterial clone contigs on marker 'scaffolds'

Where a high density map of well-characterized markers (such as STSs) exists for a chromosome, it may be advantageously used as a scaffold on which to assemble a contig of large-insert bacterial clones. BAC or PAC clones from a whole-genome library are screened by hybridization with probes derived from the markers, and overlaps between clones thus identified are established by fingerprinting. *Protocol 5* describes the production of filters carrying bacterial clones (initially arrayed in microtitre plates) suitable for hybridization with a variety of probes. Many protocols (in this book and elsewhere) describe the production of labelled probes from a variety of sources, and hence no specific protocol is given here. However, *Protocol 7* (Section 5), which describes the production and use of *Alu* PCR products as probes, can easily be modified to generate labelled probes from any STS, by substituting the appropriate PCR primers, annealing temperatures, and template (typically 20–50 ng of genomic DNA).

Depending upon the disposition of the STSs, and on the success in identifying clones containing them, it will normally be necessary to fill many gaps after the initial round of clone identification and fingerprinting. This is conveniently done by deriving end-sequences from the clones flanking such gaps (Chapter 11) and deriving new STSs from these. These STSs can then be used to identify further clones by repetition of the procedure, in a chromosome walking approach.

We believe this method to be potentially faster and more versatile than that described in Section 4.1, provided that suitable maps already exist. The identification of larger-insert clones can result in faster coverage, can be applied regionally or to a whole chromosome, and produces material which can be used as an immediate, efficient sequencing resource. An initial map having a density of one marker per 70–100 kb will yield a bacterial contig with a minimum amount of chromosome walking required for gap closure.

Protocol 5. Preparation of bacterial clone filters for hybridization

Equipment and reagents

- Sterile 8 cm × 12 cm nylon filters (Amersham—Hybond N+)[a]
- Sterile 8 cm × 12 cm Petri dishes, containing 2 × TY agar with the appropriate selective antibiotic; and lids (Hybaid—colony picker plates)[a]
- Denaturation solution: 0.5 M NaOH, 1.5 M NaCl
- 20% SDS solution (Merck)
- Neutralizing solution I: 0.5 M Tris–HCl pH 7.4, 1.5 M NaCl

- Neutralizing solution II: 50 mM Tris–HCl pH 7.4, 1.5 M NaCl)
- 50 mM Tris–HCl pH 7.4
- 20 × SSC: 3 M NaCl, 0.3 M trisodium citrate pH 7.0
- 3MM paper (Whatman)
- Trays, e.g. Biohazard trays (Scotlab)
- 96-pin replicating tool (Denley; see *Protocol 1*)[b]
- UV transilluminator (360 nm)

Method

A. Bacterial clone gridding

1. Thaw working stocks of bacterial clones in microtitre trays (stored in 7–10% glycerol).

Protocol 5. *Continued*

2. With a suitable pen, label nylon filters in the top left-hand corner (corresponding to the A1 position of the microtitre plate). Lay the filter onto the 2 × TY agar plates, being careful not to produce any air bubbles.

3. Transfer the bacterial clones from microtitre trays to the filters, either manually (using the 96-pin replicating tool) or robotically (14).

4. Incubate for 18 h at 37°C.

B. *Bacterial clone filter lysis*

1. Place two large sheets of 3MM on two large biohazard trays, saturate one tray with 10% SDS and the second with denaturation solution. Using a glass pipette, roll out the paper to get rid of air bubbles, and drain away excess solution. Place the filters colony-side up on the 3MM paper with SDS for 4 min, taking care not to allow solution to wash onto the upper side of the filters. Lift and relay the filters during the 4 min to remove any air bubbles.

2. Transfer the filters onto the tray with denaturation solution for 10 min. Ensure that there are no air bubbles: the SDS tends to foam. As with step 1 lift and replace the filters to remove any air bubbles.

3. Transfer the filters to a dry piece of 3MM paper for 10–20 min.

4. Submerge the filters, colony-side up, in 500 ml of neutralizing solution I for 5 min. (This solution can be reused.)

5. Repeat step 4 using fresh neutralizing solution I.

6. Submerge the filters in 500 ml of neutralizing solution II for 5 min.

7. Rinse the filters in 500 ml of 2 × SSC/0.1% SDS for 5 min.

8. Rinse the filters in 500 ml 2 × SSC for 5 min.

9. Rinse the filters twice in 500 ml of 50 mM Tris–HCl pH 7.4 for 5 min.

10. Air dry on 3MM paper, colony-side uppermost.

11. Expose the membrane face down on a UV transilluminator (312 nm) for 2 min.

[a]This protocol is intended for use with bacterial clones stored in microtitre format, but of course can be adapted to other formats. Petri dishes and lids can be reused if the agar is removed, the plates washed with detergent, and sterilized with 70% isopropanol.
[b]A robotic workstation with a suitable replicating tool can be used as an alternative.

4.3 Regional fingerprinting

When considering a region already spanned by a single YAC or by a minimal set of YACs, we have adopted a regional fingerprinting strategy. This involves the hybridization of either the whole YAC DNA or *Alu* PCR products from a YAC to arrays of bacterial clones. Filters of bacterial clones for hybridization may be prepared as described in the preceding section; protocol 7 (section 5) can be used to generate and hybridize *Alu* PCR products from YACs, by substituting the YAC clone for the bacterial template.

Bacterial clones identified by hybridization are placed according to the information associated with the YACs, or are further screened for the presence of other genetic and physical markers. This can give extensive information as to the relative positions and orientations of the bacterial clones, which can then be combined with fingerprinting data to assemble contigs. The YAC contig is valuable in that it can bridge gaps in the overlying bacterial contig, and indicate the best course of action for directed efforts to close these gaps (Section 5).

It has been our experience that bacterial clone coverage from a regional fingerprinting approach gives rise to approximately 60% coverage from initial screening of two bacterial libraries containing seven complexities of the chromosome or region of interest. The failure to achieve complete coverage can be attributed to either the lack of representation within the bacterial libraries (which may contain deletions relative to the YACs), or to failures of some regions of the YAC DNA to hybridize, for example if they are particularly repeat-rich.

We have found that because of the problems inherent in the YAC cloning system it is more expedient to achieve bacterial clone contiguation using the above method rather than by subcloning YACs into bacterial vector systems such as cosmids or fosmids.

5. Gap closure

Contigs produced by the strategies discussed in the previous section, whether applied regionally or to whole chromosomes, will almost inevitably contain gaps. Such gaps may result from fingerprinting an insufficient complexity of bacterial clones, or may be due to the lack of representation of specific regions within the library.

Underrepresentation in a library can be caused by sequences that are unstable in the cloning host (such as long head-to-head palindromic sequences or tandem repeats in an *E. coli* host); by the host being damaged by transcription and translation of the insert material; by a cloning bias when the library was made; or by the structural instability of the native DNA.

Many of the techniques used for gap closure are similar to those used in chromosome walking to cover targeted regions of the genome in a stepwise approach. Chromosome walking itself is discussed more fully in Chapter 11, but we present here two protocols which we have used successfully for the closure of gaps in clone contigs. The first of these, cosmid-to-cosmid/PAC hybridization (*Protocol 6*), can be used to rapidly identify bacterial clones which overlap with a cosmid lying at the end of a contig (i.e. on one side of the gap to be bridged). Fingerprinting of the clones thus identified may (if the gap does not represent a region which is 'uncloneable' in the screened library) reveal one or more which extend the contig and may bridge the gap. The use of PAC clones as probes is more problematic. The second technique, *Alu* PCR walking (*Protocol 7*) overcomes this problem by using *Alu* PCR products from PACs (again, lying adjacent to the gap to be bridged) as probes to identify overlapping clones which, again, can be fingerprinted to establish if they extend the contig or bridge the gap.

Protocol 6. Cosmid-to-cosmid or cosmid-to-PAC hybridization

The method outlined below is used routinely to hybridize cosmids, located at the ends of contigs, to gridded arrays of either cosmid or PAC clones. A standard alkaline lysis preparation is used to isolate DNA from a cosmid to use as the probe.

Equipment and reagents
- $T_{0.1}E$: 10 mM Tris–HCl, 0.1 mM EDTA pH 8.0
- Klenow DNA polymerase (5 U/μl; Boehringer)
- Oligolabelling kit (Pharmacia)
- BSA (Sigma), 10 mg/ml
- [α-^{32}P]dCTP (600 Ci/mmol; Amersham)[a]
- 20 × SSC: 3 M NaCl, 0.3 M trisodium citrate pH 7.0
- Cosmid DNA (5–10 ng/μl; *Protocol 1*)
- Filters carrying gridded bacterial (PAC or cosmid) clones (*Protocol 5*)

- Sarkosyl hybridization mix: 6 × SSC, 10 × Denhardt's reagent,[b] 50 mM Tris pH 7.4, 1% Sarkosyl (Denhardt's reagent is prepared and stored at −20°C as a 50 × concentrate: 10% (w/v) polyvinylpyrrolidone, 10% (w/v) Ficoll 400, 10% (w/v) bovine serum albumin)
- Vector DNA (identical to that in the library being screened), 120 μg/ml, sonicated
- Sonicated human placental DNA (Sigma), 10 mg/ml
- X-ray film and autoradiography cassettes

Method

1. In a 1.5 ml screw-top microcentrifuge tube add 10–15 ng of probe DNA to 17 μl of $T_{0.1}E$. Place in a boiling water-bath for 5 min, then chill on ice.

2. Add 5 μl reagent mix (oligolabelling kit), 1 μl 1/100 BSA, 1 μl [α-^{32}P]dCTP, and 1 μl Klenow. Incubate at room temperature for a minimum of 3 h.

3. Pre-hybridize gridded filters for at least 2 h in Sarkosyl hybridization mix at 65°C.

4. To the probe add 8 μl sonicated vector DNA, 20 μl sonicated human DNA, 125 μl 20 × SSC, and make up to 500 μl with $T_{0.1}E$. Incubate for 5 min in a boiling water-bath.[c]

5. Incubate the competition reaction for 20 min at 65°C, then snap-chill on ice.

6. Add the probe to 50 ml of hybridization solution pre-warmed to 65 °C in a plastic box, and mix well. Transfer the filters into this solution (ensuring that each filter is covered by solution before adding the next), and hybridize for 18 h at 65 °C.

7. Wash the filters as in *Protocol 7*C. Expose the filters to X-ray film for 8 h at room temperature.

[a]The appropriate radiation safety procedures should be followed throughout.
[b]Some of the constituents of Denhardt's solution are hazardous to health by inhalation, ingestion, and by direct contact with skin or eyes. Appropriate eye and hand protection should be worn when using Denhardt's.
[c]If it is desired to pool two or more probes, they should be pooled at this stage, the volume of $T_{0.1}E$ being reduced to accommodate the extra probes.

Protocol 7. Alu—PCR walking from PAC clones

Equipment and reagents

- Thermocycler and compatible reaction tubes
- Incubating shaker (Innova 4000; New Brunswick Scientific)
- Sterile toothpicks
- A4 clear plastic envelopes, slit open along one long edge and the bottom
- 10 cm × 15 cm and 20 cm × 20 cm plastic sandwich boxes
- Screw-top 1.5 ml microcentrifuge tubes
- $T_{0.1}E$: 10 mM Tris–HCl, 0.1 mM EDTA pH 8.0
- 10 × PCR buffer: 670 mM Tris-HCl pH 8.8, 166 mM enzyme grade $(NH_4)_2SO_4$, 67 mM $MgCl_2$
- 10 × dNTPs: 5 mM each dATP, dCTP, dGTP, dTTP (Pharmacia) in $T_{0.1}E$
- Bovine serum albumin (BSA) 0.5 mg/ml (Sigma)
- 700 mM β-mercaptoethanol (BDH Laboratory Supplies)
- *Taq* DNA polymerase, 5 U/μl ('Amplitaq', Perkin-Elmer)
- 9.4 M ammonium acetate
- 10 × ALE1 (100 ng/μl; 5′GCCTCCCAAAGT-GCTGGGATTACAG3′)
- 10 × ALE3 (80 ng/μl; 5′CCA[C/T]TGCACTCCAGCCTGGG3′)
- Ethanol
- Oligolabelling kit (Pharmacia—27-9250-01)
- $[\alpha\text{-}^{32}P]dCTP$ (600 Ci/mmol; Amersham)
- Klenow DNA polymerase (Boehringer—5 U/μl)
- Human placental DNA (Sigma—hybridization quality), 10 mg/ml
- 20 × SSC solution: 3 M NaCl, 0.3 M trisodium citrate pH 7.0
- Double distilled water
- 20% Sarkosyl solution (Merck)
- Sarkosyl hybridization mix (*Protocol 6*)
- Dextran sulfate (Pharmacia)
- Petri dishes of 2 × TY supplemented with 30 μg/ml kanamycin
- PCR-grade mineral oil (Sigma)
- Glycerol loading dyes: 10% (v/v) glycerol, 0.1% (w/v) bromophenol blue
- DNA size standards (e.g. 1 kb ladder, Pharmacia)
- Wash solution A: 0.5 × SSC, 1% Sarkosyl
- Wash solution B: 0.2 × SSC, 1% Sarkosyl

A. *Primary Alu PCR*

1. Streak the relevant PAC clone on a 2 × TY/kanamycin plate and grow overnight at 37 °C. Select a single colony and, using a sterile toothpick, transfer to 500 μl $T_{0.1}E$ in a microcentrifuge tube, vortexing well.

Protocol 7. *Continued*

2. Prepare the reaction pre-mix for *Alu* PCR and mix well. The recipe below is for one reaction. Prepare one reaction for each PAC clone, plus two spare:
   - 2.5 μl 10 × PCR buffer
   - 2.5 μl 10 × dNTPs
   - 0.8 μl BSA
   - 0.35 μl β-mercaptoethanol
   - 0.3 μl Amplitaq
   - 2.5 μl 10 × ALE1
   - 2.5 μl 10 × ALE3
   - 8.5 μl $T_{0.1}E$

3. To a thermocycler reaction tube add 20 μl of pre-mix and 5 μl of resuspended PAC colony (step 1), overlay with a drop of mineral oil, and cap firmly. In addition, prepare a negative control with water in place of the PAC suspension.

4. Thermocycle as follows:
   - 94°C × 5 min

   followed by 30 cycles of:
   - 93°C × 1 min
   - 65°C × 1 min
   - 72°C × 5 min

   followed by:
   - 72°C × 5 min

5. Precipitate the PCR products by adding 8.5 μl 9.4 M ammonium acetate and 75 μl ethanol to each 25 μl PCR reaction. Place the reaction tube on ice for 15 min then spin in a microcentrifuge at 14 000 r.p.m. (16 000 *g*) for 15 min. Pour off the supernatant and resuspend the precipitate in 25 μl of $T_{0.1}E$.[b]

6. Remove a 5 μl aliquot for gel analysis. Add 1 μl of glycerol loading buffer, and resolve by electrophoresis in a 2% agarose minigel in 1 × TBE, containing 0.4 μg/ml ethidium bromide in the gel and gel buffer, with suitable size standards. *Figure 3* shows typical amplification products from a number of PAC clones.

7. The remainder of the primary PCR product is stored at −20°C until required.

B. *Random oligolabelling of Alu PCR products*

1. Pre-hybridize bacterial clone filters (see *Protocol 5*) in the Sarkosyl hybridization mix for 3 h at 65°C, in a sandwich box, during which time perform steps 2–5.

2. In a screw-top 1.5 ml microcentrifuge tube boil 2 μl of PCR product with 15.5 μl of $T_{0.1}E$ for 5 min. Snap-chill the tube on ice until you are ready to use it.

3. Prepare the oligolabelling mix by adding to the boiled PCR probe:
   - 5 μl reagent mix (oligolabelling kit)
   - 1 μl Klenow
   - 1 μl [α^{32}P]dCTP

   Incubate at room temperature for 3 h.

4. To compete out the repetitive sequences in the probe, add 125 μl human placental DNA, 125 μl 20 $\times$ SSC, and 225 μl ddH$_2$O to the labelled probe, and boil for 5 min before snap chilling on ice for 5 min.

5. Add the probe to 50 ml of hybridization solution at 65°C in a sandwich box and mix well.

6. Transfer the filters from the pre-hybridization box into this hybridization solution. Each filter should be clone-side uppermost and should have the hybridization solution washed across its surface before the addition of the next filter. Once all the filters have been added, fit the lid onto the box and place in a perspex safety box in a 65°C incubating shaker.[c]

7. Hybridize the filters for a minimum of 3 h, or overnight.

C. *Washing regime for hybridization filters*

   We routinely wash under the following stringencies, but the durations and/or temperatures of the washes can be altered if necessary to increase the signal or reduce background.

1. Place 1 litre each of wash solutions A and B in a 65°C waterbath to pre-warm.

2. Pour off the hybridization solution and transfer the filters one at a time to a 20 cm $\times$ 20 cm plastic box containing 1 litre of 2 $\times$ SSC. Wash the filters at room temperature on a shaking platform for 5 min.

3. Repeat step 2.

4. Pour off the solution and add the pre-warmed wash solution A. Wash at 65°C for 30 min in an incubating shaker.

5. Pour off the solution and add the pre-warmed wash solution B. Wash at 65°C for 30 min in an incubating shaker.

6. Pour off the solution and wash the filters at room temperature for 5 min in 0.2 $\times$ SSC.

Protocol 7. *Continued*

7. Drain the filters lightly on 3MM paper before placing five filters colony-side up on one side of the opened A4 pocket, and close. Expose at −70 °C in a cassette with intensifying screens to pre-flashed X-ray film for 1–3 h, or to non-flashed film overnight.

---

[a]Some of the constituents of Denhardt's reagents are hazardous by inhalation, ingestion, and by direct contact with skin or eyes. Appropriate eye and hand protection should be worn.
[b]We have found that precipitating the PCR products in this way removes the background 'smear' from the reaction when run on an agarose gel and subsequently produces a better probe for hybridization.
[c]A water-bath tends to maintain its temperature better than an incubator, which can lose its temperature with constant opening.

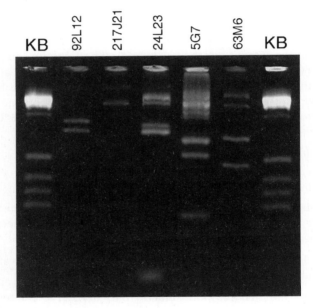

Figure 3. *Alu* PCR products of five arbitrarily named PACs resolved by electrophoresis on a 2% agarose gel containing 0.5 μg/ml of ethidium bromide. DNA size standards (1 kb ladder) are indicated by KB.

6. *Alu* PCR fingerprinting of YACs

The necessity of separating YACs from host yeast chromosomes by pulsed-field gel electrophoresis makes the purification of large quantities of YAC DNA technically inconvenient. This fact, and the large size of their inserts (0.2–2 Mb), means that the restriction fingerprinting strategies used for bacterial clones cannot be readily applied to the ordering of random sets of YACs.

Instead, we use a rapid and convenient method first described by Coffey *et al.* (15), which relies upon *Alu* PCR to amplify from each YAC a number of *Alu* element-flanked sequences. Approximately one million *Alu* elements occur in the human genome, either singly or as tandem repeats in either orientation, but they are absent from the yeast genome. Although there is some evidence to suggest that they occur more frequently in the gene-rich Giemsa-light bands of chromosomes, their frequency is enough to ensure that most YACs will yield one or more *Alu* PCR products. After the primary PCR (similar to that in *Protocol 7A*), products are labelled in a secondary re-action, and analysed using the same gel system and software (Section 7) as for restriction fingerprinting. The pattern of such amplimers, resolved by electrophoresis, acts as a fingerprint which can be analysed and compared in much the same way as the restriction fingerprint of a bacterial clone. We have applied the *Alu* PCR fingerprinting technique to a collaborative X chromosome fingerprinting project (9) as well as to several other large scale physical mapping projects; descriptions of these, and detailed protocols, will be found in refs 16–19.

7. Data entry and analysis

The entry, analysis, and storage of fingerprinting data and related informa-tion is a major part of any large scale contig assembly project. The following sections describe the software which we use. The programs are available by anonymous ftp, at `ftp.sanger.ac.uk`, where more detailed documenta-tion will also be found. The *Image* software (Section 7.1) and additional doc-umentation is in `pub/contigC`. The *FPC* software (Section 7.2) and documentation is in `pub/fpc` (contact `cari@sanger.ac.uk`). Further information is also available on the WWW (`http://www.sanger.ac.uk`).

The Sanger Centre uses ACeDB (A *C. elegans* DataBase) (20) for the final repository of physical map data. The chromosome 22 ACeDB database is in `pub/human/chr22` (contact `idl@sanger.ac.uk`). Of the many features that ACeDB provides, the ones relevant to this section are that it has a stan-dard graphics library that can present data in a variety of formats; it allows the development of various 'models' which can be adapted as necessary according to the nature of the project; and data can be exchanged with other programs in standardized textual 'ACeDB formatted' files.

7.1 Autoradiograph scanning and data entry

The autoradiograph of the fingerprint gel is read by an Amersham Film Reader scanner which converts the hard copy data into an 8-bit greyscale image.

This greyscale file is then processed using *Image* software (21) to convert the data from the image into a format that enables us to edit and enter the fingerprints from a set clones. *Image* was originally written in Fortran for the VAX, but has recently been automatically translated into C, and runs under Unix using X-windows. The ACeDB graphic library has been used for the graphical interface.

Image accepts the greyscale autoradiograph file, which can then be cropped as necessary. There are four separate functions ('pre-features', 'pre-map', 'features', and 'standard features') which must be executed before editing the data. The 'pre-features' routine identifies the parameters of the gel within which the data exists; 'pre-map' locates the position of the marker lanes; 'features' allows editing of the fingerprint lanes (giving the opportunity to edit out bands arising from incomplete digestion or from gel background); and 'standard features' aligns the marker lanes of the autoradiograph with the standardized marker set stored within the program.

In edit mode, one clone is edited at a time. Clones are identified according to standard naming conventions; a clone name is usually prefixed with the library source and library plate location. The user has the ability to enter or delete fingerprint bands, or to identify multiple co-migrating bands, with reference to the original autoradiograph.

The output of the *Image* software is a 'bands' file, representing in integer form the sizes of the bands for each clone as determined by interpolation between the λ size standards. A file containing a gel trace for each sample lane is also written. The gel trace facility is a recent valuable addition which allows the user to view the normalized fingerprint image from within the FPC software subsequently used for contig assembly.

7.2 Overlapping and analysis

The *FPC* (fingerprinted contigs) software (22) takes as input the 'band' files from *Image*, aids the user in constructing contigs, and displays them relative to landmarks (such as STSs known to lie within certain clones) and other map information. *FPC* uses the same graphical interface as ACeDB and, although it is a self-contained program and can operate without an external database, it does support the reading of ACeDB files containing information on the marker content of clones, and can create files for loading into ACeDB. *FPC* has recently replaced *Contig9* (20); although the two programs share many features and principles of operation, *FPC* greatly reduces the amount of manual intervention which is required for rapid contig assembly.

Overlap between two fingerprinted clones is identified on the basis of the number of bands which their fingerprints have in common. Naturally, any two clones (overlapping or not) may share one or more fragment sizes—within the limits of resolution of the gel system—by chance alone. *FPC* enables the identification of putative overlaps based on the probability that fragments which they have in common have arisen by chance alone. The smaller the calculated probability of coincidence, the greater the likelihood of overlap between two clones; we normally adopt a probability threshold of 10^{-6} for determining overlap between two clones in a chromosome-wide project. We initiate contig construction by performing a pairwise comparison of clones based on the likeliest matching, check the overlap of fingerprint bands with reference to the gel traces, and then establish the overlap within the database using the clone manipulation facilities in *FPC*.

There are many functions within *FPC* which speed the identification of overlaps. The *FPC* function which merges an ACeDB file of markers with the fingerprint clones provides an indication as to which clones overlap, based on shared marker data (see Section 7.3), and a variety of clone manipulation functions allow rapid addition of matching clones to existing contigs. *FPC* allows fingerprinting and contig data to be viewed in a variety of ways to facilitate contig construction. The 'fingerprint' option (*Figure 4A*) displays the fingerprints for any number of clones in the form of a simulated gel based on their fragment sizes. 'Gel image', in contrast, displays fingerprint lanes for one or more clones taken from the original autoradiographs. In the 'consensus band' (CB) display (*Figure 4B*), the fingerprint bands of a region of the contig are displayed in the physical order in which they have been deduced to lie, and the presence or absence of each of these bands in each clone is indicated. This display is similar in principal to a restriction map but, because our fingerprinting protocol identifies only a subset of the restriction fragments of each clone, a real restriction map cannot be created.

FPC has also adopted the *Contig9* scheme of identifying 'canonical clones'. If one clone, A, shares all of its fingerprint bands with a second clone, B, then clone A can be 'buried' (i.e. is contained) within clone B. All clones which cannot be so buried are termed 'canonical clones', and are the minimum set of clones needed to represent the contig. The canonical state of a clone provides evidence about the validity of the clone and its fingerprint. *FPC* allows the user to either display the 'buried' clones or to hide them when viewing the contig. *Figure 5* shows a typical display of part of a contig by *FPC*, in which the 'buried' clones are visible.

Editing functions are available for adding and removing markers from clones, manually positioning clones, identifying clones for sequencing, burying clones, adding remarks to clones, and manipulating contigs. Search commands allow the user to select a subset of clones based on their creation date, the gel on which their fingerprints were analysed, etc. Building contigs solely through interactive editing is too time-consuming, and so a suite of automatic functions has been written. These functions allow the rapid incorporation of new clones into contigs, and the merging of contigs as bridging clones are identified. Additional confirmation of overlaps (particularly of 'weak' joins, in which the apparent overlap may be due to chance alone) may be made by incorporating information on the marker content of the clones. Details of these features are provided in the *FPC* manual.

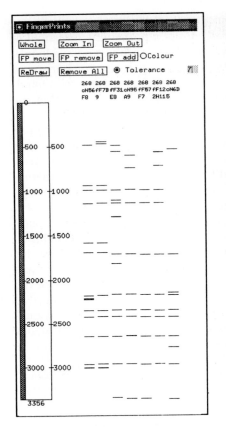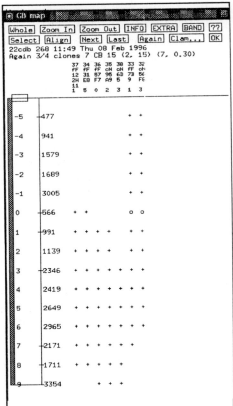

Figure 4. Example of *FPC* displays. *Left*: 'Fingerprint' display, in which fingerprint bands from several clones (268cN56F8, 268fF7B9, etc.) are displayed in the form of an idealized electrophoretogram, with the smallest bands at the top (sizes in base pairs are indicated on the left). *Right*: 'Consensus band' display representing the same clones as in the fingerprint display. Restriction fragments (numbered arbitrarily in the left-most column, and with consensus sizes indicated in the second column) are shown in the order in which they are inferred to lie. The presence of each fragment in each clone is indicated by a '+' in the columns beneath the clone names. Fragments which are expected to occur but were not detected are indicated by 'o', reflecting either an incorrect alignment of clones, a cloning artefact (e.g. deletion), or a fingerprinting error. The numbers immediately beneath the clone names are the numbers of fingerprint bands whose location within the contig could not be inferred from the data.

7.3 Incorporation of marker data by *FPC*

As mentioned in the foregoing section, information on the marker content of clones (derived, for example, by hybridization) can be combined with fingerprint data. In the case of our chromosome 22 project, a partially ordered set of 600 markers (STSs, probes, etc.) were assembled with the aid of SAM

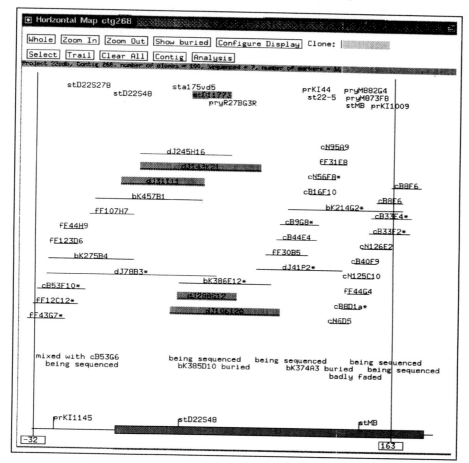

Figure 5. Example of an *FPC* contig display, showing integration of marker and clone data. The names of single copy markers (stD22S278, etc.), which hybridize to one or more of the clones (horizontal lines with clone name above each) appear along the top of the main window. Asterisks indicate canonical clones (see text). Along the bottom of the display are named three markers (prKI1145, etc.) chosen as anchor markers. Any one of these anchors may be selected, causing all clones containing it to be highlighted; in this example, anchor marker stD22S48 has been selected, and clones containing it are indicated by shaded bars in main window.

(System for Assembling Markers) (23, 24), and the results stored and displayed in our chromosome 22 ACeDB database. Hybridization results for all other markers that have been tested, even those that have not been globally positioned, are also stored in ACeDB. Files giving details of the markers and (where known) their positions are created, and imported by *FPC*.

FPC reads these marker files, and annotates each clone with the markers which it contains. It then identifies the position of each marker within a contig, by finding the location at which the greatest number of clones have been found to contain that marker. Selecting a particular marker within *FPC* allows one to display all clones which contain it (*Figure 5*); if two or more of these clones lie in different regions of the contig, there is a conflict between the fingerprinting and the marker content data, and both sources of data should be re-evaluated.

7.4 Regional assembly

A special case applies when sets of bacterial clones have been identified, prior to fingerprinting, by hybridization to single copy markers or to YACs. Some information is already available on the overlaps between these clones, and it is expected that all of the clones thus identified will fall into a single contig. Consequently, less stringent probability thresholds are required in determining overlap between two clones: we use a cut-off of 10^{-4}, rather than the much more stringent 10^{-6} threshold used for chromosome-wide contig assembly.

7.5 Complications

Complications in contig assembly can arise from a number of causes:

(a) False negatives: weak bands may be missed from the fingerprint of a clone.

(b) False positives: artefactual bands in the fingerprint.

(c) Deletions in clones, relative to genomic DNA or to other clones.

(d) Extra bands due to incomplete restriction digestion.

(e) Co-migrating bands which appear as a single band on the fingerprint

An algorithm which explored all of these possibilities, and all of the possible clone overlaps, would take an unacceptable amount of time to execute. Consequently, *FPC* uses a 'greedy' algorithm which, at each step in contig assembly, makes the best possible choice of clone (and fingerprint data) to add to the contig. Consequently, the solution is an approximation whose accuracy decreases with an increasing number of fingerprinting errors.

Though this technique does not provide exact positioning of clones, it does reliably cluster clones and gives an approximate ordering. We have tested the reliability of the algorithm by analysing 400 clone fingerprints derived from three different chromosomes, which *FPC* correctly resolved into three groups with no intermixing of clones from the different chromosomes. A further indication of FPC's reliability is the almost invariable agreement between fingerprint and marker data.

8. Conclusions

Our chromosome 22 fingerprinting project, at the time of writing, has achieved an estimated 90% coverage of the 56 Mb chromosome in 836 contigs. Approximately 82% coverage of the whole chromosome has been achieved by fingerprinting 10 200 cosmid clones from a high-purity chromosome 22-specific library. This has been complemented by the regional hybridization of YACs and single copy probes to bacterial clone arrays to identify 4000 clones in the distal half of the chromosome. In this region the combined coverage is estimated to be about 90%.

The adaptation of fluorescent technology (25, 26) to our fingerprinting methods (unpublished data) is envisaged to enhance our procedures in much the same way as it has those of sequencing. Real-time data collection will improve the speed of analysis and contig construction, and current dye technology will enable us to multiplex up to three samples per gel lane.

Acknowledgements

The authors wish to thank Gareth Howell for his assistance with the cosmid-to-PAC hybridization, and the Fingerprint group who aided in the compilation of the protocols, *FPC* suggestions, and bug reports. Fred Wobus and Richard Durbin get credit for the *Image* port to the Sun and the gel image files. Ian Longden implemented the *FPC* code for the contig, gel image, and fingerprint displays, editing functions, and search routines. Ian Dunham collaborated with C. A. S. to specify the ACeDB <--> *FPC* interface.

This work has been supported by the Wellcome Trust and the M.R.C.

References

1. Coulson, A. and Sulston, J. (1988). In *Genome analysis: a practical approach* (ed. K. E. Davies), p. 19. IRL Press, Oxford.
2. Waterston, R. and Sulston, J. (1995). *Proc. Natl. Acad. Sci. USA*, **92**, 10836.
3. Riles, L., Dutchik, J. E., Baktha, A., McCauley, B. K., Thayer, E. C., Leckie, M. P., *et al.* (1993). *Genetics*, **134**, 81.
4. Hoheisel, J., Maier, E., Mott, R., McCarthy, L., Grogoriev, A. V., Schalwyk, L. C., *et al.* (1993). *Cell*, **73**, 109.
5. Martin, C. H. (1995). *Proc. Natl. Acad. Sci. USA*, **92**, 8398.
6. Fleischmann, R. D., Adams, M. D., White, O., Clayton, R. A., Kirkness, E. F., Kerlavage, A. R., *et al.* (1995). *Science*, **269**, 496.
7. Iaonnou, P. A., Amemiya, C. T., Garnes, J., Kroisel, P. M., Shizuya, H., Chen, C., *et al.* (1994). *Nature Genet.*, **6**, 84.
8. Shizuya, H. (1992). *Proc. Natl. Acad. Sci. USA*, **89**, 8794.
9. Roest Crollius, H. and Ross, M. T. (1996). *Genome Res.*, **6**, 943.
10. Taylor, K. (1996). *Genome Res.*, **6**, 235.

11. Gibson, T. J. and Sulston, J. E. (1987). *Gene Anal. Technol.*, **4**, 41.
12. Birnboim, H. C. and Doly, J. (1979). *Nucleic Acids Res.*, **7**, 1513.
13. Coulson, A. (1986). *Proc. Natl. Acad. Sci. USA* **83**, 7821.
14. McKeown, G. (1995). *Genome Sci. Technol.*, **1**, 56.
15. Coffey, A. (1995). In *Methods in molecular biology: YAC protocols* (ed. D. Markie,) p. 54. Humana Press.
16. Forbes, S. (1996). *Genomics*, **31**, 36.
17. Van de Vosse, E. (1995). *Eur. J. Hum. Genet.*, **4**, 101.
18. Wooster, R. (1995). *Nature*, **378**, 789.
19. Aldred, M. (manuscript in preparation).
20. Sulston, J. (1989). *CABIOS*, **5**, 101.
21. Soderlund, C. A. and Longden, I. (1996). *FPC V2.5 user's manual*. The Sanger Centre, Technical Report SC-01-96.
22. Dunham, I. (1994). In *Guide to human genome mapping* (ed. M. J. Bishop), p. 111. Academic Press.
23. Sulston, J. (1988). *CABIOS*, **4**, 125.
24. Collins, J. E. (1995). *Nature*, **377** (Supplement), 367.
25. Soderlund, C. and Dunham, I. (1995). *CABIOS*, **11**, 645.
26. Currano, A. V. (1989) *Genomics*, **4**, 129.
27. Tang, X. (1994). *BioTechniques*, **15**, 294.

11

Chromosome walking

JIANNIS RAGOUSSIS and MARK G. OLAVESEN

1. Introduction

The assembly of overlapping clones to form a 'contig' can be done in one of two ways. Where large regions (or entire genomes) are to be covered, it is often most efficient to select clones at random from a library representing the region, to characterize them, and thereby establish which clones overlap one another; this is 'shotgun' contig assembly (Chapter 10). However, when smaller regions are to be covered (for example, the gap between two genetic markers known to flank a locus of interest), a more systematic approach becomes necessary. One clone is chosen as a starting point or 'founder', and clones which partially overlap it are selected (rather than fortuitously found) from the clone library. One of these new clones is then chosen, and the process repeated. In this way, a series of partially overlapping clones is assembled progressively—a process called 'chromosome walking' (1, 2).

Chromosome walking is typically performed using YAC, BAC, PAC, or cosmid clones. Typically, a fragment from one end of the insert of the 'founder' clone is isolated and used as a hybridization probe to screen the library for overlapping clones. This approach maximizes the chance that the new clone will extend significantly beyond the founder clone. Several methods have been developed for isolating end-fragments for use as probes. For example, transcription promoters near the cloning site can be used to generate RNA probes from the ends of the insert (1, 3). Other protocols involve the circularization and cloning of end-fragments (4), subcloning into plasmid vectors or the use of PCR-based techniques (5–9).

However, if the founder clone contains a chimeric DNA fragment (as is often the case in YAC libraries), the end-probe techniques are not ideal. In such cases it is desirable to isolate and use internal fragments, to avoid the chimeric regions and to achieve deeper coverage of the region.

In this chapter we concentrate on the generation of probes (both end-fragments, and internal fragments from YACs) for walking by hybridization, and present a number of protocols that cover a broad spectrum of techniques and are applicable to the most frequently used genomic cloning vectors.

2. Vectorette PCR

Vectorette PCR is a versatile method enabling the amplification of short DNA fragments which are adjacent to a known sequence. Thus, vector sequences near to the cloning site can be used to obtain fragments from the ends of the insert. Fragments obtained by this method are ideal for chromosome walking and also for confirming previously established overlaps between clones (10, 11). The method is suitable for all types of vectors including BACs, PACs, cosmids and, with the *caveat* mentioned in the introduction, YACs.

Two main steps are involved in the procedure (*Figure 1*). In the first, the clone is digested and fragments are ligated to the 'vectorette' adapter. In the second, PCR amplification is performed using two oligonucleotides which prime on the vectorette sequence and on sequences near the cloning site of the original vector. Thus, only those ligated fragments derived from the ends

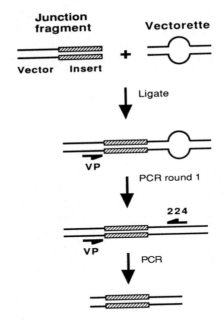

Figure 1. Vectorette PCR. Fragments of the digested clone, including a junction fragment comprising vector sequence and the end of the insert (hatched), are ligated to the vectorette, which contains a 'bubble' of non-complementary sequence. In the first round of PCR, extension from a vector-specific primer (VP) creates a fully complementary product which can now be primed by the vectorette primer (224). Subsequent rounds of PCR amplify the region between these two primers, which includes the insert end-fragment. Note that only those ligation products consisting of junction fragment + vectorette will be amplified.

of the insert will be amplified. By choosing the appropriate vector-specific primer it is possible to isolate fragments derived from one or the other end of the insert. The protocols which we describe here are designed to amplify blunt-ended fragments by ligation to a blunt-ended vectorette; this approach eliminates the unwanted PCR products which can arise when using fragments and vectorettes with 5′ overhangs.

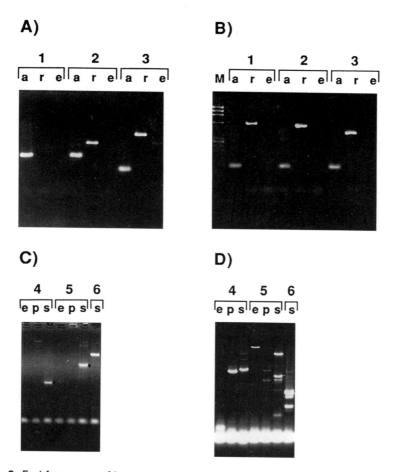

Figure 2. End-fragments of bacterial clones isolated by vectorette PCR. (A, B) End-products derived from three cosmid clones (1, 2, and 3). After digestion of the cosmid DNA with either *Alu*I (a), *Rsa*I (r) or *Eco*RV (e) and ligation of the vectorette, end fragments were amplified using vectorette primer 224 in conjunction with either the LawL primer (A; left end-fragments) or LawR primer (B; right end-fragments). Lane M contains size standards (ΦX174 DNA digested with *Hae*III). (C, D) End-products derived from three pBELOBAC clones (4, 5, 6). After digestion with *Eco*RV (e), *Pvu*II (p), or *Ssp*I (s) and ligation to the vectorette, end-fragments were amplified using the vectorette primer 224 in conjunction with either the SP6 primer (C; SP6 end-fragments) or T7 primer (D; T7 end-fragments).

2.1 Isolation of end-probes from PACs, BACs, and cosmids by vectorette PCR

For vectorette PCR, the SP6 and T7 promoter sequences of pBELOBAC, pBAC108L, and pCYPAC can be used as targets for the vector-specific primers. In Lawrist, however, the promoter sequences lie too far from the cloning site, and we instead prime on sequences closer to the insert (*Protocol 1*).

Suitable enzymes for the digestion of clones in any of these vectors include *Rsa*I, *Alu*I, *Pvu*II, *Eco*RV, and others giving blunt-ended fragments, provided that they do not cleave at or adjacent to the insert–vector junction. The best enzyme to use will depend on the disposition of sites within the insert; therefore, it is advisable to perform parallel experiments with several different enzymes to find the most suitable in any one instance. *Figure 2* gives examples of end-fragments isolated using this method.

Protocol 1. Vectorette PCR of PAC, BAC or cosmid clones

Equipment and reagents

- Thermocycler
- Sterile pipette tips
- Horizontal gel electrophoresis apparatus
- Agarose (electrophoresis grade)
- Low melting-point (LMP) agarose
- Sample buffer: 10% (v/v) glycerol, 0.1% (w/v) bromophenol blue
- Sterile deionized water
- TAE electrophoresis buffer (prepare 10 × stock: 242 g Tris base, 57.1 ml glacial acetic acid, 100 ml 0.5 M Na$_2$EDTA pH 8.0; make to 1 litre with deionized water)
- 1.5 ml microcentrifuge tubes
- T$_{0.1}$E: 10 mM Tris pH 8, 0.1 mM EDTA
- One or more of the following restriction enzymes: *Rsa*I, *Alu*I, *Hae*III, *Pvu*II, *Eco*RV; and appropriate reaction buffer (10 × concentrate)
- 3 M sodium acetate pH 5.2
- T4 ligase (United States Biochemical, 1 U/μl)
- Ligation buffer (10 ×): 500 mM Tris–HCl pH 7.6, 100 mM MgCl$_2$, 10 mM DTT, 10 mM rATP
- Perfect Match PCR specificity enhancer (Stratagene)
- dNTP mix: 5 mM each dATP, dGTP, dCTP, dTTP (Pharmacia)
- Bovine serum albumin (BSA; molecular biology grade, BRL) 5 mg/ml
- β-mercaptoethanol 0.7 M (Sigma; dilute the pure chemical 1:20 in H$_2$O)
- *Taq* DNA polymerase (Amplitaq, 5 U/μl, Perkin-Elmer)
- 10 × PCR buffer: 670 mM Tris–HCl pH 8.8, 166 mM (NH$_4$)$_2$SO$_4$, 67 mM MgCl$_2$
- Vectorette oligos, 1 μM each, as follows:

BPBI: 5'CAA GGA GAG GAC GCT GTC TGT CGA AGG TAA GGA ACGGAC GAG AGA AGG GAG AG3'
BPHII: 5'CTC TCC CTT CTC GAA TCG TAA CCG TTC GTA CGA GAA TCG CTG TCC TCT CCT TG3'

- Left and right (or SP6 and T7) vector-specific primers, appropriate to the vector in question, each at 25 μM. The sequences of the relevant primers are listed below; after each primer is given the annealing temperature, T_{ann}, to be used in step 9. For Lawrist vectors:
LawL (left) primer: 5'CGC CTC GAG GTG GCT TAT CG3' (66°C)
LawR (right) primer: 5'ATC ATA CAC ATA CGA TTT AGG TGA C3' (66°C)
For pBELOBAC:
pBELOBAC-SP6: 5'GCC AAG CTA TTT AGG TGA C3' (54°C)
pBELOBAC-T7: 5'GTA ATA CGA CTC ACT ATA GGG C3' (62°C)
For pCYPAC:
pCYPAC-SP6: 5'CGT CGA CAT TTA GGT GAC ACT G3' (62°C)
pCYPAC-T7: 5'CGC TAA TAC GAC TCA CTA TAG G3' (62°C)
For pBAC108L:
pBAC108L-SP6: 5'ATT TAG GTG ACA CTA TA3' (45°C)
pBAC108L-T7: 5'AAT ACG ACT CAC TAT AG3' (45°C)
- Vectorette primer 224 at 12.5 μM:
5'CGA ATC GTA ACC GTT CGT ACG AGA ATC GCT3'
- Ethanol, 100% and 75%
- Purified clone DNA at > 2 ng/μl, in water or T$_{0.1}$E^a

258

Method

1. In a 1.5 ml microcentrifuge tube prepare the following mixture:
   - 100ng clone DNA
   - 25 U of restriction enzyme *Rsa*I, *Alu*I, *Hae*III, *Pvu*II or *Eco*RV
   - 10 μl 10 $\times$ restriction enzyme buffer concentrate
   - H_2O to 100 μl total.

   Incubate for 1 h at 37°C. A 5 μl aliquot of the reaction should be analysed by electrophoresis on a 0.8% agarose minigel to verify that digestion is complete.

2. Add 10 μl 3 M of sodium acetate and 250 μl of ethanol to precipitate the DNA. Incubate on ice for 1 h.

3. Centrifuge at 13 000 r.p.m. in a microcentrifuge for 15 min; remove the supernatant, taking care not to lose the pellet of DNA.

4. Add 100 μl of 75% ethanol. Vortex briefly, and centrifuge at 13 000 r.p.m. in a microcentrifuge for 5 min.

5. Remove the ethanol as before and desiccate under vacuum for 5 min to remove residual ethanol. Resuspend in 20 μl of $T_{0.1}E$.

6. Prepare the vectorette by mixing equal volumes of BPBI and BPHII, heating to 65°C, and allowing to cool to room temperature over > 10 min.

7. Prepare the ligation mixture:
   - 10 μl DNA fragments from step 5
   - 2 μl 10 $\times$ ligation buffer
   - 2 μl T4 DNA ligase
   - 4 μl prepared vectorette (step 6)

   Incubate at 37°C for 1h. The mixture may be stored at -20°C after incubation.

8. Set-up PCR reactions as follows (the volumes below are for one reaction; a master mix containing the first six ingredients should be prepared and aliquotted as necessary to avoid pipetting very small volumes):
   - 2 μl 10 $\times$ PCR buffer
   - 2 μl dNTP mix
   - 0.64 μl BSA
   - 0.28 μl β-mercaptoethanol
   - 0.24 μl *Taq* polymerase
   - 0.40 μl Perfect Match
   - 10.44 μl $T_{0.1}E$
   - 1 μl appropriate vector-specific primer (left or right, or SP6, or T7)
   - 1 μl 224 primer
   - 2 μl ligation mixture (step 7)

9. Place in a thermocycler under the following conditions:
   - 94°C $\times$ 5 min[b]

Protocol 1. *Continued*

followed by 30 cycles of:

- 93°C × 1 min
- T_{ann} × 1 min[c]
- 72°C × 1 min

followed by:

72°C × 3 min

10. Add 10 μl of sample buffer. Analyse 10 μl of this on a 2% LMP agarose gel in TAE buffer containing 0.3 μg/ml ethidium bromide. After electrophoresis cut out the band and use for hybridization as described in *Protocol 7*.

[a] See, for example, Chapter 10, *Protocol 1* for details of clone DNA isolation.

[b] If non-specific PCR products are generated, they may be reduced by using a 'hot-start' procedure: omit the ligation mixture from the reaction mix (step 8), and instead add it at the end of the initial 5 min denaturation step.

[c] The annealing temperature, T_{ann}, is that appropriate to the vector-specific primer, as detailed in 'Equipment and reagents'.

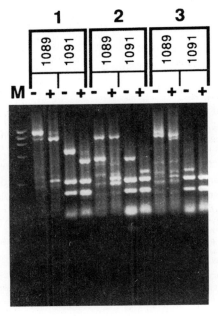

Figure 3. End-fragments of YAC clones isolated by vectorette PCR. DNA from three YAC clones (1, 2, 3) was digested with *Rsa*I, ligated to the vectorette, and amplified using the vectorette primer 224 in conjunction with vector primers 1089 or 1091. Comparison of samples of amplification products before (−) and after (+) digestion with *Eco*RI allows the correct amplification product to be identified. These true end-fragments will be re-amplified in a second PCR (*Protocol 2*). M: size standards (ΦX174 DNA digested with *Hae*III).

2.2 Isolation of end-probes from YACs by vectorette PCR

Because YAC clones cannot conveniently be purified from the complex yeast host genome, two rounds of PCR are necessary to ensure that the YAC insert end is isolated in pure form. The first digestion, ligation of the vectorette, and PCR gives an end-fragment carrying some vector sequence, and also (usually) several contaminant bands. The true end-fragment contains an *Eco*RI site at its junction with the vector sequence: it can thus be identified on an agarose gel by comparing undigested with *Eco*RI digested PCR products. Once identified, the correct band is then isolated and re-amplified using primers close to the cloning site, to give pure end-fragment almost free of vector (*Figure 3*).

As in the case of bacterial clones, *Rsa*I, *Alu*I, *Hae*III, *Pvu*II, or *Eco*RV may be used for the initial digestion; the optimal enzyme will vary from case to case.

Protocol 2. Vectorette PCR of YAC clones

Equipment and reagents

- Thermocycler
- Sterile pipette tips
- Agarose electrophoresis apparatus
- Agarose (electrophoresis grade)
- Low melting-point (LMP) agarose
- Sterile deionized H$_2$O
- TBE electrophoresis buffer (prepare 10 × stock): 108 g Tris base, 55 g boric acid, 9.3 g Na$_2$EDTA; make to 1 litre with deionized water
- TAE electrophoresis buffer (prepare 10 × stock): 242 g Tris base, 57.1 ml glacial acetic acid, 100 ml 0.5 M Na$_2$EDTA pH 8.0; make to 1 litre with deionized water
- Sample buffer: 10% (v/v) glycerol, 0.1 mg/ml bromophenol blue
- 1.5 ml microcentrifuge tubes
- T$_{0.1}$E: 10 mM Tris pH 8, 0.1 mM EDTA
- *Eco*RI restriction enzyme and manufacturer's 10 × digestion buffer
- One or more of the following restriction enzymes: *Rsa*I, *Alu*I, *Hae*III, *Pvu*II, *Eco*RV, and appropriate digestion buffer (10 × concentrate)
- T4 ligase (United States Biochemical, 1 U/μl)
- Ligation buffer (10 ×): 500 mM Tris–HCl pH 7.6, 100 mM MgCl$_2$, 10 mM DTT, 10 mM rATP

- Perfect Match PCR enhancer (Stratagene)
- dNTP mix: 5 mM each dATP, dGTP, dCTP, dTTP (Pharmacia)
- Bovine serum albumin (BSA; molecular biology grade, BRL) 5 mg/ml
- β-mercaptoethanol (Sigma): 0.7 M (dilute the pure chemical 1:20 in H$_2$O)
- *Taq* polymerase (Amplitaq, 5 U/μl, Perkin-Elmer)
- 10 × PCR buffer: 670 mM Tris–HCl pH 8.8, 166 mM (NH$_4$)$_2$SO$_4$, 67 mM MgCl$_2$
- Vectorette oligos (*Protocol 1*)
- PCR primers, each at 12.5 μM, as follows:
  1089: 5'CAC CCG TTC TCG GAG CAC TGT CCG ACC GC3'
  1091 : 5'ATA TAG GCG CCA GCA ACC GCA CCT GTG GCG3'
  Sup4–2: 5'GTT GGT TTA AGG CGC AAG AC3'
  Sup4–3: 5'GTC GAA CGC CCG ATC TCA AG3'
  224 : 5'CGA ATC GTA ACC GTT CGT ACG AGA ATC GCT3'
- Sterile wooden toothpicks
- Light mineral oil (Sigma)
- YAC clone DNA in agarose plugs, prepared as in ref. 12
- DNA size standards (e.g. ΦX174 DNA digested with *Hae*III; HT Biotech)

A. *First digestion, ligation, and amplification*

1. Equilibrate the agarose plug containing the YAC DNA in 1 ml of T$_{0.1}$E for 60 min, changing the buffer once after 30 min.

Protocol 2. *Continued*

2. Trim the block to give a volume of approximately 50 μl (50 mm³), and incubate in a 1.5 ml microcentrifuge tube at 65°C for 15 min to melt the agarose.

3. Incubate at 37°C for 10 min and add 5 μl of the appropriate 10 × restriction enzyme buffer (for *Rsa*I, *Alu*I, *Hae*III, *Pvu*II or *Eco*RV).

4. Add 20 U of the corresponding restriction enzyme and continue to incubate at 37°C for 2–15 h.

5. Resolidify the agarose on ice. Add 200 μl TE and allow the agarose to equilibrate at 4°C for 1 h.

6. Prepare the vectorette by mixing equal volumes of BPBI and BPHII, heating to 65°C, and allowing to cool to room temperature over > 10 min

7. Remove excess liquid from the resolidified agarose and melt at 65°C for 15 min.

8. Incubate at 37°C for 10 min. Add 10 μl 10 × ligation buffer, 20 U T4 DNA ligase, 10 μl of vectorette and adjust volume to 100 μl with H_2O. Incubate at 37°C for 1 h.

9. Add 400 μl $T_{0.1}E$. The mixture may be stored at −20°C at this stage if required.

10. Prepare a 'master mix' for the first round PCR. The following recipe is for one reaction; a total of two reactions will be required for each clone, plus two additional reactions:

    - 5 μl dNTP mix
    - 5 μl 10 × PCR buffer
    - 1.6 μl BSA
    - 0.7 μl β-mercaptoethanol
    - 0.6 μl *Taq* polymerase
    - 1.1 μl TE
    - 1 μl Perfect Match

11. Set-up the first round PCR reactions. It is usually convenient to set-up reactions for the left- and right-hand end-fragments at the same time.

    - 5 μl ligation mixture (step 9)
    - 35 μl master mix (step 10)
    - 5 μl vectorette primer 224
    - 5 μl of primer 1089 (for left end-fragment) or 1091 (for right end-fragment)

    In addition, negative controls (with $T_{0.1}E$ in place of ligation mixture) should be set-up with primers 1089 and/or 1091 as appropriate. Over-lay each reaction with 50 μl of mineral oil.

12. Place in a thermocycler under the following conditions:

- 94 °C × 5 min

followed by 38 cycles of:

- 93 °C × 1 min
- 65 °C × 1 min
- 72 °C × 1 min

followed by:

- 72 °C × 5 min

B. *Identification of true end-fragment*

1. Take 10 µl from each PCR reaction, and add 2 µl of 10 × *Eco*RI digestion buffer concentrate and 20 U of *Eco*RI. Add water to a final volume
of 20 µl and incubate at 37 °C for 1 h.

2. Analyse 10 µl of the undigested PCR products and controls (step A12)
and 20 µl of the *Eco*RI digested amplification products (step B1), each
mixed with 0.5 vol. of loading buffer on a 1.5% agarose gel in TBE
buffer. Load suitable size standards (covering the range ~ 300–1500
bp) on adjacent tracks. Each sample lane (except the negative
controls) will contain several bands.

3. Compare the undigested and *Eco*RI digested left end amplification
products. They should be identical except for one band, which will be
287 bp smaller in the *Eco*RI digested sample than in the undigested
sample; this band, in the undigested sample, is the correct left end-
fragment. Similarly, the correct right end-fragment·(in the undigested
sample) will be replaced by a fragment 172 bp smaller in the *Eco*RI
digested sample.

C. *Re-amplification and purification of end-fragment*

1. Prepare a PCR reaction consisting of:

- 35 µl master mix (as for step A10)
- 5 µl vectorette primer 224
- 5 µl primer Sup4-2 (for left end-fragments) or Sup4-3 (for right end-
fragments)
- 5 µl water

2. Using a sterile wooden toothpick, stab the correct band in the
'undigested' sample lane of the gel (step B2). Swirl the end of the
toothpick in the PCR reaction mixture and overlay with 50 µl of
mineral oil.

3. Place in a thermocycler under the following conditions:

- 94 °C × 5 min

Protocol 2. *Continued*

followed by 20 cycles of:

- 93°C × 1 min
- 59°C × 1 min
- 72°C × 1 min

followed by:

- 72°C × 5 min

4. Analyse 5 μl samples of the amplification products (mixed with an equal volume of sample buffer) on a 2% LMP agarose minigel with suitable size standards. The PCR products should be 287 bp (left end-fragment) or 172 bp (right end-fragment) smaller than the bands which were toothpicked from the previous gel.

5. Excise the band from the gel for use as a probe (*Protocol 7*).

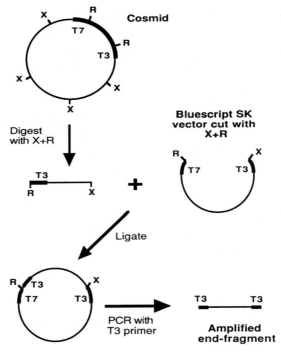

Figure 4. Bluescript-mediated isolation of end-fragments. The cosmid is digested with *Xba*I (X) and *Eco*RI (R), and the fragments ligated into Bluescript SK⁺ cut with the same enzymes. Of all possible ligation products, only that shown (containing the cosmid T3 promoter and corresponding insert end-fragment) can be amplified by the PCR with the T3 primer, to yield an end-fragment carrying two partial T3 sequences. The T7 end of the cosmid can be isolated instead by digesting cosmid and Bluescript with *Hind*III, and amplifying ligation products with the T7 primer.

3. Bluescript-mediated isolation of end-fragments

This method is ideal for isolating insert end-sequences from vectors containing T3 and T7 promoter sequences flanking their cloning sites (9), such as the pWE and 'Supercos' cosmid vectors (Stratagene).

The cosmid is digested with *Eco*RI and another restriction enzyme. This releases insert end-fragments abutting either the T3 or T7 vector sequences (*Figure 4*). The fragments are then ligated directionally into Bluescript SK$^+$ (Stratagene), which also has T3 and T7 promoter sequences flanking its cloning site. By amplifying the ligated material with *either* a T7 *or* a T3 primer alone, the desired end-fragment can be isolated.

Protocol 3. Bluescript-mediated end-fragment isolation from cosmids

Equipment and reagents

- Thermocycler
- Sterile pipette tips
- Agarose gel electrophoresis apparatus
- Low-melting point agarose (electrophoresis grade)
- Sterile deionized H$_2$O
- TBE electrophoresis buffer (prepare 10 × stock): 108 g Tris base, 55 g boric acid, 9.3 g Na$_2$EDTA; make to 1 litre with deionized water
- TAE electrophoresis buffer (prepare 10 × stock): 242 g Tris base, 57.1 ml glacial acetic acid, 100 ml 0.5 M Na$_2$EDTA pH 8.0; make to 1 litre with deionized water
- Sample buffer: 10% (v/v) glycerol, 0.1 mg/ml bromophenol blue
- 1.5 ml microcentrifuge tubes
- T$_{0.1}$E buffer: 10 mM Tris pH 8, 0.1 mM EDTA
- Restriction endonucleases *Eco*RI, *Hind*III, and *Xba*I, and manufacturer's 10 × *Eco*RI digestion buffer[a]
- pBluescript SK$^+$ vector (Stratagene)

- T4 ligase (1 U/μl, United States Biochemical)
- Ligation buffer (10 ×): 500 mM Tris–HCl pH 7.6, 100 mM MgCl$_2$, 10 mM DTT, 10 mM rATP
- dNTP mix: 5 mM each dATP, dGTP, dCTP, dTTP (Pharmacia)
- Bovine serum albumin (BSA; molecular biology grade, BRL) 5 mg/ml
- β-mercaptoethanol (Sigma) 0.7 M (dilute the pure chemical 1:20 in H$_2$O)
- *Taq* polymerase (Amplitaq, 5 U/μl, Perkin-Elmer)
- 10 × PCR buffer: 670 mM Tris–HCl pH 8.8, 166 mM (NH$_4$)$_2$SO$_4$, 67 mM MgCl$_2$
- Primers, at 20 μM each
  T3 primer: 5'ATT AAC CCA CTC TAA AG3'
  T7 primer: 5'AAT ACG ACT CAC TAT AG3'
- Phenol/chloroform, 1:1
- 3 M sodium acetate pH 5.2
- Ethanol, 100% and 75%
- Cosmid DNA, 5–10 ng/μl[b]

Method

1. Digest three 100 ng aliquots of the cosmid as follows:
   - 100 ng cosmid DNA
   - 25 U *Eco*RI
   - 25 U of either *Hind*III (for T7 end-fragments) or *Xba*I (for T3 end-fragments), or neither (control)
   - 3 μl 10 × *Eco*RI digestion buffer
   - water, to 30 μl total volume

   Incubate at 37 °C for 2 h, or overnight.

Protocol 3. *Continued*

2. Inactivate the restriction enzymes by incubating at 65 °C for 10 min.

3. Analyse 5 μl aliquots of each reaction by electrophoresis on a 1% agarose gel to verify that digestion is complete.

4. Add 25 μl of H_2O to the remaining 25 μl of each reaction. Add 100 μl of phenol/chloroform, vortex briefly and spin in a microcentrifuge at 13 000 r.p.m. for 15 min. Transfer the upper (aqueous) phase to a clean microcentrifuge tube, avoiding the interface.

5. Precipitate the DNA by adding 5 μl 3 M sodium acetate and 125 μl of ethanol.

6. Centrifuge at 13 000 r.p.m. in a microcentrifuge for 15 min. Remove the supernatant, taking care not to lose the pellet of DNA.

7. Add 100 μl 75% ethanol. Vortex briefly, and centrifuge at 13 000 r.p.m. for 5 min.

8. Remove the ethanol as before and desiccate under vacuum for 5 min to remove residual ethanol. Redissolve the DNA in 17 μl H_2O.

9. Digest two 100 ng aliquots of pBluescript SK$^+$ DNA as follows:
   - 100 ng pBluescript SK$^+$ DNA
   - 25 U *Eco*RI
   - 25 U of either *Hind*III (for T7 end-fragments) or *Xba*I (for T3 end-fragments)
   - 1 μl 10 × *Eco*RI digestion buffer
   - water, to 10 μl total volume

   Incubate at 37 °C for 2 h, or overnight, then inactivate the restriction enzymes at 65 °C for 10 min.

10. Prepare the two ligation reactions, using the cosmid and pBluescript DNAs digested with either *Eco*RI + *Hind*III (for isolation of T7 end-fragments); or with *Eco*RI + *Xba*I (for isolation of T3 end-fragments) as follows:
    - 10 μl digested pBluescript SK$^+$ (step 9)
    - 7 μl digested, purified cosmid DNA (step 8)
    - 2 μl 10 × ligation buffer
    - 1 μl T4 ligase

    Incubate at 16°C overnight.

11. Prepare the two PCR reactions for amplification of the ligated end-fragment:
    - 1 μl ligation mixture (step 10)
    - 2 μl 10 × PCR buffer
    - 2 μl dNTP mix

- 0.2 μl T7 or T3 primers (for amplification of T7 end or T3 end insert fragments respectively)
- 0.5 μl *Taq* polymerase
- 14.3 μl H$_2$O

12. Place in a thermocycler under the following conditions:

- 94°C × 1 min

followed by 30 cycles of:

- 94°C × 1 min
- 44°C × 1 min
- 72°C × 2 min

followed by:

- 72°C × 3min

13. Add 10 μl of sample buffer to each reaction and resolve each entire sample on a 1% gel made of low-melting point agarose in TAE buffer. Excise the gel slice containing the PCR product from each lane and proceed to *Protocol 7.*[a]

[a] The size of the PCR products will depend on the distribution of restriction sites within the insert. Fragments of > 5 kb do not amplify well in the PCR, whilst very small fragments may be of limited use as probes. If either of these is the case, alternative enzymes may be used for digestion of the cosmid and pBluescript DNAs. *Hind*III may be replaced by *Hind*II, *Acc*I, *Xho*I, *Kpn*I, *Sal*I or *Asp*718 for isolation of T7 end-fragments; *Xba*I can be replaced with *Pst*I, *Sma*I, *Sac*I or *Spe*I for isolation of T3 end-fragments.

[b] Chapter 10, *Protocol 1* may be used to purify cosmid DNA.

4. Identification of end-fragments by hybridization with oligonucleotides.

The Supercos/pWE and Lawrist cosmid vectors, as well as BAC and PAC vectors, all have *Eco*RI sites outside the T3 (or SP6) and T7 promoters. Therefore, an *Eco*RI digest will release insert end-fragments carrying these promoter sequences. Hybridization of *Eco*RI digests with end-labelled oligos complementary to the promoter sequences therefore allows the identification of such fragments (*Figure 5*), which may then be isolated by gel electrophoresis, as described in *Protocol 4*.

Such an approach has the advantage of being applicable on a large scale, for end-fragment identification of many clones at once. This technique is also useful for confirming overlaps between clones: once end-fragments have been identified it is easier to compare the rest of the fragments and estimate possible overlaps.

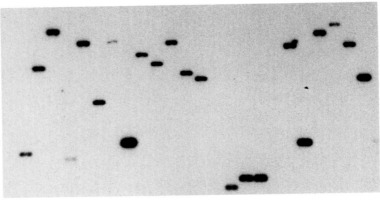

SP6

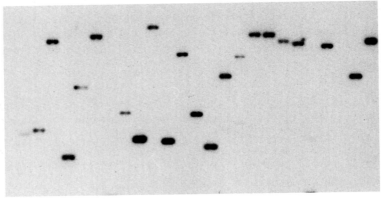

T7

Figure 5. Identification of end-fragments by oligonucleotide hybridization. 25 different Lawrist 4 cosmids were digested with *Eco*RI, the fragments resolved by electrophoresis, and blotted. Insert end-fragments carrying vector sequences were detected by auto-radiography after probing with labelled oligonucleotides complementary to either the SP6 (upper panel) or T7 (lower panel) promoter sequences. The clones are shown in the same order in each panel.

Protocol 4. Detection of end-fragments by vector-specific oligo hybridization

Equipment and reagents

- Sterile pipette tips
- Agarose gel electrophoresis apparatus
- Low-melting point agarose (electro-phoresis grade)
- *Eco*RI restriction enzyme and manu-facturer's 10 × digestion buffer

- Sterile deionized H_2O
- TBE electrophoresis buffer (prepare 10 × stock): 108 g Tris base, 55 g boric acid, 9.3 g Na_2EDTA; make to 1 litre with deion-ized water
- 1.5 ml microcentrifuge tubes

- TAE electrophoresis buffer (prepare 10 × stock): 242 g Tris base, 57.1 ml glacial acetic acid, 100 ml 0.5 M Na$_2$EDTA pH 8.0; make to 1 litre with deionized water
- Sample buffer: 10% (v/v) glycerol, 0.1 mg/ml bromophenol blue
- T$_{0.1}$E buffer: 10 mM Tris pH 8, 0.1 mM EDTA
- Hybridization oven and bottles
- T4 kinase (10 U/µl; Boehringer Mannheim) and manufacturer's 10 × buffer concentrate
- 20 mM EDTA pH 8.0
- [γ-^{32}P]ATP(Amersham; 3000 Ci/mmol)
- P4 Superfine Sephadex (Pharmacia)
- TE buffer: 10 mM Tris–HCl pH 8, 1 mM EDTA
- Washing solution: 2 × SSC, 0.1% SDS (20 × SSC stock solution is 3 M NaCl, 0.3 M sodium citrate, adjusted to pH 7.0 with NaOH)

- Glass wool
- SaranWrap plastic film
- Hybridization solution: 6 × SSC, 10 × Denhardt's reagent, 50 mM Tris–HCl pH 7.4, 1% (w/v) Sarcosyl, 10% (w/v) dextran sulfate. Denhardt's reagent is prepared as a 50 × stock (10% (w/v) Ficoll 400, 10% (w/v) polyvinylpyrrolidone, 10% (w/v) BSA) and stored at −20°C
- Oligonucleotide at 5 µM—for pWE and Supercos vectors use either T7 or T3 oligonucleotides (depending on which end of the insert is sought); for Lawrist, BAC, and PAC vectors use either T7 or SP6 oligonucleotides:

T7: 5'AAT ACG ACT CAC TAT AG3'

T3: 5'ATT AAC CCA CTC TAA AG3'

SP6: 5'ATT TAG GTG ACA CTA TA3'
- Bacterial clone DNA at >10 ng/µl[a]
- Equipment and reagents as for Chapter 12, *Protocol 8*

Method

1. Digest 100 ng of clone DNA with 25 U of *Eco*RI in 10 µl of the manufacturer's digestion buffer for 2 h at 37°C.

2. Add 5 µl of sample buffer to a 5 µl aliquot of the digest and resolve the fragments by electrophoresis in a 0.8% agarose gel, stain with ethidium and photograph. Blot the fragments onto nylon membranes (Chapter 12, *Protocol 8*), cutting one corner of the filter to ensure that it can be orientated with respect to the gel. Retain the other half of the sample.

3. In a microcentrifuge tube, prepare the oligolabelling reaction:
   - 10 µl appropriate oligonucleotide
   - 2 µl H$_2$O
   - 2 µl 10 × kinase buffer
   - 5 µl [γ-^{32}P]ATP
   - 1 µl T4 kinase

   Incubate at 37°C for 30 min.

4. Add another 1 µl kinase and incubate again at 37°C for 30 min.

5. Add 80 µl 20 mM EDTA.

6. Prepare a gel filtration column by filling the barrel of a 1 ml syringe, its nozzle plugged with glass wool, with P4 Superfine Sephadex in TE buffer. Overlay the Sephadex with the reaction mixture, place the syringe in a 15 ml centrifuge tube and spin at 1 *g* for 1 min. Retain the eluted liquid.

7. Place the column in a clean 15 ml tube, add 100 µl of TE buffer to the column, and spin as before, retaining the eluted liquid.

Protocol 4. *Continued*

8. Repeat step 8 a further four times, giving a total of six eluted fractions.

9. Count 2 μl samples of each of the six fractions using a hand-held monitor. The labelled oligo will normally be eluted in the first and/or second fraction(s), samples of which should register > 1000 c.p.m. Retain the fraction with the greatest activity, or pool the two most active samples. Later fractions should contain little radioactivity.

10. Add the probe to a sufficient quantity of hybridization solution (typically 10 ml, depending on the size of the filter and of the hybridization bottles). Place, with the filter, in a hybridization bottle or other container and hybridize at 42°C overnight.

11. Pour off the hybridization solution. Wash the filter for 30 min with gentle agitation in washing solution at room temperature for 30 min.

12. Pour off the wash solution, replace with fresh washing solution and wash for a further 10 min at room temperature.

13. Wrap the filter in SaranWrap, and expose overnight to X-ray film.

14. Resolve the remaining half of the digested clone (step 1) on a 0.8% low-melting point agarose gel in TAE buffer and stain with ethidium. Comparison with the previous gel photograph and the autoradiograph should enable you to identify the clone end-fragment which hybridizes with the probe.

15. Excise the band containing the desired fragment, and proceed to *Protocol 7*.

16. Filters may be stripped for re-probing with a different oligonucleotide, provided that they are not allowed to dry completely at any time. Strip by immersing in a solution of 1% SDS at 90°C for several minutes, with gentle agitation.

ªSee Chapter 10, Protocol 1

5. End-fragment rescue by vector religation

Religation of bacterial clones after restriction digestion enables fragments to be recovered simultaneously from each end of an insert in cosmids, PACs, or BACs. This method, described in *Protocol 5*, has the advantage of giving cloned end-fragments for storage, and of requiring no knowledge of the vector sequences flanking the insert.

The clone is first digested with any restriction enzyme which does not cleave within the vector or at the insert–vector junctions (*Figure 6*). The digestion fragments are recircularized and transformed into competent

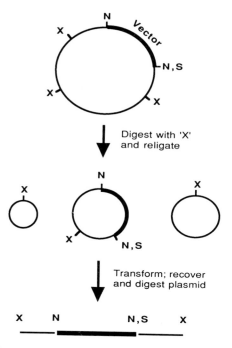

Figure 6. Isolation of end-fragments by vector religation. The bacterial clone (vector shown as heavy line) is digested with restriction enzyme X, which cleaves only within the insert. The fragments are recircularized by ligation and transformed into competent cells. Only the circular fragment consisting of the vector carrying the two insert end-fragments will be viable. Digestion with enzymes X + S or X + N will release one or both insert end-fragments from the purified fragment, respectively.

cells. The recircularized vector sequence, carrying fragments from either end of the original insert, will be propagated, whilst internal fragments from the insert will not. End-fragments can then be recovered, if necessary, from the recircularized vector.

5.1 Choice of restriction enzymes

Any restriction enzymes that do not cleave within the vector sequence or at the cloning site can be used. For example, *Sca*I, *Sst*I, or *Xba*I can be used with inserts cloned into the pCYPAC vector (13). For recovery of end-fragments from the recircularized vector, this first enzyme is used in combination with one which cleaves at one or both of the vector–insert boundaries (for recovery of one or both end-fragments). In the case of pCYPAC, for example, *Not*I can be used to cleave at both insert boundaries whilst *Sal*I cleaves close to only one of the insert boundaries.

Protocol 5. Vector religation of cosmid, BAC and PAC clones

Equipment and reagents

- Thermocycler and compatible reaction vessels
- Horizontal electrophoresis apparatus
- Agarose (electrophoresis grade)
- TBE buffer (prepare 10 × stock): 108 g Tris base, 55 g boric acid, 9.3 g Na₂EDTA; make to 1 litre with deionized water
- $T_{0.1}E$: 10 mM Tris–HCl pH 8, 0.1 mM EDTA
- 3 M sodium acetate pH 6.5
- Phenol/chloroform/isopropanol (25:24:1)
- 10 × ligation buffer: 500 mM Tris-HCl pH 7.5, 200 mM dithiothreitol (DTT), 100 mM MgCl₂, 5 mM spermidine, 1 mg/ml BSA
- T4 DNA ligase (United States Biochemical)
- *E. coli* competent cells (HB101, Promega)
- LB agar plates containing the selective antibiotic appropriate to the vector

- Low melting-point (LMP) agarose
- Clone DNA at 20–50 ng/μl[a]
- Restriction enzymes and appropriate 10 × buffer concentrates, chosen with reference to the vector sequence (see text). At least three enzymes (A, B, C) will be needed:
  (A) should not cleave within the vector or vector-insert junctions.
  (B) should cleave at or close to both vector-insert junctions, but not elsewhere within the vector.
  (C) should cleave at or close to only one vector-insert junction, but not elsewhere within the vector.
- 3M sodium acetate pH 5.2
- Ethanol, 100% and 70%
- Equipment and reagents for Chapter 10, *Protocol 1*

Method

1. Digest 200 ng of clone DNA with 5 U of restriction enzyme A in a 20 μl reaction for 2 h under conditions recommended by the enzyme manufacturer.

2. Remove a 5 μl aliquot of the reaction products and analyse on 0.8% (w/v) agarose gel in 1 × TBE to check that digestion is complete.

3. Add 20 μl of phenol/chloroform/isopropanol, vortex for 30 sec, and spin at 13 000 r.p.m. in a microcentrifuge for 5 min. Transfer the upper (aqueous) phase to a clean 0.5 ml microcentrifuge tube.

4. Precipitate the DNA by adding 1.5 μl 3 M sodium acetate and 40 μl of ethanol. Incubate at −20°C for 15 min, and pellet the DNA at 13 000 r.p.m. for 10 min in a microcentrifuge.

5. Decant the ethanol, wash the pellet with ∼ 100 μl 70% (v/v) ethanol, centrifuge as in step 4, remove the remaining ethanol, and allow the pellet to air dry for 10 min.

6. Resuspend the DNA in the following mixture:
   - 2 μl 10 × ligation buffer
   - 2 μl 5 mM rATP
   - 2 μl T4 DNA ligase (2.5 U/μl)
   - 14 μl H₂O

   Incubate at 14°C for 4–6 h.

7. Transform *E. coli* competent cells with 5 μl of ligation reaction, following the manufacturer's instructions, and plate onto LB agar containing the appropriate antibiotic.

8. Prepare DNA from the recombinant clones (Chapter 10, *Protocol 1*)
   and digest 100 ng samples of clone DNA with:
   (a) enzyme A.
   (b) enzymes A + B.
   (c) enzymes A + C.
   In each case, digestion should be performed in 20 µl of the enzyme
   supplier's recommended buffer with 25 U of enzyme (or of each
   enzyme) for 2 h at the appropriate temperature.

9. Analyse the products of these digestions on a 0.8% LMP agarose gel.
   (a) Should give a single linear product (the vector carrying both end-
   fragments). (b) Should give three products: the vector (identified by
   its known size) and each end-fragment. (c) Will give two products: one
   free end-fragment; plus the other end-fragment attached to the vector.
   Comparison of digests (b) and (c) allows one to determine which
   insert end corresponds to each of the two end-fragments in digest (b);
   excise the appropriate bands from the gel and proceed to *Protocol 7*.

*ª*See Chapter 10, *Protocol 1*

6. Isolation of internal sequences from YAC clones

As discussed in the introduction, it is often desirable to isolate several internal sequences from inserts cloned in YACs. Given the difficulty of purifying
YAC DNA from the host yeast genome, a convenient way to accomplish this
is to generate and subclone inter-*Alu* repeat PCR products from human-
derived inserts. This is done by amplifying the YAC DNA (in the presence of
the yeast genomic background) using one or more primers complementary to
the ends of human *Alu* interspersed repeat sequences (14). The orientation
of the primers is such that they will selectively amplify any regions (of less
than a few kilobases) which are flanked by *Alu* sequences in the cloned
insert. The distribution of *Alu* elements in the human genome is such that
most YAC inserts will yield one or more inter-*Alu* PCR products. Similar
approaches can be used on inserts from other species, provided that suitable
species-specific repeats exist to act as priming sites.

PCR products from *Taq* polymerase amplification contain a single
deoxyadenosine at the 3′ ends of each strand, which reduces the efficiency of
cloning into blunt-ended vectors, unless ends are made blunt using Klenow
or T4 polymerase. Using a vector whose ends have single 3′ deoxythymidine
overhangs, individual PCR products can be subcloned and isolated. Examples of such vectors include pCRII from the TA Cloning Kit (Invitrogen),
and pGEM-T (Promega). Recombinant clones are then analysed by second
PCR to verify that they contain inter-*Alu* sequences and to determine their
insert sizes, as described in *Protocol 6* (see *Figure 7*).

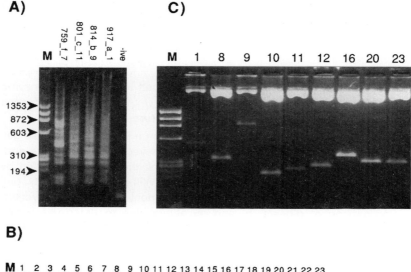

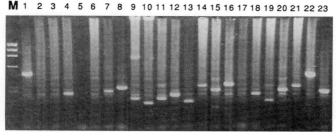

Figure 7. Isolation of *Alu* PCR products from YACs. (A) Primary *Alu* PCR products of four YAC clones (759_f_7, 801_c_11, 814_b_9, and 917_a_1) and negative control (-ive). (B) Products from a fifth clone (912_g_9; not shown) were subcloned into pCRII, and DNA from 23 recombinant clones (1–23) was amplified by a secondary *Alu* PCR and analysed by electrophoresis. (c) The inserts from nine of the subclones were excised and purified, and analysed by electrophoresis. M: DNA size standards (ΦX174 DNA digested with *Hae*III; sizes in base pairs are indicated in A).

Protocol 6. Isolation of inter-*Alu* sequences from YAC clones

Equipment and reagents

- Thermocycler
- Horizontal electrophoresis apparatus
- Low melting-point (LMP) agarose
- Agarose (electrophoresis grade)
- TBE buffer (*Protocol 5*)
- TE: 50 mM Tris–HCl pH 8.0, 1 mM EDTA
- T$_{0.1}$E: 50 mM Tris–HCl pH 8.0, 0.1 mM EDTA

- 10 × PCR buffer: 500 mM KCl, 100 mM Tris–HCl pH 9.0, 1.0% Triton X-100, 75 mM MgCl$_2$
- *Taq* DNA polymerase (Amplitaq, 5 U/µl; Perkin-Elmer)
- dNTP mix: 2 mM each of dATP, dCTP, dGTP, dTTP
- 1 ml syringe

6. Incubate at 65°C for 30–60 min and add to hybridization buffer (the volume of hybridization buffer will depend on the size of the filter).

7. Hybridize overnight at 65°C and remove the hybridization buffer.

8. Wash the filter once in 2 × SSC, 0.1% SDS for 10 minutes at room temperature. Then once in 2 × SSC, 0.1% SDS for 20 minutes at 65°C. Finally twice in 0.1 × SSC, 0.1% SDS for 10 min at 65°C.[c]

9. Wrap the filter in SaranWrap and expose to X-ray film overnight, or as necessary. The developed autoradiograph should reveal the locations of colonies containing inserts matched by the probe.

[a] Pure β-mercaptoethanol is approximately 14M.
[b] Or other species' DNA, as appropriate.
[c] The temperature of the final wash can be increased to 68°C to reduce background if necessary.

Acknowledgements

We would like to thank Ruth Mason, Ferina Ismail, Richard Stephens, and Nicos Tripodis for verifying protocols and providing data, and Dr David Bentley for providing initial vectorette protocols. The work was supported by the MRC-HGMP and EEC Biomed-1 grants.

References

1. Cross, S. H. and Little, P. F. (1986). *Gene*, **49**, 9.
2. Evans, G. A. and Wahl, G. M. (1987). In *Methods in enzymology* (ed. S. L. Berger and A. R. Kimmel) Vol. 152, p. 604. Academic Press, London.
3. Wahl, G. M., Lewis, K. A., Ruiz, J. C., Rothenberg, B., Zhao, J., and Evans, G. A. (1987). *Proc. Natl. Acad. Sci. USA*, **84**, 2160.
4. Shero, J. H., McCormick, M. K., Antonarakis, S. E., and Hieter, P. (1991) *Genomics*, **10**, 505.
5. Ochman, H., Ajioka, J. W., Garza, D., and Hartl, D .L. (1990). *Bio/Technology*, **8**, 759.
6. Triglia, T., Peterson, M. J., and Kemp, D. J. (1988). *Nucleic Acids Res.*, **16**, 8186.
7. Riley, J., Butler, R., Ogilvie, D., Finniear, R., Jenner, D., Powell, S., *et al.* (1990). *Nucleic Acids Res.*, **18**, 2887.
8. Devon, R. S., Porteous, D. J., and Brookes, A. J. (1995) *Nucleic Acids Res.*, **23**, 1644.
9. Haberhausen, G. and Muller, U. (1995). *Nucleic Acids Res.*, **23**, 1441.
10. Anand, R., Ogilvie, D. J., Butler, R., Riley, J. H., Finniear, R. S., Powell, S. J., *et al.* (1991). *Genomics*, **9**, 124.
11. Coffey, A. J., Roberts, R. G., Green, E. D., Cole, C. G., Butler, R., Anand, R., *et al.* (1992). *Genomics*, **12**, 474.
12. Ragoussis, J. (1995). In *YAC protocols* (ed. D. Markie), p. 69. Humana Press.

13. Ioannou, P. A., Amemiya, C. T., Garnes, J., Kroisel, P. M., Shizuya, H., Chen, C., *et al.* (1994). *Nature Genet.*, **6**, 84.
14. Cole, C. G., Dunham, I., Coffey, A. J., Ross, M. T., Meier-Ewert, S., Bobrow, M., *et al.* (1992). *Genomics*, **14**, 256.
15. Carter, M. J. and Milton, I. D. (1993). *Nucleic Acids Res.*, **21**, 1044.

12

Long-range restriction mapping of genomic DNA

WILFRIED BAUTSCH, UTE RÖMLING, KAREN D. SCHMIDT,
AKHTAR SAMAD, DAVID C. SCHWARTZ,
and BURKHARD TÜMMLER

1. Principle of long-range restriction mapping

Long-range restriction maps are important tools to determine the arrangement and distance of gene markers, and a prerequisite before committing oneself to costly and time-consuming cloning and sequencing projects in order to identify gene organization and rearrangements. For the construction of a low-resolution physical map of a genome or chromosomal region, the intact DNA is digested with a restriction endonuclease that cleaves infrequently and the fragments are separated by pulsed-field gel electrophoresis (PFGE) (1, 2). Restriction fragment patterns of small genomes, such as those of prokaryotes or lower eukaryotes, can be visualized directly by gel staining. With larger genomes, the complex mixture of fragments is blotted onto membranes after electrophoresis, and hybridized with suitable probes to detect fragments of interest. The order of probes and of fragments is established from the combinatorial analysis of partial and complete digests with one or more restriction enzymes. In the simplest case, two probes will hybridize to the same restriction fragment, thus revealing physical linkage between them. However, independent confirmation must always be sought to exclude hybridization to distinct but co-migrating bands. This may be accomplished by, for example, hybridization of the two probes in question to the fragments produced by *partial* digestion with the same enzyme as was used in the complete digest: both probes should detect the same bands (and with the same relative intensities) in this more complex mixture of fragments. Combined usage of several different enzymes should then allow construction of a macrorestriction map around both loci, which may be refined by double digestion analysis.

Similarly, genomic rearrangements (deletions, insertions, chromosomal translocations) may be mapped by looking for an altered electrophoretic mobility of defined fragments in affected individuals. Again, independent confirmation is required to exclude other explanations for such alterations, such as differential methylation or point mutations which create or destroy sites. Such confirmation may come, for example, from the investigation of several tissues, cell lines and somatic cell hybrids; or the demonstration of probe deletion or abnormal physical linkage of probes known to have originated from different chromosomal positions. The optimal strategy depends on the genome size and on the frequency and spacing of restriction sites in the genome being studied. For a more detailed overview the reader is referred to a recent monograph (2) and the original literature.

This chapter provides protocols for all practical aspects fundamental to the long-range restriction mapping of genomic DNA by PFGE: DNA isolation, restriction digestion, one- and two-dimensional electrophoresis, blotting, and hybridization with labelled DNA probes. The generation of suitable DNA probes (cloned genes, sequence-tagged sites, primer pairs defining unique PCR products, etc.) is not covered, but many aspects of probe generation are discussed in other chapters of this book and elsewhere.

2. Preparation of agarose-embedded genomic DNA

Before restriction digestion is performed, intact chromosomal DNA must be prepared. Cells are embedded in agarose and then treated with detergents and enzymes which lyse the cell wall and allow proteins and other molecules to diffuse out. The long DNA molecules remain trapped in—and protected by—the agarose. Although the basic method is always the same, the choice of detergent and enzymes depends on the constituents of cells and their extracellular matrix. We supply protocols for the most common applications: the processing of mammalian cells and tissues (*Protocol 1*), plants (*Protocol 2*), and bacteria (*Protocol 3*). *Protocol 3* is also applicable to the processing of yeast and fungi, but pretreatment with cell wall-lysing enzymes prior to proteolytic digestion is necessary. We routinely encapsulate cells into agarose plugs of $10 \times 6 \times 1$ mm in size. Alternatively, the cells can be embedded in agarose microbeads, which are essentially minute agarose blocks. The use of microbeads is rarely described in the literature, as they tend to stick to plastic and produce fuzzy bands and more background than DNA embedded in plugs. However, these shortcomings have been overcome by improved protocols which have been successfully applied to bacterial, yeast, and plant cells (3, 4) (*Protocol 4*).

The cell walls make it difficult to prepare DNA from Gram-positive bacteria (see footnote in *Protocol 3*) and particularly from plant tissues without prior treatment. Hence, most groups working with plants first isolate protoplasts using cell wall hydrolases such as cellulase, pectolyase Y-23, or macerase, and then embed the protoplasts in agarose. The inherently large volume of plant cells (due to the internal vacuole) causes problems in obtaining sufficiently high DNA concentrations. Therefore, cell wall digestion is performed in incubation media of high osmolarity to reduce the cell volume. Furthermore, it is often advantageous to starve the plants for a few days in the dark before harvesting, to decrease the number of starch granules. *Protocol 2A* describes the necessary steps for processing tomato plants (6). Types and amounts of cell wall hydrolases and enzyme action conditions vary for different plant species and need to be empirically optimized (see the review by Wu *et al.* in ref. 2 for further details).

Protocol 1. Preparation of unsheared genomic DNA from mammalian cells

Equipment and reagents

- Perspex block mould, preferably with slots of 10 × 6 × 1 mm
- Histopaque-1077 (Sigma)
- Percoll (Sigma)
- High quality low-melting point (LMP) agarose (Sigma, type VII)
- Haemocytometer
- Glass or Potter homogenizer (for fresh or frozen solid tissues)

Glassware, plastic disposables, and all buffers listed below are to be autoclaved

- ES: 0.5 mg/ml Proteinase K, 1% (w/v) *N*-lauroylsarcosine, 0.5 M EDTA pH 9.5

- Percoll-1095: 69% (v/v) Percoll, 10% (v/v) 1.5 M NaCl, 21% (v/v) H_2O
- PBS: prepare by making solutions A and B (below), autoclaving them separately, and mixing after cooling to room temperature. Solution A: 0.13 g $CaCl_2.2H_2O$, 0.1 g $MgCl_2.6H_2O$ in 500 ml H_2O. Solution B: 0.2 g KCl, 0.2 g KH_2PO_4, 2.9 g $Na_2HPO_4.12H_2O$, 8 g NaCl in 500 ml H_2O
- SE: 75 mM NaCl, 25 mM EDTA pH 7.5
- 10 × TE: 10 mM EDTA, 0.1 M Tris pH 7.7 (measured at 20°C; the pH of Tris buffers is strongly temperature-dependent)
- Mortar and pestle

A. Cells in culture

1. Harvest early stationary phase cells from suspension, or subconfluent cells from plates (e.g. by trypsination or mechanical treatment).

2. Pellet the cells at 500 *g* for 10 min at room temperature and wash twice with PBS.

3. Resuspend the cells at 3 × 10^7/ml in SE buffer at 37°C.

4. Melt 2% (w/v) LMP agarose in SE, cool to 50°C in a water-bath, and mix gently with an equal volume of the cell suspension.

5. Immediately dispense the mixture into the slots of the mould.

6. Let the agarose blocks solidify for 15 min at 4°C.

7. Gently remove the solidified blocks into 1 ml ES in an Eppendorf tube, placing five blocks in each tube.

8. Incubate at 56°C for 15 h. Tubes should be gently mixed by inverting a few times during incubation. The agarose blocks will be transparent by the end of this period.

9. Remove SE with a Pasteur pipette and equilibrate the agarose blocks in TE, using four changes of buffer at intervals of at least 30 min.

10. Cut each block into halves (i.e. to 5 × 6 × 1 mm) and store at 4°C in TE. The encapsulated DNA is stable for at least four years.

Protocol 1. *Continued*

B. *Fresh blood*

1. Layer 10 ml of Percoll-1095 under 10 ml of Histopaque-1077 in a 50 ml conical centrifuge tube.
2. Dilute 5 ml freshly drawn K–EDTA blood with 15 ml PBS.
3. Carefully layer the diluted blood onto the step gradient.
4. Spin at 400 g for 30 min at room temperature.[a]
5. Aspirate off all but 5 ml of the PBS.
6. Collect the two fractions of granulocytes and lymphocytes at the interfaces using a Pasteur pipette, and transfer them both into another 50 ml centrifuge tube.
7. Dilute the cell suspension to 50 ml with PBS.
8. Centrifuge at 500 g for 20 min at room temperature.
9. Resuspend the cell sediment in 0.2 ml PBS. Count an aliquot in a haemocytometer and adjust to 3×10^7 cells/ml in PBS.
10. Proceed as in Part A, steps 4–10.

C. *Fresh solid tissue*

1. Mince the tissue into $\leqslant 2$ mm pieces using sterile scissors and scalpel.
2. Transfer into PBS and break into single cells by several strokes with a tight-fitting glass or Potter homogenizer.
3. Count an aliquot in a haemocytometer; if large cell aggregates are present, repeat the homogenization. Adjust to 3×10^7 cells/ml in SE.
4. Proceed as in Part A, steps 4–10.

D. *Frozen tissue*

1. Pre-chill a mortar and pestle with dry ice.
2. Grind the tissue under liquid nitrogen to a fine powder. Wear a face mask when handling human or other potentially infectious material.
3. Transfer the powder into PBS using a sterile spatula (≈ 5 ml PBS per gram of mammalian tissue).
4. Count an aliquot in a haemocytometer; if cell clumps are present, homogenize with a few strokes of a glass or Potter homogenizer. Adjust to 3×10^7 cells/ml in SE.
5. Proceed as in Part A, steps 4–10.

[a] The two-step gradient centrifugation (5) separates lymphocytes (banding at the PBS–plasma/Histopaque-1077 interface) and granulocytes (banding at the Histopaque-1077/Percoll-1095 interface) from erythrocytes sedimenting at the bottom. For optimum separation, freshly drawn K–EDTA blood should be immediately processed to avoid haemolysis which reduces the yield of nuclear cells.

As protoplast isolation on a large scale is costly and tedious, straight-forward alternatives have been developed. Leaves, flour or seeds are ground to a fine powder and the crushed tissue is directly embedded in agarose plugs (7) (*Protocol* 2B). In a more time-consuming method (*Protocol* 2C), which has been successfully applied to many plant species (4), the plant cells walls are physically broken and the nuclei are isolated from the homogenate and embedded in agarose plugs or microbeads.

Protocol 2. Megabase DNA preparation from plant tissue

Equipment and reagents

All solutions are to be autoclaved unless otherwise stated.

- CPW-salts solution: 27.2 mg/litre KH_2PO_4, 0.16 mg/litre KI, 0.025 mg/litre $CuSO_4.5H_2O$, 0.101 g/litre KNO_3, 0.246 g/litre $MgSO_4.6H_2O$ (8)
- Enzyme medium: 9% (w/v) mannitol, 3 mM 2-(*N*-morpholino)-ethanesulfonic acid (MES)–KOH pH 5.8, 1% (w/v) cellulase (EC 3.2.1.4, ICN Pharmaceuticals), 0.2% (w/v) macerase (Calbiochem), 0.05% (w/v) pec-tolyase Y-23 (ICN Pharmaceuticals), in CPW-salts solution—prepare fresh medium for each use and filter-sterilize
- Wash medium: 3 mM MES–KOH pH 5.8, 2% (w/v) KCl in CPW-salts solution

- 10 × HB: 100 mM Tris, 800 mM KCl, 100 mM EDTA, 10 mM spermidine, 10 mM spermine, pH 9.4 adjusted with NaOH
- HB–SMT: 1 × HB, 0.5 M sucrose, 0.15% (v/v) β-mercaptoethanol, 0.5% (v/v) Triton X-100
- ES, SE, and TE buffers (*Protocol 1*)
- Nylon mesh (80µm and 30-40µm pore sizes; Millipore)
- Miracloth polyester mesh (Calbiochem, Cat. No. 475855)
- Haemocytometer
- Mortar and pestle
- Crucible

A. *Protoplast isolation*[a]

1. Harvest 2-4 g of young, fully expanded leaves from four- to six-week old shoot cultures grown in a greenhouse.

2. Dissect the leaves with a sterile scalpel into 2 mm slices in 10 ml of enzyme medium in a Petri dish.

3. Add 30 ml of enzyme medium and incubate in the dark at room tem-perature for 15 h without shaking. Check the efficiency of protoplast formation by microscopic examination of an aliquot.

4. Remove residual leaf pieces and rinse them by shaking in a small volume of wash medium to release all protoplasts, pooling with the previous medium. Filter protoplasts through 80 µm and then 30–40 µm nylon mesh. Rinse Petri plate and filters thoroughly with wash medium to increase protoplast yield, filter the washes, and pool.

5. Pellet the protoplasts (10 min, 200 *g*, room temperature).

6. Discard the supernatant and wash the pellet twice (pelleting for 5 min, 200 *g*) with 2 ml wash medium.

7. Count an aliquot of the resuspended protoplast suspension using a haemocytometer and adjust to 5×10^7 cells/ml in wash medium.

8. Proceed as in *Protocol 1A*, steps 4–10.

Protocol 2. *Continued*

B. *Encapsulation of crushed plant tissue in agarose (7)*

1. Pre-chill a mortar and pestle with dry ice.
2. Grind about 0.4 g of young green leaves under liquid nitrogen to a fine powder.[b]
3. Transfer the powder to a crucible pre-heated at 50°C.
4. Mix with 2 ml of 0.7% (w/v) LMP agarose in SE at 42°C and gently stir with a sterile spatula to obtain a homogeneous mixture.
5. Dispense the mixture into the mould, shaking gently while pouring to maintain homogeneity.
6. Turn the mould on its side (to avoid deposition of debris at the bottom of each plug). Let the agarose blocks solidify for 20 min at 4°C.
7. Proceed as in *Protocol 1*A, steps 7–10.

C. *Preparation of megabase DNA from plant nuclei (4)*

1. Pre-chill a mortar and pestle with dry ice.
2. Grind about 20 g of frozen plant tissue under liquid nitrogen to a fine powder.[c]
3. Transfer the powder into 200 ml ice-cold HB–SMT.
4. Stir the contents for 10 min on ice and filter the homogenate into an ice-cold centrifuge bottle through two layers of cheesecloth and one layer of Miracloth by squeezing with gloved hands.
5. Centrifuge at 1800 *g* for 20 min at 4°C.
6. Gently resuspend in 30 ml ice-cold HB–SMT and filter through two layers of Miracloth by gravity into a 50 ml centrifuge tube.
7. Remove tissue residues and intact cells by centrifugation at 60 *g* (2 min, 4°C).
8. Transfer the supernatant to another centrifuge bottle and pellet the nuclei (1800 *g*, 15 min, 4°C).
9. Wash the pellet by resuspension in ice-cold HB–SMT and centrifugation (1800 *g*, 15 min, 4°C) one to three times.
10. Count an aliquot of the nuclei using a haemocytometer and adjust to 5×10^7 cells/ml in ice-cold HB.
11. Pre-warm the suspension of nuclei to 40°C in a water-bath.
12. Proceed as in *Protocol 1*A, steps 4–10.

[a] Adapted from a protocol by van Daelen *et al.* (6) for tomato leaves. Further protocols for protoplast isolation from rice, wheat, barley, and several dicotyledonous plants are described in an article by Wu *et al.* in ref. 2.
[b] Coarse flour or crushed seeds can be used instead. Seeds are crushed in a mortar and pestle without liquid nitrogen (7).
[c] Fresh plant tissue may be used instead. Cut it into pieces and homogenize in ice-cold $1 \times$ HB plus 0.15% β-mercaptoethanol with a blender. Add Triton X-100 to a final concentration of 0.5% (v/v) and continue with step 4.

Protocol 3. Preparation of unsheared genomic DNA from bacteria[a]

Reagents

• SE, ES, and TE buffers (as *Protocol 1*)

Method

1. Transfer a single bacterial colony into culture broth and incubate for ~ 18 h at the appropriate temperature.

2. Dilute 0.1 ml of this culture with 0.9 ml of SE. Measure OD_{570} (0.6 $OD_{570} \approx 1 \times 10^9$ cells/ml).[b]

3. Harvest the cells by centrifugation for 10 min at 1400 g.

4. Resuspend the pellet in the same volume of SE and centrifuge again.

5. Adjust the cell suspension to 3×10^9 cells/ml in SE.

6. Proceed as in *Protocol 1A*, steps 4–10.[c]

[a] This protocol works with many Gram-negative and Gram-positive bacteria. However, the procedure needs to be modified for bacteria with unusual cell walls (Nocardia, Mycobacteria) or with highly flexible genomes (Neisseria). Some species have to be grown on agar slants.
[b] Often it is necessary to empirically determine the relation between OD_{570} and cell density; the stated figure is correct for *E. coli*, staphylococci, *Pseudomonas* and *Burkholderia* species.
[c] This procedure is sufficient for Gram-negative bacteria. Prior to treatment with ES, agarose-embedded Gram-positive bacteria should be incubated for 2 h at 37°C with lysozyme buffer (10 mg/ml lysozyme in 100 mM NaCl, 25 mM Tris pH 5.0). Some species such as staphylococci need an even more aggressive treatment, such as an additional overnight incubation with 10–60 mg/ml lysostaphin. Yeasts such as *Saccharomyces cerevisiae* and *Schizosaccharomyces pombe* are encapsulated at a concentration of 6×10^8 cells/ml in SE buffer. Cell wall lysis requires pre-treatment for 3 h at 37°C with 20 mM dithiothreitol, 1 mg/ml zymolyase 20T in SE buffer.

Protocol 4. Preparation of genomic DNA in agarose microbeads

This protocol is based on those described in refs 3 and 4.

Reagents

• Light mineral oil (Fluka, Cat. No. 76235)
• Buffers (*Protocol 1*)
• 10 × EX: 0.01% (v/v) Triton X-100, 0.5 M EDTA pH 9.0
• TEX: 0.01% (v/v) Triton X-100 in TE
• EX: 0.01% (v/v) Triton X-100, 50 mM EDTA pH 8.0
• LMP agarose (Sigma, type VII), 1% in SE buffer

Method

1. Harvest mammalian, protozoan, yeast, or bacterial cells or plant nuclei and adjust cell concentrations as described above in the *Protocols 1, 2C* and *3*, respectively.

Protocol 4. *Continued*

2. Pre-warm light mineral oil at 42 °C.

3. Melt 1% (w/v) LMP agarose in SE and keep it at 42 °C in a water-bath.

4. Pre-warm 1 ml of suspended cells or nuclei to room temperature and mix with an equal volume of the agarose. Pour into a pre-warmed 500 ml flask.

5. Add 20 ml of the pre-warmed light mineral oil, shake the contents of the flask vigorously by hand for 2–3 sec, and pour the emulsion into 150 ml of ice-cold SE buffer vigorously stirred with a magnetic bar.

6. Stir the solution for 10 min on ice, during which agarose microbeads solidify.

7. Harvest the microbeads by centrifugation at 1000 *g* for 20 min at 4 °C in a swinging bucket centrifuge.[a]

8. Discard the supernatant and resuspend the pelleted microbeads in 5–10 vol. of ES.[b]

9. Incubate at 56 °C for 15 h with gentle shaking.

10. Wash once in 10 × EX for 1h at 50 °C and once in EX for 1 h on ice.[c]

11. Store the microbeads in TEX at 4 °C.

---

[a] If the microbeads do not pellet well, centrifugation at up to 1800 *g* can be used (4).
[b] If yeast or other Gram-positive bacteria are used, pre-incubate with other lysing enzymes as appropriate (*Protocol 3*) at this step.
[c] To change a buffer, pellet the microbeads by centrifugation (1000 *g*, 4 °C, 20 min) and pour the buffer off the pellet. Add the new buffer and completely resuspend the pelleted microbeads by vigorous vortexing. Clumps of microbeads that do not separate by vortexing are broken by pipetting up and down through a disposable pipette. The consistency should be similar to that of heavy oil. The Triton X-100 prevents the microbeads from sticking to themselves and to the plastic and glassware.

3. Choice of restriction endonuclease

3.1 General guidelines

Enzymes that recognize sequences longer than 6 bp are potentially useful for long-range restriction mapping, as they tend to cut infrequently. *Table 1* lists the restriction endonucleases and intron-encoded endonucleases with recognition sequences > 6 bp that are commercially available. Moreover, due to the non-random arrangement of base pairs in a genome, certain endonuclease recognition sequences may be substantially underrepresented in the genome of interest. Useful criteria for the selection of enzymes are the GC content, codon usage, degree of methylation, nearest neighbour data of dinucleotide frequencies, and nucleotide sequence data. Reliable predictions can be made for bacteria, yeast, and mammals (see below), but for inverte-

Table 1. Restriction and intron-encoded endonucleases with recognition sequences longer than 6 bp[a]

Name	Recognition or homing sequence
7 base recognition sequence	
*Rsr*II, *Csp*I, *Cpo*I	CG⇓GWCCG
8 base recognition sequence	
*Sgr*AI	CR⇓CCGGYG
*Fse*I	GGCCGG⇓CC
*Sgf*I	GCGAT⇓CGC
*Not*I	GC⇓GGCCGC
*Sfi*I	GGCCNNNN⇓NGGCC
*Asc*I	GG⇓CGCGCC
*Srf*I	GCCC⇓GGGC
*Sse*8387I	CCTGCA⇓GG
*Pme*I	GTTT⇓AAAC
*Swa*I	ATTT⇓AAAT
*Pac*I	TTAAT⇓TAA
Intron-encoded nucleases[b]	
I-*Ceu*I	TAACTATAACGGTC ↑ CTAA⇓GGTAGCGA
PI-*Tli*I	GGTTCTTTATGCGG ↑ ACAC⇓TGACGGCTTTATG
I-*Ppo*I	ATGACTCTC ↑ TTAA⇓GGTAGCCAAA
PI-*Psp*I	TGGCAAACAGCTA ↑ TTAT⇓GGGTATTATGGGT
VDE	ATCTATGTCGG ↑ GTGC⇓GGAGAAAGAGGTAAT-GAAATGGCA
I-*Sce*I	TAGGG ↑ ATAA⇓CAGGGTAAT

[a]For enzymes recognizing degenerate sequences, only those whose recognition sequence is the equivalent of more than six full bases are listed. R: A or G. W: A or T. Y: C or T. ⇓: strand break. ↑: break of complementary strand (if not symmetrical).
[b]Intron-encoded nucleases tolerate degeneracies in their homing sequence. According to the enzyme supplier, the sequence specificity under carefully controlled conditions is about 8–10 bases for I-*Ppo*I, PI-*Tli*I, and PI-*Psp*I; that of VDE is > 11 bases. I-*Ceu*I tolerates some sequence degeneracy, but minimal tolerance was found within the core sequence of 19 bases extending from −12 to +7 relative to the top strand break (⇓) (9).

brates and particularly plants the usefulness of endonucleases has to be tested on a case-to-case basis.

3.2 Bacteria

Restriction endonucleases with a 7 bp or 8 bp recognition sequence are expected to cleave bacterial chromosomes into only a few fragments. Enzymes recognizing 6 bp sites composed of A and T (e.g. *Dra*I, *Ssp*I, *Asn*I) may cut rarely in GC-rich genomes, whilst those recognizing hexamers composed of G and C (e.g. *Sma*I, *Nae*I, *Eag*I, *Bss*HII, *Bgl*I, *Sac*II, *Apa*I) cut rarely in AT-rich genomes. The sequence CTAG is underrepresented in most bacteria (10) and at least one of the four enzymes *Spe*I, *Xba*I, *Nhe*I, *Avr*II that include this tetranucleotide in their recognition sequence should be useful for bacterial genome mapping. In many bacteria, some palindromic sequences are counterselected; for example *Pvu*I, *Bam*HI, *Bgl*II and *Bcl*I cleave infrequently in archaebacterial DNA, regardless of its GC content.

Of particular interest is the intron-encoded endonuclease I-*Ceu*I (11) which recognizes a 19–24 bp large consensus sequence in the 23S rDNA gene which is conserved among chloroplast, mitochondrial, and prokaryotic genomes (11). Hence, genomic digests with I-*Ceu*I allow the evaluation of the number of *rrn* operons in a given bacterial strain.

3.3 Yeast

The genomes of *Saccharomyces cerevisiae* and of *Schizosaccharomyces pombe* (though not necessarily those of other fungi) are very AT-rich, and enzymes with GC-rich recognition sequences therefore cut rarely. *Asc*I, *Not*I, *Sfi*I, *Srf*I and *Sse*I, produce an average fragment size of more than 100 kb from yeast DNA.

3.4 Mammals

The nuclear genomes of mammals are all approximately 40% GC and the dinucleotide 5′-CpG is fivefold more rare than would be expected from the GC content alone. Most CpG dinucleotides are methylated at the cytidine residue, but methylation at a specific site may be influenced by cell type, growth, and differentiation.

Bulk mammalian DNA is interspersed with 0.5–2 kb stretches of GC-rich DNA called 'CpG islands' in which CpG dinucleotides are not underrepresented. Moreover, CpG dinucleotides within these islands are generally not methylated (with a few exceptions, such as the inactivated X chromosome and some cell lines). CpG islands account for $\approx 2\%$ of the genome, and occur predominantly in the GC-rich R-bands of chromosomes. They are associated with the 5′ end of housekeeping genes and of a large fraction of genes that are expressed in a tissue-specific manner, about 60% of human genes being associated with CpG islands (12).

3.4.1 Impact of CpG islands and cytidine methylation on long-range mapping

Almost all restriction enzymes with one or more CpG dinucleotides in their recognition sequence are prevented from cleaving if the cytidine residue is methylated. Partial methylation at such sites amongst a population of DNA molecules can hence result in incomplete digestion by methylation-sensitive enzymes, giving a complex pattern of fragments which may differ between cell lines and tissues and which can be exploited to confirm physical linkage of DNA probes (see Section 1). Non-methylated CpG dinucleotides are abundant in the CpG islands, and therefore enzymes with GC-rich recognition sequences will predominantly cleave in the islands at the 5′ end of genes. Those 'CG-rich' enzymes that generate large restriction fragments have been divided into groups based on two parameters: the proportion of all recognition sites that are within CpG islands; and the average number of sites per island (13).

(a) Group I: (*Asc*I, *Not*I). Recognition sites for these enzymes are located almost exclusively in CpG islands, but only about 30% of islands contain a site. Both enzymes are very useful for long-range mapping and for the identification of chromosome breakpoints arising from deletions or translocations (causing a shift in fragment size compared to non-affected chromosomes).

(b) Group II: (*Bss*HII, *Eag*I, *Sac*II). About 75% of sites for these enzymes reside in CpG islands. Clusters of these sites on the restriction map are strong evidence for a CpG island and hence for a nearby gene.

(c) Group III: (*Nae*I, *Nar*I, *Sma*I). These methylation-sensitive enzymes have recognition sites within nearly every island, but there are also many sites outside the islands which are subject to variable methylation.

(d) Group IV: (*Mlu*I, *Nru*I, *Pvu*I, *Spl*I). The recognition sequences of these enzymes each contain two CpG dinucleotides but, in contrast to those of groups I–III, they also contain two AT bases. More than 90% of sites of sites for these enzymes reside outside CpG islands and are therefore subject to variable methylation. These enzymes are particularly useful for extending restriction maps beyond CpG islands and for establishing physical linkage of markers. Unequivocal mapping of rearrangements, however, is jeopardized by donor-to-donor and allele-to-allele variability of methylation.

These rules are of course only applicable to genomic DNA isolated directly from mammalian cells; cloned DNA is not methylated at CpG dinucleotides.

4. Restriction digestion of agarose-embedded DNA

Most restriction endonucleases function in the presence of agarose. Since diffusion within agarose is more limited than in liquid reactions, higher concentrations of enzymes and longer incubation times are generally required for complete digestion. *Protocols 5* and *6* describe typical experimental conditions for total and partial digests, respectively. For most restriction enzyme buffers, one should follow the recipes of the suppliers. But we have found that certain enzymes work better in the following buffers: *Dpn*I—10 mM MgCl$_2$, 150 mM NaCl, 10 mM Tris–HCl pH 7.5 (37°C); *Pac*I—10 mM magnesium acetate, 100 mM potassium glutamate, 25 mM Tris–acetate pH 7.6 (37°C); *Spe*I—10 mM MgCl$_2$, 50 mM NaCl, 10 mM Tris–HCl pH 7.5 (37°C). DNA cut with restriction enzymes having recognition sites of at least 8 bp or with cloned enzymes sometimes requires Proteinase K treatment after digestion, to remove non-covalently bound enzyme. Some sources recommend treating the agarose plugs with phenylmethylsulfonyl fluoride (PMSF) after DNA isolation, in order to inactivate residual amounts of Proteinase K which might destroy restriction enzymes. In our experience, loss of restriction enzyme activity is usually not significant. Thus PMSF treatment seems to be only appropriate if small amounts of restriction enzyme (< 1 U) are used, as may be the case in partial digests.

Protocol 5. Total restriction digestion of embedded DNA

Reagents

- Bovine serum albumin (BSA), 10 mg/ml: add 0.1 g BSA to 9 ml autoclaved water in a sterile tube; adjust volume to 10 ml, and store 1 ml aliquots at −20°C.)
- Restriction enzyme and appropriate buffer
- TE (*Protocol 1*)

- Dithiothreitol (DTT), 0.5M (Add 0.77g DTT to 9 ml autoclaved water in a sterile tube. Adjust volume to 10 ml, and store 1 ml aliquots at −20°C.)
- TEX (*Protocol 4*)

A. *Digestion in agarose plugs*

1. Equilibrate half a block (i.e. 5 × 6 × 1 mm) three times for 30 min with 1 ml of the restriction buffer in an Eppendorf tube. Remove the buffer each time using a Pasteur pipette.

2. Add 60 μl restriction buffer, 1 μl DTT, 1 μl BSA and 10 U restriction enzyme.

3. Incubate overnight at the appropriate temperature.

4. Stop the reaction by adding 1 ml TE and store the tubes at 4°C.[a]

B. *Digestion in agarose microbeads*

1. Pipette 20–30 μl of microbeads into an Eppendorf tube.[b]

2. Wash the beads once with 0.2 ml TEX and then twice with 0.2 ml restriction buffer.[c] Resuspend in a total volume of 0.2 ml of restriction buffer.

3. Add 1 μl BSA and 5U restriction enzyme.

4. Incubate for 1 h at the appropriate temperature.

5. To stop the reaction, wash with 0.2 ml TEX.[d]

[a] When fragments of < 20 kb are to be analysed, use the plugs immediately as small fragments may diffuse out of the plug.
[b] Microbead suspensions should be pipetted with disposable tips, from which ~ 5 mm of the end has been removed with a razor blade.
[c] To wash microbeads, pipette strongly into the tube to disperse the microbeads, centrifuge for 1 min in an Eppendorf centrifuge at maximum speed, and remove the supernatant.
[d] Small fragments may be lost by diffusion more rapidly from agarose microbeads than from plugs.

Protocol 6. Optimization of a partial restriction digestion

Reagents
• As for *Protocol 5*

Method

1. Use plugs of 6 × 5 × 1 mm in size containing a high cell density (e.g. > 1 × 10⁸ mammalian cells/ml). Equilibrate three times with 1 ml restriction enzyme buffer for 30 min each time at room temperature.

2. Prepare a series of enzyme solutions, each consisting of the appropriate amount of enzyme in 60 μl of reaction buffer, and add 1 μl BSA and 1 μl DTT. Incubate at the appropriate temperature for various times.[a,b]

3. Stop the reaction with 1 ml TE.

4. (a) For single-dimension gels (when determining digestion conditions), cut the plugs to a size of approximately 6 × 2 × 1 mm before loading.

   (b) When running the first dimension of a two-dimensional gel (see below), cut the plug into two halves, carry out the end-labelling reaction (optional step; see below), and load the two plugs in the gel, separated by one lane.

5. Run a 1.5% agarose gel under the appropriate separation conditions.[c]

[a] There are no reliable rules for the adequate incubation conditions and the optimal amount of enzyme. Slight variations of enzyme activity may lead to irreproducible results and may vary depending on the supplier, the enzyme batch, and its age. Hence, it is extremely important to keep one tube of enzyme for the partial digestion experiments and to carry out the tests over a short period. Digestion conditions depend on the enzyme, the supplier and the DNA template. Incubation time should exceed 45 min (as shorter digestions with larger amounts of enzyme tend to be less reproducible) and a wide range of enzyme quantities should be tested. On one occasion we varied the enzyme amounts between 0.5 and 3 U and the time between 2 and 8 h. However, when we used a batch of the same enzyme from another supplier, we found that enzyme activities from 0.008 to 0.5 units and incubation times from 45 to 90 min were appropriate.
[b] Pre-incubation on ice is not necessary.
[c] Under appropriate running conditions, the upper separation limit should correspond to the sum of the molecular weight of the two largest fragments existing in the chromosomal region of interest.

---

5. One-dimensional pulsed-field gel electrophoresis

5.1 PFGE techniques

Pulsed-field gel electrophoresis separates DNA molecules of up to 10 Mb in size, using at least two alternating electric fields which act in different directions. Migration of the DNA molecules through the gel is governed by the strength of the electric fields, the pulse time (i.e. the time between alternations of the field), the geometry of the fields, temperature, ionic strength, and the type and concentration of the agarose (2, 14). Although a variety of PFGE techniques have been developed in recent years, transverse alternating field electrophoresis (TAFE) (15), contour-clamped homogeneous electric field electrophoresis (CHEF) (16), and the programmable autonomously-controlled electrophoresis (PACE) (17) are nowadays used almost exclusively. CHEF generates uniform electric fields across the gel by using an hexagonal array of electrodes at predetermined ('clamped') electric potentials, to generate alternating fields separated by an angle of 120°. In the PACE system, the potential of point-electrodes around the gel can be individually varied to mimic a wide variety of field geometries. TAFE, CHEF, and PACE are preferred for physical genome analysis because of the high resolution and sharpness of bands in the whole size range from 1 to 10 000 kb, the straight migration path, and the almost identical electrophoretic mobility in all lanes.

Pulse times (which may vary over the course of the run) and electric field strength are the principal parameters which determine the size range of molecules separated by PFGE, and these need to be optimized on a case-to-case basis. Electrophoretic mobility of linear DNA molecules decreases monotonically with size, but the situation becomes more complex at the extremes of molecular weight range. Large fragments separated by PFGE may be difficult to interpret, as the electrophoretic mobility passes an inflection point of minimum mobility and then rises again, such that larger molecules move faster than smaller ones (14). We have also observed pulse time-dependent inversion of the relative mobility for linear DNA molecules smaller than 50 kb, because their behaviour is a superposition of that expected in continuous field electrophoresis and that caused by the field alternation. These phenomena impede the unequivocal assignment of DNA fragments run under different PFGE separation conditions.

An increase in pulse time and/or electric field strength results in larger molecules being resolved. Electric fields of > 3 V/cm, however, may lead to degradation and trapping of megabase sized DNA molecules in the agarose matrix (14). Faint or fuzzy bands of the largest fragments are indicative of this phenomenon. Hence, whereas fields of 4–7 V/cm are typically employed for separation in the 10 to 1000 kb size range, larger molecules should be separated with large pulse times at field strengths below 2 V/cm.

5.2 Size markers for pulsed-field gel electrophoresis

The chromosomes of lower eukaryotes, a ladder of bacteriophage λ oligomers, and a restriction digest of λ DNA are appropriate commercially available size markers. Chromosomes of *Saccharomyces cerevisiae*, *Candida albicans*, and *Schizosaccharomyces pombe* cover the range from 200 kb to 9

Mb. Separations are slightly distorted into an hourglass shape across the gel in CHEF and PACE. In order to account for these imperfections, the same size markers should be applied to at least the centre and the outermost lanes of the gel.

5.3 Electrophoresis

5.3.1 General hints for the preparation and loading of pulsed-field gels

(a) All solutions, glass-, and plasticware should be sterilized whenever possible to prevent contaminations by DNase.

(b) Wear gloves whenever handling agarose blocks in order to avoid DNase contamination which might lead to degradation of DNA in a subsequent restriction digest.

(c) Always prepare the agarose gel with the appropriate electrophoresis buffer.

(d) The agarose blocks should not be mechanically stressed, especially when loading the gel, as this may damage the DNA and lead to diffuse bands after electrophoretic separation.

5.3.2 PFGE running conditions

The fragments produced by complete and partial digestion of genomic DNA with one or more enzymes usually encompass two orders of magnitude or more in size. It is not usually possible to analyse the full size range effectively using a single set of electrophoresis conditions. When starting a mapping project, the expected size range should be divided into two to four overlapping regions (e.g. < 50 kb; 30-700 kb; 200-2000 kb; and 1000-7000 kb). Standardized conditions should then be established for resolving each of these ranges on the user's own PFGE apparatus. Guidelines are normally provided by the manufacturer, or reference should be made to the literature covering the relevant type of gel system.

Protocol 7. Casting, loading and staining of PFGE gels

Reagents

- 10 × TBE: 0.04 M EDTA, 0.9 M boric acid, 0.9 M Tris pH 8.4 (20 °C)
- TE (*Protocol 1*)
- Ultrapure agarose (Life Technologies)
- Pulsed-field gel apparatus

- 1000 × aqueous ethidium bromide stock solution (10 mg/ml)[a]
- Gel photography system and UV trans-illuminator

A. *Gel casting*

1. Boil the agarose in 0.5 × TBE on a heater with stirring, or in a microwave oven. The volume required can be calculated from the

---

295

Protocol 7. *Continued*

   area and thickness (0.4 cm) of the gel; allow approximately 20–50 ml extra.

2. Let the agarose cool until the flask is hand-hot.

3. Put the gel casting chamber on a level surface.

4. Place the comb about 1 cm from the edge of the frame, perpendicular to the gel plate.

5. Pour the agarose in the gel chamber to a depth of about 0.4 cm. Keep the rest of the agarose liquid in a water-bath at 45 °C.

6. When the agarose becomes cloudy, the gel is solidified (about 20 min).

B. *Sample loading*

1. Remove the comb and fill the slots with TE.

2. Cut the agarose blocks to the size of the gel slots.

3. Load the blocks into the slots with the aid of a scalpel and a plastic spatula.[b]

4. Carefully remove the TE and seal all slots with warm agarose (see step A5).

5. Equilibrate the gel for 30 min in the electrophoresis chamber.

6. Start the electrophoresis using the appropriate parameters.

C. *Ethidium bromide staining*

1. After the run, stain overnight with ethidium bromide (10 μg/ml in water).[a]

2. Destain three times for 30 min with water.

3. Photograph on an ultraviolet light box.

---

[a] Ethidium bromide stock solution should be stored in the dark. It is a carcinogen, and should be handled with gloves.
[b] To load the gel with agarose microbeads, load the TE/microbead suspension directly on a tooth of a (horizontal) comb using a cut pipette tip. Form a uniform droplet. Remove TE with a small piece of absorbent paper towel. Add 5–10 μl of warm agarose to glue the microbeads together. After 1 min carefully place the comb with the beads into the sample well. Remove the comb after the agarose has solidified (~ 30 min) and seal the slots with warm agarose (3). Continue with step 5.

6. Blotting and hybridization of pulsed-field gels

The hybridization of pulsed-field blots with markers is the core technique for long-range restriction mapping of large genomes. PFGE-separated DNA can be conveniently blotted onto nylon membranes by alkaline transfer and subsequent immobilization with UV light (*Protocol 8*). Uncharged membranes are preferable to charged membranes because of the ease with which the former can be stripped and reused. Hybridization of pulsed-field blots is feasible with both radioactively and non-radioactively labelled probes; both are described in *Protocol 9*. As the resolution of PFGE is sensitive to overloading with DNA, the high specific activity of ^{32}P- or ^{33}P-labelled probes is advantageous for the detection of single copy sequences in large genomes. In contrast, we routinely use digoxigenin labelled probes and signal detection by chemiluminescence for hybridization analysis of genomic DNA from microorganisms.

Protocol 8. Capillary transfer of DNA from pulsed-field gels to nylon membranes (Southern transfer)

Reagents

- Uncharged nylon membrane (e.g. Hybond-N, Amersham)
- Blotting paper GB003 (Schleicher & Schuell) or 3MM (Whatman)
- Neutralizing solution: 50 mM Na-phosphate pH 6.5
- Denaturation/transfer solution: 1.5 M NaCl, 0.5 M NaOH
- SaranWrap (or any similar UV-transparent film)

Method

1. Take a Polaroid print of the ethidium bromide stained gel.[a]

2. Denature the DNA by soaking the gel two times for 15 min in denaturing solution. (This step is optional; the alkalinity of the transfer solution is sufficient for denaturation.)

3. Construct a blotting 'bridge'. This consists of a tray or shallow tank of transfer solution, containing a glass plate or other horizontal surface, slightly larger than the gel, supported 1–2 cm clear of the solution. Onto this surface place, in order from bottom to top:

   - two sheets of blotting paper, hanging down into the solution at each end
   - the gel (underside uppermost)
   - nylon membrane (2 mm larger than the gel on all sides)
   - two sheets of blotting paper cut exactly to the size of the gel
   - a 15 cm stack of paper towels, slightly smaller than the blotting paper
   - a weight of ~ 1 kg

   Make sure that the transfer solution in the tray is replenished as it is absorbed through the gel. Blot for 48 h, or until at least 1.5 litres of solution has been transferred.[b]

4. Neutralize the membrane by shaking it for 5 min in neutralizing solution. Allow the membrane to air dry until damp. Wrap in SaranWrap.

Protocol 8. *Continued*

5. Immobilize the DNA by placing the blot DNA side down on a UV trans-illuminator. Irradiate for about three times the time required for a Polaroid print.[c]

6. Store the blot at room temperature in the dark.

---

[a] This exposure of the DNA to UV light is sufficient to break the nucleic acid into transferable fragments. No depurination with HCl is necessary.
[b] Use 0.4 M NaOH as the denaturation/transfer solution for blotting onto a positively charged membrane.
[c] Positively charged nylon membrane need not be irradiated.

Protocol 9. Hybridization of pulsed-field blots

Reagents

- Sephadex G-50 (Pharmacia)
- Hexanucleotide mixture: 0.5 M Tris, 0.1 M $MgCl_2$, 1 mM DTE, 2 mg/ml BSA, hexanucleotides (62.5 A_{260}U/ml)
- Dye marker: 0.8% (w/v) dextran blue (M_r 2×10^6), 0.8% (w/v) phenol red
- Klenow polymerase
- TE (*Protocol 1*)
- SDS: 10% (w/v) stock
- Pre-hybridization and hybridization solution: 7% (w/v) SDS, 1 mM EDTA, 500 mM Na-phosphate pH 7.2 (20)—for non-radioactive hybridization add 0.5% (w/v) blocking reagent (Boehringer Mannheim); dilute the blocking reagent by warming the solution to 50°C–70°C 1 h before use
- Washing solution: 1% (w/v) SDS, 40 mM Na-phosphate pH 7.2
- Probe stripping solution: 1% SDS (w/v), 0.2 M NaOH

For DIG labelling and detection with CDP-Star[TM]
- Buffer 1: 150 mM NaCl, 100 mM Tris–HCl pH 7.5

- Buffer 2: buffer 1, 0.5% (w/v) blocking reagent (Boehringer Mannheim)
- Buffer 3: 100 mM NaCl, 50 mM $MgCl_2$, 100 mM Tris–HCl pH 9.5
- DIG labelling mixture: 1 mM dATP, 1 mM dCTP, 1 mM dGTP, 0.65 mM dTTP, 0.35 mM DIG-dUTP pH 7.5 (Boehringer Mannheim supply either the complete mixture, or the individual components)
- Anti-digoxigenin antibody conjugated with alkaline phosphatase, Fab fragments (Boehringer Mannheim); dilute just before use 1:5000 with buffer 2
- CDP-Star[TM] (Tropix): prepare 1:100 and 1:1000 dilutions of the 25 mM CDP-Star stock solution in buffer 3 just before use

For radioactive labelling
- Radioactive labelling mixture: 1 mM dATP, 1 mM dTTP, 1 mM dGTP, ~ 40 mCi [α-^{32}P]dCTP (3000 Ci/mmol), total volume 2 ml per reaction (other [α-^{32}P] or [α-^{33}P]dNTPs may be used, the three other unlabelled dNTPs being present at 1 mM each)

A. *Probe labelling (18, 19)*

1. Denature the template DNA (25 ng–2 µg in 15 µl water) in a microcentrifuge tube in a boiling water-bath for 5 min.

2. Place it on ice immediately. Spin down any liquid which has condensed on the tube walls as fluid (10 sec, 13 000 *g*).

3. Add, to a microcentrifuge on ice: 20 µl denatured DNA, 2 µl hexanucleotide, 2 µl of either DIG or radioactive labelling mixture, 2 U Klenow polymerase.

4. Incubate for 20 h at 37°C.

5. Remove unincorporated dNTPs by Sephadex G-50 column chromatography. Plug the bottom of a sterile 15 cm Pasteur pipette with glass wool, and fill with Sephadex G-50 pre-swollen in TE + 0.1% SDS. Wash the column with several volumes of TE + 0.1% SDS before use.

6. Mix probe with 10 ml of dye marker solution and apply it to the column. Elute the DNA with TE + 0.1% SDS. The dextran blue runs with the labelled DNA fragments in the exclusion volume, the phenol red with the unincorporated dNTPs.

7. Radioactively labelled probes should be stored at −20°C and used within three days. DIG-labelled probes are stable at −20°C for at least two years.

B. Hybridization (20)

1. Incubate the filter in pre-hybridization solution for at least 1 h at 68°C. Use at least 0.5 ml/cm^2 membrane area.[a]

2. Denature the DNA probe by boiling in a water-bath for 10 min, then chill rapidly on ice, and mix with hybridization solution. 2 ml of hybridization solution is sufficient when using a hybridization oven.

3. Discard the pre-hybridization solution and replace it with the hybridization solution containing the probe.

4. Hybridize 24–48 h at 68°C.

5. Discard the hybridization solution.[b] Be careful not to let the blot dry out.

6. Add washing solution. Incubate at room temperature for 5 min.

7. Wash twice with washing solution at 68°C for 25 min.[c]

8. (a) For radioactively labelled probe: wrap the filter in SaranWrap and expose to an X-ray film. After a suitable exposure has been obtained, strip the membrane (*Protocol 9*D).

   (b) For DIG labelled probes: use the filter immediately for detection of signals or store at −20°C.

C. Immunological detection of DIG labelled DNA with CDP-Star^{TM}

Shake the filter in a container on a rocker platform at all times, with the exception of step 6.

1. Equilibrate the filter for 1 min in buffer 1.

2. Incubate the filter for 30 min with buffer 2 (0.5 ml/cm^2 filter area).

3. Drain off all excess buffer and incubate the filter with antibody solution (25–50 μl/cm^2 membrane) for 30 min. Make sure that the small volume soaks the whole filter.

Protocol 9. *Continued*

4. Wash the blot three times for 15 min each with 200 ml buffer 1.

5. Equilibrate the filter for 2 min in buffer 3.

6. Incubate with the 1:100 diluted CDP-Star solution for 5 min (25 μl/cm^2 filter).[d]

7. Rinse briefly with the 1:1000 dilution of the CDP-Star solution.

8. Discard the CPD-Star solution and wrap the filter in thin SaranWrap, excluding air bubbles.

9. Expose the filter to an X-ray film for 1 h, or as long as necessary to obtain a strong signal.

10. Strip the membrane (*Protocol 9*D).

D. *Stripping of hybridized nylon membranes*[e]

1. Incubate the filter two times in 0.1% (w/v) SDS/0.2 M NaOH at room temperature with shaking.

2. Wash with H$_2$O until the pH value is neutral.

3. Wrap the moist filter in SaranWrap and store at $-20\,°$C.

[a]For hybridization one can use a hybridization oven or seal the blot into plastic freezer bags and incubate them in a water-bath.
[b]DIG labelled probes in hybridization solution can be stored at $-20\,°$C for months and reused several times.
[c]Can be substituted by wash steps at lower stringency (lower temperature, higher ionic strength).
[d]The 1:100 dilution can be used three or four times if stored at 4 °C.
[e]Blots can be repeatedly used for hybridization provided that they are not allowed to dry out, as this causes the hybridized probe to bind irreversibly to the filter. After detection of signals the membrane should be stripped immediately.

7. Two-dimensional pulsed-field gel electrophoresis

Electrophoretic separation in two dimensions increases the spatial resolution and the informativeness of restriction digests for physical mapping. Applications include chromosomal regions in large genomes with repetitive DNA that otherwise are hard to map (21–23) and the complete mapping of small genomes by combinatorial analysis (24–26) (*Protocol 10*).

Long-range restriction mapping of complex genomes by one-dimensional PFGE requires a set of single copy DNA probes which are difficult to find for regions that contain repeats (23) or clusters of homologous genes (21, 22). Two-dimensional PFGE that allows redigestion and resolution of PFGE fragments with a second restriction enzyme is useful for the analysis of these multicopy DNA families, within which probes can hybridize to multiple genomic locations.

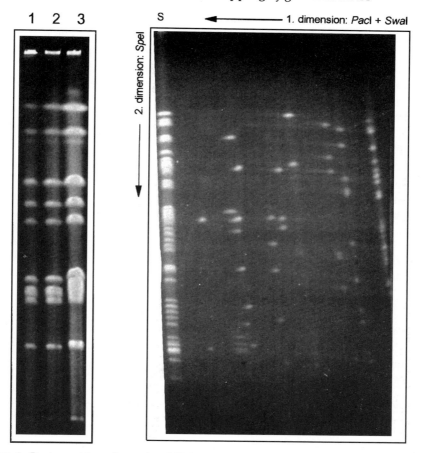

Figure 1. Reciprocal two-dimensional PFGE separation of the DNA fragments generated by a *PacI–SwaI–SpeI* triple restriction digest of *Pseudomonas aeruginosa* C19 chromosomal DNA. (A) PFGE separation of *PacI–SwaI* digested DNA to determine the appropriate DNA concentration for two-dimensional gel electrophoresis (cell density within the agarose plugs increases from lane 1 through 3) and to optimize the separation conditions. Running conditions were 3.8 V/cm and 1% agarose in 0.5 × TBE; pulse times were linearly increased in two ramps from 40 to 80 sec for 40 h and from 60 to 340 sec for 40 h. (B) Ethidium bromide stain of one reciprocal two-dimensional gel. After separation of the *PacI–SwaI* digested DNA in the first dimension the whole lane was cut out, digested with *SpeI*, and separated in the second dimension. A complete *PacI–SwaI–SpeI* triple digest was applied as size marker (lane S). The running conditions of the first dimension were the same as in (A), but the first ramp was reduced to 20 h. The separation conditions of the second dimension were 6 V/cm and 1.5% agarose in 0.5 × TBE; pulse times were increased in four linear ramps from 8 to 45 sec for 20 h, 12 to 25 sec for 22 h, 1 to 14 sec for 14 h, and 1 to 4 sec for 4 h. PFGE was conducted in a Bio-Rad CHEF DR II or DR III apparatus.

The smaller genomes of bacteria and lower eukaryotes can be mapped by two-dimensional PFGE without any need for supplementary genetic data. The fragment order is established by two strategies: comparison of partial versus complete digestion ('partial–complete mapping'), or of consecutive digests with two different enzymes, in each order ('reciprocal digest mapping') (*Figures 1–3*). The mapping strategy requires high quality agarose-embedded DNA and optimal separation conditions (see the example shown in *Figure 1*).

7.1 Partial–complete mapping

In this technique, a partial restriction digest is first separated by PFGE in one dimension, then redigested to completion with the same enzyme, and subsequently resolved in the second orthogonal dimension. After the partial digestion, the DNA is usually labelled in the agarose plug with ^{32}P-nucleotides by Klenow DNA polymerase (see *Protocol 11*). The partial digest is separated by PFGE, the entire lane is cut out, and the fragments in the agarose are redigested to completion within the agarose using the original enzyme. The lane is oriented in the second agarose gel, with the width of the

---

Figure 2. Hints for the analysis of two-dimensional gels of small genomes. Typical combinations of fragments that are difficult to interpret: reciprocal two-dimensional pulsed-field gels of complete double digests with two enzymes A and B. The linear DNA fragment (top) harbours four recognition sites for enzyme A (fragments designated a–d) and three sites for enzyme B (fragments designated *a–d*). Double digestion fragments are numbered 1 to 8. Co-migrating fragments of similar size are indicated by the same number (fragments 6). Fragment migration was from left to right in the first dimension and from top to bottom in the second dimension. Closed circles indicate end-labelled fragments. To resolve fragment order from the two-dimensional gels consider the following points. All gels must contain the same number of digestion fragments irrespective of the digestion order (this may help to detect fragment clustering). The sum of the sizes of a double digestion lane must be equal to the sum of the original first dimension fragment (e.g., fragment a is cleaved by enzyme B into fragments 1, 5, 6 [bottom left]; the size of [1 + 5 + 6] must be equal to the size of fragment a). Fragment clustering (like c′, c″) may be detected by mass law and increased signal intensity. Tiny double digestion fragments (like fragment 8) may run out of the gel or get lost by diffusion and may therefore not be visible on the two-dimensional gel. However, their presence is revealed by a slight reduction of the size of the original fragment (e.g. c″ is cleaved by enzyme B into fragments 4 and 8; although 8 is not visible on the gel because of its small size, its presence is indicated by the small, but significant reduction of fragment 4 as compared to fragment 3 [bottom left]). Digestion fragments that are not cleaved in the second dimension are located on the contour lane (indicated by the dashed diagonal). On the corresponding reciprocal gel these fragments are located in lanes with three or more double digestion fragments (unless they are located at the end of a linear genome), because they are generated by internal cleavage of the first dimension fragment with the second dimension enzyme. Such internal fragments are not end-labelled and are only visible in the ethidium bromide stain (e.g. fragment 1, bottom left). More than two internal restriction fragments cannot be ordered unambiguously, as permutation of their order would yield the same pattern on the two-dimensional gel. The reader is advised to deduce the fragment order of the hypothetical genome shown above from the fragment patterns on the reciprocal gels. End-labelling is necessary to completely resolve the fragment order, otherwise the sequence remains ambiguous at the termini.

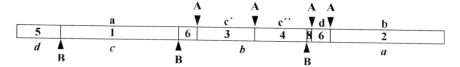

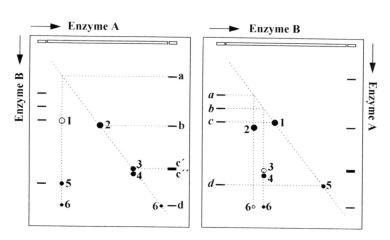

lane of the first dimension becoming the height of the second dimension. The complete digest is separated in the second dimension by PFGE, and the two-dimensional gel is stained with ethidium bromide. Subsequently, the gel is blotted for autoradiographic detection of any end-labelled fragments. (Blotting gives higher resolution than drying the gel for autoradiography; transfer of fragments is efficient provided that the gel has been exposed to UV for the time required to take one or two Polaroid pictures.) All fragments which are generated by complete digestion of the initial partial digest fragments are visualized by ethidium bromide staining, with the fluorescence intensity being proportional to their length. In contrast, the autoradiogram only detects those fragments which corresponded to the ends of the partial-digest fragments, with the same signal intensity irrespective of size. Hence, comparative evaluation of stained gel and the autoradiogram will facilitate the determination of fragment order. Fragments that are linked to each other are directly read off from the two-dimensional gel (see *Figure 3*) (25, 26).

7.2 Reciprocal digest mapping

A complete restriction digest with enzyme A is performed. The products are usually end-labelled with ^{32}P-nucleotides by Klenow DNA polymerase (see *Protocol 11*). These products are then separated by electrophoresis in the first dimension. The gel lane is excised and the fragments it contains are re-

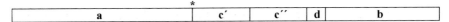

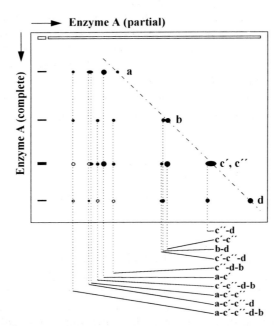

Figure 3. Hints for the analysis of two-dimensional gels of small genomes. Typical combinations of fragments that are difficult to interpret: two-dimensional gel of a partial–complete single digest. Closed circles indicate end-labelled fragments. Fragments cleaved to completion in the partial digest are located on the contour lane (indicated by the dashed diagonal line). Always check the consistency of your established links by determining the size of the partial digestion fragment. Extrapolate the second dimension lane onto the contour lane where the completely digested fragments provide convenient size markers, e.g. the link c′–c″ is verified by the size of the original partial digestion fragment which roughly matches that of fragment b. Slight deviation from the diagonal indicates link to a small fragment (c″–d; b–d). Ellipsoids instead of circles indicate more than one partial fragment in close proximity (a–c′–c″, a–c′–c″–d). More than one partial digestion fragment per lane may be differentiated by intensity, mass law, or slight parallel shifts (b, c′–c″, b–d, c′–c″–d). An intensity that is higher for the partial digestion fragment than for its two constituents indicates that the two fragments flank a hard-to-cut site (indicated by an asterisk in the map): compare the relative intensities of fragments a and b in the partials and on the contour lane.

digested (to completion) with enzyme B, and then separated in the second orthogonal direction. On a separate series of gels the order of restriction digestions is reversed. Again, blotting and autoradiography is performed after visualization with ethidium, to identify end-labelled fragments. The corresponding spots on both two-dimensional gels are identified by their identical molecular weights. 'Linking' fragments are those which carry the recognition site for enzyme A at one end and for enzyme B at the other. These linking fragments will therefore be end-labelled in both two-dimensional gels (i.e. regardless of the order of digestion) and may be identified by autoradiography. The assignment of a linking fragment to its two parental fragments in the single digestions with enzyme A or B establishes the fragment overlap. Note that, if necessary, one or both of the enzymes in a double digest can be replaced by a mixture of suitable enzymes, to give a 'compound' set of cleavage sites (see, for example, *Figure 1*).

Protocol 10. Preparation of two-dimensional gels

Equipment and reagents
- Pulsed-field gel apparatus
- Perspex incubation chamber : 20 × 1 × 1.2 cm (l × w × h) internal dimensions, with lid
- Reagents: see *Protocols 7* and *9*
- Perspex blocks: 1.5 × 1 × 1 cm (to adjust the effective length of the incubation chamber)
- Sterile sewing thread
- Ethanol, 70%

Method

1. Prepare a 1.5% agarose gel in 0.5 TBE buffer (*Protocol 7*).[a]

2. Load a 6 × 2 × 1 mm agarose plug, (containing DNA digested partially in the case of partial–complete mapping, or to completion with enzyme 'A' in the case of reciprocal digest mapping, and optionally end-labelled according to *Protocol 11*) in the bottom of a gel slot.[b]

3. Load a second plug (treated identically to the first) into the adjacent slot to check the quality and reproducibility of digestion, loading and electrophoresis.

4. Run the first dimension.[c] No size standard is required.

5. Use a taut, sterile sewing thread to make grooves on either side of the lane which will be used in the second dimension. Cut along these grooves with a scalpel, using the edge of a glass plate as a guide. Store this lane on a sterile glass plate at 4°C during staining and examination of the rest of the gel.[d]

6. Stain the rest of the gel with ethidium bromide solution, photograph, and check the quality of the control lane. If the signals are weak, do not proceed.

7. Prepare the Perspex chamber for redigestion of the agarose lane. Rinse the tub several times with 70% high quality ethanol, and finally soak the tub and the Perspex blocks in 70% ethanol for 5 min. Wipe the lid with a soft tissue soaked in 70% ethanol. Pour the alcohol out of the tub and allow to dry. Rinse with 5 ml of restriction enzyme buffer.

8. Trim the excised gel lane to remove parts at the top and bottom which do not contain DNA fragments, using the photograph of the

---

Protocol 10. *Continued*

control lane as a guide. Slide the trimmed lane into the incubation chamber by pushing it gently with the edge of a scalpel from the glass plate. Fill the whole tub with restriction enzyme buffer. Incubate for 30 min without shaking. Repeat this procedure a further three times.

9. After removing the washing buffer, insert the Perspex blocks, if necessary, to minimize the 'dead space' at either end of the lane. Fill the chamber with restriction enzyme buffer until the lane is fully covered. For economic use of enzyme, no more than 3 ml restriction enzyme buffer should be used.[e]

10. For every millilitre of restriction enzyme buffer added to the chamber, add 13.2 μl each of 0.5 M DTT and 10 mg/ml BSA. Add restriction enzyme (the same as the first enzyme in the case of a partial–complete digest, or enzyme 'B' in the case of a reciprocal double digestion) to a concentration of 66 U/ml. Mix thoroughly with a Pasteur pipette.[f,g]

11. Cover the tub with the lid and incubate overnight (incubation time may be extended to 48 h). If necessary one can add additional enzyme after 24 h.

12. Remove enzyme buffer with a Pasteur pipette. Fill the tub to the brim with TE and leave for 30min at room temperature to wash out enzyme. Change the buffer one or two times during this period.

13. Cast a 1.5% agarose gel using a preparative comb with a width at least equivalent to the length of the redigested gel lane. Fill the slot with TE.

14. Insert the gel lane into the TE filled slot, turning it on edge so that what was previously its width is now its height.

15. Pour 1.5% agarose around the gel lane until the surface is level. The agarose should be close to gelling temperature.[h]

16. Use a scalpel to make slots at either end of the inserted lane and load them with suitable size standards.[i]

17. Run the second dimension at the desired conditions.

18. Stain the gel in 10 μg/ml ethidium bromide solution overnight, destain it three times for 30 min each time with water whilst shaking, and photograph the gel using UV illumination.

19. (a) To detect end-labelled fragments, blot the gel onto a nylon membrane over 48 h (see *Protocol 8*) and expose to X-ray film.

    (b) Alternatively, blot the gel (*Protocol 8*) and hybridize with suitable DNA probes (see *Protocol 9*).

[a] The gel should be as thin as possible for maximum cooling during electrophoresis. No temperature gradient must be generated which otherwise leads to a 'rocket-like' running of DNA fragments within the gel. To ensure results displaying clear signals use 1.5% agarose gels whenever possible.

[b] Avoid rough handling of the plug as mechanical stress will cause shearing of the DNA.

[c] The fragments should be resolved over as great a distance as possible.

[d] Try to cut out the lane as close to its edges as possible because the width of the lane is the height of the gel in the second dimension. The thicker the gel in the second dimension, the more intensity is lost.

[e] In order to minimize the amount of enzyme used, try to use just enough restriction enzyme buffer to cover the lane for digestion.

[f] The concentration 66 U/ml is an average value which may be adjusted on the basis of prior one-dimensional PFGE. Bear in mind that the enzyme concentration may vary according to the supplier and the batch, e.g. we used concentrations from 24 U/ml (*Spe*I, Eurogentech) to 100 U/ml (*Swa*I, Boehringer Mannheim).

[g] Several factors may influence the quality of the digestion. Acetylated BSA may inhibit the restriction enzymes; certain agaroses may inhibit digestion, and test digests should be performed on DNA encapsulated in the relevant agarose. We use Ultra Pure Agarose (Life Technologies) for two-dimensional gels. All tested enzymes worked well in this agarose with the exception of *Swa*I. In this case, the fragments of the first dimension were separated in 1% low-melting agarose (Sigma, type VII).

[h] Gel and lane should form an even surface. For standard two-dimensional gels we use 1.5% agarose for both directions. The agarose concentration for the second dimension should be the same as or higher than that for the first dimension. If this is not the case, the spots will appear distorted.

[i] Analysis of complex genomes: apply bacteriophage λ DNA oligomers and yeast chromosomes as standards. Analysis of small genomes: use a total digest as standard for partial-complete digests; in reciprocal gels we recommend using a single digest of the enzyme used for the second digestion, and also a double digest with both enzymes, as standards in two adjacent lanes.

Protocol 11. End-labelling of DNA in an agarose plug (applicable to small genomes only)[a]

Reagents

- Klenow polymerase (Boehringer Mannheim)
- [α-^{32}P]dNTP (> 3000 Ci/mmol)[b]

- Klenow end-labelling buffer: 10 mM MgCl$_2$, 50 mM NaCl, 10 mM Tris–HCl pH 7.5 (37°C)
- TE (*Protocol 1*)

Method

1. Digest the DNA in the plug (size 6 × 5 × 1 mm) with the appropriate restriction enzyme in microcentrifuge tube.

2. Remove buffer plus restriction enzyme, equilibrate twice for 30 min each with 1 ml TE at room temperature.

3. Adjust the plug to the final size needed (6 × 2 × 1 mm) for the subsequent electrophoresis and transfer it back into the microcentrifuge tube.

4. Equilibrate the plug twice for 30 min with 1 ml Klenow end-labelling buffer at room temperature.

Protocol 11. *Continued*

5. Thoroughly remove all buffer.

6. Add 20 μl Klenow end-labelling buffer, 2 μCi [α-^{32}P]dNTP (> 3000 Ci/mmol), 1 U Klenow polymerase. Incubate for 30 min at room temperature.[c]

7. Remove the buffer.

8. Wash the plug twice with 1 ml TE for 20 min at room temperature without mixing.

9. Check that the nucleotide is incorporated by holding the microcentrifuge tube with the plug against a Geiger counter. The signal should be around 100 c.p.s.[d]

[a] We use the end-labelling procedure for two purposes: detection of linking fragments or end-fragments in two-dimensional gels, and visualization of fragments below 5 kb in one-dimensional gels.
[b] Klenow polymerase incorporates nucleotides only at 5' protruding ends. Thus, for fragment end-labelling use the dNTP complementing the first base in the 5' overhang of your restriction enzyme site. If the restriction enzyme creates blunt ends or 5' recessive ends, nucleotides will be incorporated throughout the length of the fragment. This procedure is useful for increasing the sensitivity of fragment detection when compared to the ethidium bromide stain.
[c] For the success of the procedure it is strongly recommended that neither the incubation time nor the enzyme concentration be increased. If the signal intensity is too low, the amount of radioactivity should be increased.
[d] We strongly recommend that you evaluate the quality of the end-labelling before proceeding with a two-dimensional gel. This can be done by running the labelled fragments in a single dimension, blotting for 48 h, and exposing membrane for X-ray film; signals should be visible after 24 h exposure. For two-dimensional gels, strongly labelled fragments will give a signal after one day and weakly labelled ones after one to two weeks. Size standards should also be labelled in two-dimensional gels.

8. Alteration of recognition specificity for long-range restriction mapping

8.1 Use of DNA methyltransferase to modify restriction digestion

The sensitivity of many restriction endonucleases to methylation of the recognition sequence may be exploited to generate rare cutting restriction sites. Four different approaches have been tested.

(a) Sequential two-step DNA 'cross-protection'. If the recognition sequence of DNA methyltransferase and a methylation-sensitive restriction endonuclease partly overlap, it is possible to block a subset of the restriction sites. A list of overlapping methylase/endonuclease targets is compiled in refs 27 and 28. As an example, the recognition sequences of *Alu*I methylase (5'-AGCT) and endonuclease *Nhe*I (5'-GCTAGC) overlap by three base pairs. Any *Nhe*I site that is preceded by an A or followed by a

T residue will be protected from cleavage by methylation with *Alu*I methylase. One would predict that 7 out of 16 recognition sites are thus blocked in a genome with 50% G + C content. This 'cross-protection' procedure has been applied to bacterial genome analysis to generate rare restriction sites (29) and to whole-chromosome mapping (30).

(b) Methyl-dependent cleavage using adenine methylase and *Dpn*I. The restriction enzyme *Dpn*I (5'-G^{6m}ATC) is unique in its requirement for N6-methylated adenosines on both strands in order to cleave. Adenosine methylation of 5'-GATC is rare in prokaryotes and absent in eukaryotes. Methylation with an adenine methylase and subsequent cleavage with *Dpn*I generates 8 to 14 bp recognition sequences suitable for megabase mapping (27). For example, methylation of DNA with M-*Cla*I (5'-ATCG^{6m}AT) generates the 10-bp bimethylated recognition site ATCG^{6m}ATCG^{6m}AT susceptible to digestion with *Dpn*I. The procedure has been demonstrated to work with various methylases for megabase cutting of yeast and bacterial chromosomes.

(c) Partial internal methylation. As an alternative to *Protocol 6*, partial digestion patterns may also be produced by combined treatment of the genomic DNA with a restriction endonuclease in competition with a cleavage blocking methyltransferase. For instance, *Bsp*RI methylase (5'-GGCC) methylates the internal cytosine in its recognition site, thus blocking cleavage by the restriction enzyme *Not*I (5'-GCGGCCGC). Simultaneous incubation of template with both enzymes in a suitable buffer results in a reproducible partial digestion pattern sufficient for long-range restriction mapping (31, 32).

(d) Creation of a unique (non-methylated) cleavage site. The engineering of a single cleavage site within several megabases of genomic DNA would be another useful tool for long-range restriction mapping. This goal may be accomplished by insertion or transposon mutagenesis, whereby the inserted sequence carries the recognition site for an intron-encoded endonuclease (33, 34). Alternatively, a unique site can be created by protection of a single site from methylation by either a DNA binding protein (Achilles' heel cleavage) (35–37) or oligonucleotide-directed triple helix formation (38, 39) while methylating all other sites.

Although these methods are attractive for large scale physical mapping, sequential methylation and cleavage has only sporadically been applied to the analysis of complex genomes. Reasons for this are the low number of commercially available methyltransferases, insufficient purity of the enzyme preparations, the difficulty of achieving complete methylation at all sites, and the inherent limitations in creating atom-specific chemistry on DNA (40). Therefore no protocol is given here. Detailed assays are described by McClelland (2, 27, 31, 32), Szybalski (35–37), Dervan (38, 39) and co-workers.

8.2 CpG methylation: use of 5-azacytidine to reveal cryptic restriction sites

The methylation of cytosine, which modulates eukaryotic gene expression, is inhibited by the pyrimidine analogue 5-azacytidine (azaC). The incorporation of azaC into DNA inhibits the DNA cytosine methyltransferase by covalent trapping of the enzyme (41) which also is the underlying mechanism for the high toxicity of the drug (42). Treatment of cells with azaC (*Protocol 12*) can be exploited to render normally methylated CG sequences susceptible to cleavage with ^{m}CG-sensitive restriction endonucleases (see Section 3.4.2). The activation of cryptic restriction sites particularly improves the resolution of long-range mapping of heavily methylated regions in mammalian genomes that are devoid of CpG islands (43–45). Southern analysis of partially demethylated sites can be used as alternative to that of a partial restriction digest of nonmethylated DNA. Moreover, demethylation enables one to distinguish differential methylation from restriction fragment length polymorphisms. This feature is especially useful for the mapping of regions with amplified DNA and for the identification of chromosome breakpoints with group IV enzymes (see Section 3.4.2).

Protocol 12. Treatment of cultured cell lines with 5-azacytidine (azaC) to generate hypomethylated DNA[a]

Reagents
- 1 mM stock solution of 5-azacytidine (azaC; Sigma/Calbiochem) in water—store at 4°C for no longer than three weeks (43)[b]

Method

1. Add azaC to the cell growth medium to a final concentration of 0.2–5 μM. The sensitivity of cell lines to azaC varies considerably, and a range of concentrations should be tested to titrate the yield of demethylated restriction sites.

2. Grow the cells in the presence of azaC for up to 14 days, though in most cases exposure for 24–72 h will suffice; the optimum period must be determined on a case-to-case basis.

3. Harvest the cells and prepare unsheared genomic DNA as described in *Protocol 1A*.

[a] 5-aza-2'deoxycytidine may be used instead of azaC. A further less toxic alternative is *S*-adenosyl-L-homocysteine (46), but experience with this agent is as yet limited.
[b] azaC is toxic. Wear gloves at all times.

---

9. Perspective: Optical Mapping

Ordered restriction maps provide precise genomic distances between physical landmarks, assist in contig formation for gene mapping and for sequencing efforts, and can assist in preliminary characterization of genetic alterations involving specific loci. The associated techniques for the generation of restriction maps have changed little over the last decade, primarily because they use electrophoresis as the central technology. However, recently the first practical non-electrophoretic genomic physical mapping approach, termed 'Optical Mapping' (47–50) has been developed in the laboratory of one of the authors (D. S.).

Optical Mapping is a single-molecule technique for the rapid production of ordered restriction maps from single DNA molecules. In the original method (47), individual fluorescently-labelled DNA molecules were elongated in a flow of molten agarose generated between a coverslip and microscope slide. The restriction endonuclease (previously added to the molten gel) was triggered by addition of magnesium ions, and the resulting cleavage events were recorded by fluorescence microscopy as time-lapse digitized images. Restriction enzyme sites were visible as gaps that appeared as the DNA was cleaved and the fragments relaxed. However, improvements in the lower limit of size resolution and in throughput were required if a wide range of cloning vectors (cosmid, bacteriophage, yeast artificial chromosome [YAC]) were to be analysed by Optical Mapping. Given the usefulness of such clones in contig formation for mapping and sequencing projects, a second-generation Optical Mapping approach has been developed. This dispenses with agarose, and instead involves fixing elongated DNA molecules onto polylysine-treated glass surfaces by sandwiching them between a treated coverslip and glass slide. A cooled charge-coupled device (CCD) camera is used to image molecules ranging in size from 28 kb down to 0.8 kb (49), representing a major improvement from the prior gel-based Optical Mapping approach, with the lower limit of resolution improved to ~ 1 kb (47). Recent experiments have lowered the limit of size resolution to 300–500 bp. Since the accuracy of restriction fragment sizing is a function of the number of fragment molecules analysed, an accuracy comparable to, and in some cases better than, conventional gel electrophoresis has been achieved. Further refinements of the Optical Mapping technology include the use of silane derivatives for charge modification of the glass surface, and increases in throughput towards the future goal of whole-genome analyses. Generally, 100 molecules of each clone (i.e. cosmid) were analysed by the optical mapping software, and five to ten molecules selected for map construction based on their map content, the latter defined from histogram bins containing the maximum number of restriction fragments. A final map is then constructed from molecules having the same restriction cleavage pattern, and stated fragment sizes reflect an averaging of the observed restriction fragments.

Although a large fraction of the human genome is presently covered in YAC contigs (51), for which low resolution maps have been reported, limitations such as a high frequency of cloning artefacts among YACs, and the low complexity of fingerprints generated by hybridization, have hampered generation of extensive YAC restriction maps. Ordered restriction maps of YACs have been generated at the New York laboratory by Optical Mapping (50), with overall relative sizing accuracies that are comparable to routine PFGE analysis. Moreover, large scale restriction maps of loci in the human and mouse genomes were constructed by Optical Mapping. Automation and miniaturization of Optical Mapping should yield dramatic increases in throughput and reductions in cost. It is hoped that the advantages of Optical Mapping will accelerate both the completion of the initial goals of the Human Genome Project, and facilitate the mapping and analysis of disease genes on a population scale.

References

1. Schwartz, D. C. and Cantor, C. R. (1984). *Cell*, **37**, 67.
2. Burmeister, M. and Ulanovsky, L. (ed.) (1992). *Pulsed-field gel electrophoresis*. Humana Press, Totowa, NJ.
3. Koob, M. and Szybalski, W. (1992). In *Methods in enzymology* (ed. R. Wu), Vol. 216, p. 13. Academic Press, London.
4. Zhang, H.-B., Zhao, X., Ding, X., Paterson, A. H., and Wing, R .A. (1995). *Plant J.*, **7**, 175.
5. Boyum, A. (1968). *Scand. J. Clin. Lab. Invest.*, **21** (Suppl. 97), 77.
6. Van Daelen, R. A. J., Jonkers, J. J., and Zabel, P. (1989). *Plant Mol. Biol.*, **12**, 341.
7. Guidet, F. and Langridge, P. (1992). In *Methods in enzymology* (ed. R. Wu), Vol. 216, p. 1. Academic Press, London.
8. Frearson, E. M., Power, J. B., and Cooking, E. C. (1973). *Dev. Biol.*, **33**, 130.
9. Marshall, P. and Lemieux, C. (1992). *Nucleic Acids Res.*, **20**, 6401.
10. McClelland, M., Jones, R., Patel, Y., and Nelson, M. (1987). *Nucleic Acids Res.*, **15**, 5985.
11. Liu, S. L., Hessel, A., and Sanderson, K. E. (1993). *Proc. Natl. Acad. Sci. USA*, **90**, 6874.
12. Cross, S. H. and Bird, A. P. (1995). *Curr. Opin. Genet. Dev.*, **5**, 309.
13. Bickmore, W. A. and Bird, A. P. (1992). In *Methods in enzymology* (ed. R. Wu), Vol. 216, p. 224. Academic Press, London.
14. Gemmill, R. M. (1991). *Adv. Electrophoresis*, **4**, 1.
15. Gardiner, K., Laas, W., and Patterson, D. (1986). *Som. Cell Mol. Genet.*, **12**, 185.
16. Chu, G., Vollrath, D., and Davies, R. W. (1986). *Science*, **234**, 1582.
17. Clark, S. M., Lai, E., Birren, B. W., and Hood, L. (1986). *Science*, **241**, 1203.
18. Feinberg, A. P. and Vogelstein, B. (1983). *Anal. Biochem.*, **132**, 6.
19. Feinberg, A. P. and Vogelstein, B. (1984). *Anal. Biochem.*, **137**, 266.
20. Church, G. M. and Gilbert, W. (1984). *Proc. Natl. Acad. Sci. USA*, **81**, 1991.
21. Woolf, T., Lai, E., Kronenberg, M., and Hood, L. (1988). *Nucleic Acids Res.*, **16**, 3863.

22. Walter, M. A. and Cox, D .W. (1989). *Genomics*, **5**, 157.
23. Warburton, P. E. and Willard, H. F. (1990). *J. Mol. Biol.*, **216**, 3.
24. Bautsch, W. (1988). *Nucleic Acids Res.*, **16**, 11461.
25. Römling, U., Grothues, D., Bautsch, W., and Tümmler, B. (1989). *EMBO J.*, **8**, 4081.
26. Römling, U. and Tümmler, B. (1991). *Nucleic Acids Res.*, **19**, 3199.
27. Nelson, M. and McClelland, M. (1992). In *Methods in enzymology* (ed. R. Wu), Vol. 216, p. 279. Academic Press, London.
28. McClelland, M., Nelson, M., and Raschke, E. (1994). *Nucleic Acids Res.*, **22**, 3640.
29. Quiang, B.-Q., McClelland, M., Podar, S., Spokauskas, A., and Nelson, M. (1990). *Gene*, **88**, 101.
30. Bautsch, W. (1993). *FEMS Microbiol. Lett.*, **107**, 191.
31. Hanish, J. and McClelland, M. (1990). *Nucleic Acids Res.*, **18**, 3287.
32. Hanish, J., Rebelsky, M., McClelland, M., and Westbrook, C. (1991). *Genomics*, **10**, 681.
33. Thierry, A. and Dujon, B. (1992). *Nucleic Acids Res.*, **20**, 5625.
34. Colleaux, L., Rougeulle, C., Avner, P., and Dujon, B. (1993). *Hum. Mol. Genet.*, **2**, 265.
35. Koob, M. and Szybalski, W. (1990). *Science*, **250**, 271.
36. Kur, J., Koob, M., Burkiewicz, A., and Szybalski, W. (1992). *Gene*, **110**, 1.
37. Koob, M., Burkiewicz, A., Kur, J., and Szybalski, W. (1992). *Nucleic Acids Res.*, **20**, 5831.
38. Strobel, S. A. and Dervan, P. B. (1991). *Nature*, **350**, 172.
39. Strobel, S. A., Doucette-Stamm, L. A., Riba, L., Housman, D. E., and Dervan, P. B. (1991). *Science*, **254**, 1639.
40. Oakley, M. G., Turnbull, K. D., and Dervan, P. B. (1994). *Bioconjug. Chem.*, **5**, 242.
41. Santi, V. D., Norment, A., and Garrett, C. E. (1984). *Proc. Natl. Acad. Sci. USA*, **81**, 6993.
42. Jüttermann, R., Li, E., and Jaenisch, R. (1994). *Proc. Natl. Acad. Sci. USA*, **91**, 11797.
43. Dobkin, C., Ferrando, C., and Brown, W. T. (1987). *Nucleic Acids Res.*, **15**, 3183.
44. Heard, E. and Fried, M. (1990). *Nucleic Acids Res.*, **18**, 6147.
45. van der Hout, A. H., Hulsbeek, M. M. F. and Buys, C. H. C. M. (1995). *Cytogenet. Cell Genet.*, **70**, 134.
46. DeCabo, S. F., Santos, J., and Ferndez-Piqueras, J. (1995). *Cytogenet. Cell Genet.*, **71**, 187.
47. Schwartz, D. C., Li, X., Hernandez, L. I., Ramnarain, S. P., Huff, E. J., and Wang, Y.-K. (1993). *Science*, **262**, 110.
48. Wang, Y.-K., Huff, E. J., and Schwartz, D. C. (1995). *Proc. Natl. Acad. Sci. USA*, **92**, 165.
49. Meng, X., Benson, K., Chada, K., Huff, E., and Schwartz, D. C. (1995). *Nature Genet.*, **9**, 432.
50. Cai, W., Housman, D. E., Wang, Y.-K., and Schwartz, D. C. (1995). *Proc. Natl. Acad. Sci. USA*, **92**, 5164.
51. Cohen, D., Chumakov, I., and Weissenbach, J. (1993). *Nature*, **366**, 698.

A1

Non-commercial resources for genome mapping

RAMNATH ELASWARAPU

1. Introduction

A key issue in genome research is the cost of producing two major types of resource: biological reagents and computing resources. It is therefore recognized that the sharing of materials and data within the scientific community is essential for achieving the goals of the program in the most cost-effective way. Consequently, an increasing number of research centres are making their reagents and data accessible to other investigators. Most of the centres provide their resources free of charge (often in exchange for collaboration, acknowledgement, or feedback of data), or at a price sufficient to cover the costs involved in managing and distributing the resources.

This appendix lists those centres, worldwide, who are prepared to make their biological resources available to the genome community on a non-commercial basis. It has been compiled on the basis of information collected from publicly available data sources—including over 1000 WWW pages—and, in some cases, through personal contacts. Inevitably, some resources may have been missed, and new resources are made available at an ever-increasing rate, but the sources listed here will serve as good starting points from which to track down reagents released since the publication of this book.

The resources are listed hierarchically, with an initial subdivision into vertebrates, invertebrates, plants, and fungi. These are further subdivided by species, and then by resource type. Although the nature of resources varies from species to species, the general scheme has been to list them, where possible, in the following order: whole organisms; cell lines and genomic DNAs (including mutant cells, linkage mapping panels, etc.); hybrid or transformed cell lines; genome-wide clone libraries; chromosome-specific or regional clone libraries; cDNA and EST libraries; other clone libraries; probes and PCR primers.

Only a minimum of descriptive data have been included with each entry; details of the nature and availability of each resource should be obtained from the sources listed.

2. Vertebrates

2.1 Human

2.1.1 Cell lines

A variety of cell cultures (and DNA derived from them) are provided which include mutant cell lines, recombinant DNA material, abnormal and ageing cells and material from the CEPH reference families. Extensive catalogues and detailed information will be found at:

- American Type Culture Collection (ATCC), Rockville—
  `http://www.atcc.org`
- European Collection of Animal Cell Cultures (ECACC), Salisbury—
  `http://www.gdb.org/annex/ecacc/HTML/ecacc.html`
- The Institute of Physical and Chemical Research (RIKEN), Saitama—
  `http://www.riken.go.jp`
- The Galliera Genetic Bank—fax. (+39) 10 5632628
- Coriell Cell Repositories—`ccr@arginine.umdnj.edu`

2.1.2 Cell hybrid panels

Somatic cell hybrids, produced by fusion of human and rodent cells, are useful sources of large regions of human genomic material. A hybrid cell line may contain several human chromosomes, a single chromosome, or fragments of one or more chromosomes. 'Radiation hybrids', produced by irradiation of the human 'donor' cells prior to fusion, typically contain many small fragments of one or more human chromosomes, and panels of such hybrids allow mapping of markers.

Chapters 4 and 6 deal, respectively, with radiation hybrids and with somatic cell hybrids. Details of available resources, together with information on their properties and applications, will be found in these chapters.

2.1.3 Total genomic libraries

i. YAC libraries

Yeast artificial chromosome (YAC) libraries are constructed using pYAC-series vectors, and are propagated in yeast cells. The superiority of these clones lies in the fact that they can carry inserts of over 2 Mb. Their drawback is that many YAC clones are chimeric, containing DNA fragments which are not colinear with the genomic DNA. The degree of chimerism varies from library to library. The available libraries also differ in their complexity, in the type of donor DNA used, and in the size range of the cloned fragments. YAC libraries are distributed either as PCR pools (combinatorial pools of clone DNA, which allow positive clones to be identified with the minimum number of PCR screens for the target sequence), or as high density arrays on filters for screening by hybridization. Distributors of various human

YAC libraries are:

- CEPH (Fondation Jean Dausset), Paris—http://www.cephb.fr
- UK HGMP Resource Centre (HGMP-RC), Hinxton—
  http://www.hgmp.mrc.ac.uk
- Baylor College of Medicine (BCM), Houston—
  http://www.bcm.tmc.edu
- Resource Centre/Primary Database of the German Human Genome
  Project, Berlin—http://www.rzpd.de
- YAC Screening Centre (YSCL), Leiden—
  yscl@ruly46.medfac.leidenuniv.nl
- DIBIT-HSR and IGBE-CNR, Milano—salac@dibit.hsr.it
- National Centre for Human Genome Research (NCHGR)—
  http://www.nih.gov
- The Institute of Physical and Chemical Research (RIKEN), Saitama—
  http://www.riken.go.jp

ii. BAC libraries

Bacterial artificial chromosomes, or BACs, are constructed in plasmid-based vectors propagated in bacteria. They can retain inserts of up to 300 kb, though the average insert size in a typical library will usually be 120–150 kb. BACs are easier to handle than YACs (for example, the cloned DNA can easily be isolated from the host background), and exhibit a very low rate of chimerism. Human BAC libraries are available from:

- CEPH (Fondation Jean Dausset), Paris—http://www.cephb.fr
- Cedars-Sinai Medical Centre (CSMC), Los Angeles—
  http://www.csmc.edu

iii. P1/PAC libraries

These libraries are constructed in bacterial-based pCYPAC vectors which can carry inserts of 80–150 kb. They are propagated in an *E. coli* host and hence, like BACs, are easy to grow and to isolate DNA from. P1 clones differ from PACs in that the former are produced by transfection of the host cells with packaged vector–insert constructs, whereas the latter are produced by transformation with unpackaged DNA. The main problem with these libraries is that many of the clones (up to 25% or more) are without human inserts. Sources of human PAC and P1 libraries are:

- UK HGMP Resource Centre (HGMP-RC), Hinxton—
  http://www.hgmp.mrc.ac.uk
- Resource Centre/Primary Database of the German Human Genome
  Project, Berlin—http://www.rzpd.de

- The European Pancreatic Cancer Reference Library System (EPCRLS), Ulm—http://www.uni-ulm.de
- National Centre for Human Genome Research (NCHGR)— http://www.nih.gov
- Cedars-Sinai Medical Centre (CSMC), Los Angeles— http://www.csmc.edu

2.1.4 Chromosome-specific libraries

i. Cosmid libraries

Chromosome-specific cosmid libraries, with typical insert sizes of 40–50 kb, are available for several chromosomes. The following centres are the distributors of various such libraries:

- Resource Centre/Primary Database of the German Human Genome Project, Berlin—http://www.rzpd.de
- Lawrence Livermore National Laboratory (LLNL), Livermore— http://www-bio.llnl.gov
- The Institute of Physical and Chemical Research (RIKEN), Saitama— http://www.riken.go.jp
- UK HGMP Resource Centre (HGMP-RC), Hinxton— http://www.hgmp.mrc.ac.uk

ii. Bacteriophage libraries

A variety of bacteriophage libraries are available for most of the human chromosomes from the following sources:

- American Type Culture Collection (ATCC), Rockville— http://www.atcc.org
- Lawrence Livermore National Laboratory (LLNL), Livermore— http://www-bio.llnl.gov

2.1.5 cDNA libraries

i. Tissue- and organ-specific cDNA libraries

A variety of cDNA libraries is available from the sources listed below. Detailed descriptions of the content and construction of these libraries will be found at the sites listed:

- Resource Centre/Primary Database of the German Human Genome Project, Berlin—http://www.rzpd.de
- The European Pancreatic Cancer Reference Library System (EPCRLS), Ulm—http://www.uni-ulm.de
- National Centre for Human Genome Research (NCHGR)— http://www.nih.gov

- Japanese Collection of Research Bioresources (JCRB), Tokyo—
  `http://www.nihs.go.jp`
- The Institute of Physical and Chemical Research (RIKEN), Saitama—
  `http://www.riken.go.jp`

In addition, several cDNA libraries are available from the UK HGMP Resource Centre. These have been made from a variety of tissues and organs, and are available either as 'full-length' or as 'normalized fragmentary' libraries. The former were generated from cDNAs of which the first strand synthesis was primed by oligo-dT; those of > 500 bp were then cloned non-directionally into the *Bst*XI site of the consitutive mammalian expression vector pCDM8. These libraries are available as insert–vector ligation mixtures ready for transformation into *E. coli* MC1061/p3; selection with ampicillin and tetracyclin is mediated by a suppressor tRNA encoded in the pCDM8 vector.

The 'normalized fragmentary' libraries were generated by a novel procedure in which restriction fragments of cDNA (derived from a mixture of tissues) are subdivided into 256 subsets before cloning, such that the combination of subsets represents the entire cDNA population. Adaptors were ligated to the fragment ends to allow PCR amplification, followed by ligation into the pCRII vector. These libraries are available from:

- UK HGMP Resource Centre (HGMP-RC), Hinxton—
  `http://www.hgmp.mrc.ac.uk`

ii. IMAGE consortium cDNA libraries

The IMAGE (Integrated Molecular Analysis of Genomes and their Expression) consortium was initiated by four academic groups (led by Drs C. Auffray, G. Lennon, M. H. Polymeropoulos, and M. B. Soares) on a collaborative basis, with the aim of providing a common set of shared resources for efficient gene sequencing, mapping, and expression studies. Researchers using these libraries return the information, which is then placed in the public domain as quickly as possible, thus ensuring that data are integrated efficiently and with the minimum of duplication.

At the time of writing, over 500 000 distinct oligo-dT primed, directionally cloned plasmid cDNAs are arrayed at the Lawrence Livermore National Laboratory, each clone having both a Genome Database (GDB) accession number and an IMAGE number. These libaries are available from:

- UK HGMP Resource Centre (HGMP-RC), Hinxton—
  `http://www.hgmp.mrc.ac.uk`
- Resource Centre/Primary Database of the German Human Genome Project, Berlin—`http://www.rzpd.de`
- American Type Culture Collection (ATCC), Rockville—
  `http://www.atcc.org`
- Lawrence Livermore National Laboratory (LLNL), Livermore—
  `http://www-bio.llnl.gov`

2.1.6 CpG island library

CpG islands are short, dispersed regions of unmethylated DNA with a higher frequency of CpG dinucleotides than the rest of the genome. These islands are associated with 5′ ends of 'housekeeping' genes and of some tissue-specific genes. A CpG island library is available as a lyophilized *E. coli* 'XL1-BlueMRF' culture from :

- UK HGMP Resource Centre (HGMP-RC), Hinxton—
  `http://www.hgmp.mrc.ac.uk`

2.1.7 Probes and PCR primers

There are a number of sources of human probe DNAs and PCR primers, including those which detect or amplify polymorphic loci. Catalogues of available resources will be found at:

- UK HGMP Resource Centre (HGMP-RC), Hinxton—
  `http://www.hgmp.mrc.ac.uk`
- Co-operative Human Linkage Centre (CHLC), Iowa—
  `http://www.chlc.org`
- Japanese Collection of Research Bioresources (JCRB), Tokyo—
  `http://www.nihs.go.jp`
- European Collection of Animal Cell Cultures (ECACC), Salisbury—
  `http://www.gdb.org/annex/ecacc/HTML/eccac.htm`
- Lawrence Berkeley National Laboratory (LBNL), Berkeley—
  `http://rmc-www.lbl.gov`

2.2 Non-human primates

2.2.1 Cell lines

Cell lines derived from kidney cells of African green monkey are available from:

- American Type Culture Collection (ATCC), Rockville—
  `http://www.atcc.org`

2.2.2 Genomic libraries

Cosmid libraries are available for gorilla, African green monkey, Formosa monkey *(Macaca cyclopisis)*, gibbon, and orangutan, from:

- Japanese Collection of Research Bioresources (JCRB), Tokyo—
  `http://www.nihs.go.jp`

2.3 Pig

2.3.1 Reference family DNA stocks

65 sets of DNA from reference families are available for linkage mapping from:

- Department of Animal Science, Iowa State University, Ames— mfrothsc@iastate.edu

2.3.2 Genomic libraries

A pig YAC library is available from:

- Resource Centre/Primary Database of the German Human Genome Project, Berlin—http://www.rzpd.de

2.3.3 PCR primers

About 300 pairs of PCR primers which amplify microsatellite-containing sequences are distributed from the following centre:

- Department of Animal Science, Iowa State University, Ames— mfrothsc@iastate.edu

2.4 Sheep
2.4.1 PCR primers

More than 100 PCR primer pairs which amplify polymorphic microsatellite markers are available from:

- Utah State University, Logan—fanoelle@cc.usu.edu

2.5 Mouse
2.5.1 Live animals and embryos

These banks contain a range of valuable mutant, transgenic, and inbred mouse lines, supplied as frozen embryos or as breeding nuclei of live animals. The banks also offer services for the collection and storage of stocks.

- MRC Experimental Embryology and Teratology Unit, Tooting—fax: (+44) 0181 767 9109
- MRC Radiobiology Unit, Didcot—fax: (+44) 01235 835 691

2.5.2 Cell cultures

A vast collection of cell lines, assembled from the original developers, is available from:

- Japanese Collection of Research Bioresources (JCRB), Tokyo— http://www.nihs.go.jp
- The Institute of Physical and Chemical Research (RIKEN), Saitama— http://www.riken.go.jp

2.5.3 Somatic cell hybrid panel

A panel of 27 mouse × Chinese hamster polychromosomal somatic cell hybrids has been produced, and should provide a useful tool for comparative mapping and chromosomal localization. DNA from this panel is available from:

- UK HGMP Resource Centre (HGMP-RC), Hinxton—
  `http://www.hgmp.mrc.ac.uk`

2.5.4 Genomic DNA stocks

DNAs from inbred, hybrid, wild, recombinant inbred, mutant, congenic, transgenic, and karyotypically abnormal mice are supplied for use in Southern blotting and PCR amplification by:

- The Jackson Laboratory, Maine—`http://www.jax.org`

2.5.5 Backcross DNA panel mapping resource

The EUCIB *Mus domesticus* × *M. spretus* backcross mapping panel has been produced to facilitate high-resolution genetic mapping (Chapter 2). The following centres distribute DNA from this panel:

- UK HGMP Resource Centre (HGMP-RC), Hinxton—
  `http://www.hgmp.mrc.ac.uk`
- The Jackson Laboratory, Maine—`http://www.jax.org`

2.5.6 Genomic libraries

i. YAC libraries
Mouse YAC libraries have been produced from both male and female mouse spleen, and are available from:

- Resource Centre/Primary Database of the German Human Genome Project, Berlin—`http://www.rzpd.de`
- UK HGMP Resource Centre (HGMP-RC), Hinxton—
  `http://www.hgmp.mrc.ac.uk`

ii. P1 library
A high density gridded P1 library is available from the following centre for screening by hybridization:

- Resource Centre/Primary Database of the German Human Genome Project, Berlin—`http://www.rzpd.de`

2.5.7 cDNA libraries

i. Murine cDNA libraries
Libraries from various tissues are available from:

- UK HGMP Resource Centre (HGMP-RC), Hinxton—
  `http://www.hgmp.mrc.ac.uk`

- Resource Centre/Primary Database of the German Human Genome Project, Berlin—http://www.rzpd.de

ii. IMAGE consortium cDNA libraries
45 IMAGE consortium murine cDNA libraries from various tissues are currently available. Details of the IMAGE consortium, and of availability, are as for item 2.1.5(ii).

2.6 Zebrafish
2.6.1 Mutant strains
Over 70 mutant strains, produced in a collaborative effort, are available from:

- Institute of Neuroscience, University of Oregon, Eugene— http://www.uoregon.edu

2.6.2 Genomic libraries
A variety of zebrafish genomic libraries is available from:

- Resource Centre/Primary Database of the German Human Genome Project, Berlin—http://www.rzpd.de

- Department of Molecular and Cell Biology, University of California, Berkeley—ses@mendel.berkely.edu

2.6.3 cDNA libraries
cDNA libraries made from staged embryos are available from:

- Eccles Institute, University of Utah, Salt Lake City— grunwald@msscc.med.utah.edu

2.6.4 Cosmid library
- Resource Centre/Primary Database of the German Human Genome Project, Berlin—http://www.rzpd.de

2.7 Pufferfush (*Fugu*)
2.7.1 Cosmid libraries
Cosmid libraries made from genomic DNA are available from:

- UK HGMP Resource Centre (HGMP-RC), Hinxton— http://www.hgmp.mrc.ac.uk
- Resource Centre/Primary Database of the German Human Genome Project, Berlin—http://www.rzpd.de
- YAC Screening Centre (YSCL), Leiden—e-mail: yscl@ruly46.medfac.leidenuniv.nl

2.7.2 cDNA libraries

- Resource Centre/Primary Database of the German Human Genome Project, Berlin—http://www.rzpd.de

3. Invertebrates

3.1 *Drosophila*
3.1.1 Mutant and wild strains
Drosophila melanogaster mutant and wild stocks are available from:

- National Institute of Genetics, Mishima Shizuoka— http://www.grs.nig.ac.jp
- Tübingen *Drosophila* stock collection— http://www.mpib-tuebingen.mpg.de

3.1.2 P-element lethal lines
Drosophila stocks carrying mapped single P-element lethal insertions are available from the following centre:

- Department of Biology, Indiana University—matthewk@indiana.edu

3.1.3 Genomic libraries
i. YAC library
A YAC library of onefold mean coverage, providing 90% coverage of the euchromatic portion of the genome, has been constructed. YAC clones are available from:

- Department of Biology, WU, St. Louis—duncan@biodec.wustl.edu

ii. P1 library
The following 16 laboratories have volunteered to make *Drosophila* P1 genomic libraries:

- University of Wisconsin-Madison, Laboratory of Molecular Biology, Madison—http://www.bocklabs.wisc.edu
- Carnegie Institute, Baltimore—http://www.ciwemb.edu
- Baylor College of Medicine, Houston—http://www.bcm.edu
- Fred Hutchinson Cancer Research Centre, Seattle— http//:www.fhcrc.org
- Indiana University, Bloomington—http://www.bio.indiana.edu
- University of California, Los Angeles—http://www.ucla.edu
- University of Utah, Salt Lake City—http://www.utah.edu
- Kyushu University, Fukuoka (Japan)—http://www.kyushu-u.ac.jp

- Yale University, New Haven—http://www.yale.edu
- Salk Institute, San Diego—http://www.salk.edu
- Harvard University, Cambridge (USA)—http://www.harvard.edu
- McGill University, Montreal—http://www.mcgill.ca
- European Molecular Biology Laboratory, Heidelberg—http://www.embl-heidelberg.de
- University of California, Berkeley—http://www.berkely.edu
- Stanford University School of Medicine, Stanford—http://www.stanford.edu
- Department of Genetics, Cambridge University (UK)—http://www.cam.ac.uk

iii. Cosmid library

- Resource Centre/Primary Database of the German Human Genome Project, Berlin—http://www.rzpd.de

iv. Mapped cosmid clones

Mapped cosmids are provided for use in physical mapping of the genome as a European collaborative project.

- The European Drosophila Genome Mapping Project—inga@nefeli.imbb.forth.gr

3.2 *Caenorhabditis elegans*

3.2.1 Genetic stocks

Over 2400 strains of wild and mutant types are available from:

- Caenorhabditis Genetics Centre (CGC): stier@molbio.cbs.umn.edu

3.2.2 Cosmid transgenics

Cosmid transgenics are produced by microinjection into the gonads of cosmids (containing *C. elegans* DNA) coupled to a marker plasmid. They have been constructed in association with the *C. elegans* Genome Sequencing laboratories in St. Louis, Missouri and The Sanger Centre, Hinxton. More than 100 transgenic strains are available from:

- Institute of Molecular Biology and Biochemistry, Simon Fraser University, Burnaby—djanke@darwin.mbb.sfu.ca

4. Plants

4.1 Rice

4.1.1 Plant cell cultures

- American Type Culture Collection (ATCC), Rockville—http://www.atcc.org

4.1.2 Genomic YAC library

A YAC library is available as clones and as high density filters for screening by hybridization, from:

- STAFF Institute, Ibaraki—http://www.staff.or.jp

4.1.3 cDNA clones

A large number of cDNA clones are available from the following centres:

- STAFF Institute, Ibaraki—http://www.staff.or.jp
- Cornell University, Ithaca—http://greengenes.cit.cornell.edu

4.1.4 PCR primers

PCR primers amplifying microsatellite-containing sequences are available from:

- Cornell University, Ithaca—http://greengenes.cit.cornell.edu

4.2 Maize

4.2.1 Probes

A vast collection of probes for use in genome mapping is available from:

- USDA-ARS, University of Missouri, Columbia—
  ed@teosinte.agron.missouri.edu

4.3 Wheat

4.3.1 Recombinant inbred line populations

- Department of Plant Breeding and Biometry, Cornell University, Ithaca—
  http://greengenes.cit.cornell.edu

4.3.2 Genomic libraries

Clones are available from the following centre:

- Department of Plant Breeding and Biometry, Cornell University, Ithaca—
  http://greengenes.cit.cornell.edu

4.3.3 Probes

- USDA-ARS, Albany, California—oandersn@pw.usda.gov

4.4 Oats

4.4.1 Recombinant inbred line populations

- Department of Plant Breeding and Biometry, Cornell University, Ithaca—
  http://greengenes.cit.cornell.edu

4.4.2 Genomic libraries

- Department of Plant Breeding and Biometry, Cornell University, Ithaca—
  http://greengenes.cit.cornell.edu

4.5 *Arabidopsis thaliana*

4.5.1 Plant cell cultures

- American Type Culture Collection (ATCC), Rockville—
  http://www.atcc.org

4.5.2 Mapping lines

These include recombinant inbred lines, transposon, and T-DNA tagged lines.

- Arabidopsis Biological Resource Centre (ABRC), Columbus Ohio—
  arabidopsis+@osu.edu1

4.5.3 Mutant lines

Over 8000 unique lines of *Arabidopsis thaliana* are maintained at the following centres. These include seed stocks associated with mapping or cloning of genes.

- Nottingham Arabidopsis Stock Centre (NASC), Nottingham—
  http://nasc.life.nott.ac.uk

- Arabidopsis Biological Resource Centre (ABRC), Columbus Ohio—
  arabidopsis+@osu.edu

4.5.4 Genomic libraries

i. YAC libraries

Arabidopsis YAC libraries are available as clones and filters from:

- Arabidopsis Biological Resource Centre (ABRC), Columbus Ohio—
  arabidopsis+@osu.edu

ii. P1 library

An *Arabidopsis* P1 library is available as clones and as filters from:

- Arabidopsis Biological Resource Centre (ABRC), Columbus Ohio—
  arabidopsis+@osu.edu

iii. BAC library

An *Arabidopsis* BAC library is available as clones or on filters from:

- Arabidopsis Biological Resource Centre (ABRC), Columbus Ohio—
  arabidopsis+@osu.edu

4.5.5 cDNA libraries

- Arabidopsis Biological Resource Centre (ABRC), Columbus Ohio—
  arabidopsis+@osu.edu

4.5.6 EST library

Approximately 20 000 cloned ESTs are available from:

- Arabidopsis Biological Resource Centre (ABRC), Columbus Ohio—
  `arabidopsis+@osu.edu`

5. Fungi

5.1 *Saccharomyces cerevisiae*

5.1.1 Genomic libraries

Genomic libraries in plasmid, bacteriophage, and cosmid vectors are available from:

- American Type Culture Collection (ATCC), Rockville—
  `http://www.atcc.org`

5.1.2 cDNA library

Clones from a cDNA library constructed in bacteriophage vector can be obtained from:

- American Type Culture Collection (ATCC), Rockville—
  `http://www.atcc.org`

5.2 *Schizosaccharomyces pombe*

5.2.1 Genomic libraries

i. Cosmid library
- Resource Centre/Primary Database of the German Human Genome
  Project, Berlin—`http://www.rzpd.de`

ii. YAC library
- Resource Centre/Primary Database of the German Human Genome
  Project, Berlin—`http://www.rzpd.de`

iii. P1 library
- Resource Centre/Primary Database of the German Human Genome
  Project, Berlin—`http://www.rzpd.de`

Acknowledgements

I wish to thank the very large number of people who provided details of their resources for inclusion in this appendix. I would also like to thank my colleagues in the Centre, particularly Steve Gamble for his help in setting-up and maintaining my computer link with the rest of the world, and our Director, Dr Keith Gibson, for his support and encouragement. Finally and most importantly, I wish to thank my wife Rekha and daughters Richa and Ritu, for their tolerance and support during the many long weekends and late evenings that went into compiling this appendix.

Bioinformatics for genome mapping

PETER M. WOOLLARD and GARY WILLIAMS

1. Introduction

The term 'bioinformatics' describes the application of computing in many aspects of biology. It includes the provision of biological data and analytical tools via computers, ranging from databases of sequences and map information to tools for performing sequence analysis and literature searches. For the researcher involved in genome analysis, bioinformatics offers many advantages, including:

(a) Efficient access to biological data of all sorts, including 'search' facilities.

(b) Automation of repetitive tasks, such as PCR primer design or sequence alignment.

(c) Access to software and computing resources for performing complex operations, such as large scale linkage analysis.

This chapter cannot cover, in detail, all aspects of bioinformatics relevant to genome analysis. Not only is the scope of the field vast, but new data and software are added daily. A more extensive treatment will be found in ref. 1; newly added resources can usually be found by using the sites detailed here as starting points. We hope, in this chapter, to introduce the reader to the principal aspects of bioinformatics, and to provide a listing of the major resources currently available.

Readers new to bioinformatics are strongly urged to seek advice from more experienced users and local computing services, who should be able to help in setting up the basic hardware and software, and in arranging access to the Internet. Bioinformatics courses (often available through the Online Bioinformatics Centres discussed in Section 4.3) can offer more specific advice relevant to biologists. In addition, many of the computer sites described throughout this chapter offer detailed advice and assistance to novice users.

2. The Internet and World Wide Web

The Internet is a global network of interconnected computers. It is not owned or run by a single entity, though there is international co-ordination in the assignment of the Internet 'addresses' which identify each site or server. A bewildering array of software and data transfer protocols exist to allow communication between these computers. One of the most important protocols is 'FTP', or file transfer protocol, a universally agreed standard which allows any computer file to be passed from computer to computer. Many software packages exist to implement FTP; your local computing centre will advise on the most appropriate one.

In recent years, access to bioinformatics has been immensely simplified by the development of the 'World Wide Web' or WWW. This is not an alternative computer network, but a much simpler and friendlier system for communicating via the Internet. Web sites, each identified by a unique WWW address, present their data in a standard format known as 'Hypertext Markup Language' (HTML). Browser programs (such as *Netscape* or *Mosaic*) present this data to the user as pages carrying text, images and icons. Highlighted keywords and icons, when selected with the computer mouse, act as links to call up and display further pages of information either at the same web site or at another. In this way, navigation between web sites is greatly simplified. Most web browser programs also provide other features, such as the ability to download selected information to disc.

Increasingly, web sites are becoming interactive, allowing the user to enter data and to use search facilities and programs implemented on the host computer. Most of the sites detailed in this chapter provide access via the WWW, as indicated by an address starting with 'http://'. Some of the larger and more heavily used databases are duplicated at several 'mirror sites'. Many sites (such as the UK HGMP Resource Centre) allow limited public access via the WWW; greater access (including file access to their own data, and the ability to run software) is available to registered users via the WWW. Telnet, which offers more 'traditional' terminal-based access to computing facilities, is still supported at many centres (including the HGMP Resource Centre), but is fast becoming redundant. Recent developments, such as the 'Java' language, are making the WWW utilities even more powerful.

3. Hardware and software requirements

Your local computing centre should be able to advise you on suitable hardware and software for Internet access, and to assist with its installation and use. Your requirements will, naturally, be influenced by exactly what you want to do: graphic-intensive work, for example, demands a fast computer and a high-resolution monitor. A realistic minimum requirement, however, comprises:

(a) A computer with suitable operating system, having at least 1 Gb of disc space, 16 Mb of memory, a 17 inch colour monitor, and a Pentium processor or equivalent; OR a graphic terminal (e.g. X-terminal) connected to a local server.

(b) Basic Internet communication software and (for WWW access) a suitable browser program such as *Netscape* or *Mosaic*.

(c) Physical connection to the Internet

(d) For home use, subscription to an Internet access provider.

The price/performance ratio of hardware continues to improve dramatically. The longer you hold out, the better the system you will get for your money, but you do have to buy sometime! There are many academic discounts available for both hardware and software: your local computing centre will advise. If you are in the UK look at the CHEST deals on the NISS (National Information Services and Systems) server (http://www.niss.ac.uk/).

Recently, vendors have been producing cheap and simple 'plug and play' systems which give Internet access. Such systems are competitive, but may be limited in their speed of access, or in their ability to download and store data.

In addition to the basic minimum of software mentioned above, software for handling electronic mail (e-mail), Telnet communication and FTP are extremely useful. The Internet also offers access to the 'newsgroups' or electronic bulletin boards; again, your local computing centre will advise on software for this. X-windows (which allows the display of graphical windows via the Internet) is also very useful. Users with some experience of computing will find a 'C' compiler useful, as it allows source code for many applications to be modified and compiled for use on their own machine.

If Internet access is not already provided on site, there are many commercial access providers and network suppliers. A useful set of addresses is available from:

- http://www.niss.ac.uk/it/nw-suppliers.html
- http://home.netscape.com/home/internet-directory.html

Some of the most well known Internet access providers are:

- America Online (http://www.aol.com/)
- CompuServe (http://www.compuserve.com/)
- Pipex (http://www.pipex.com/)

Finally, it is impossible to over-emphasize the importance of keeping regular backups of software and data. This may be done conveniently using tape streamers or writable high-capacity optical drives, or your local computing centre may provide facilities for this.

4. Starting points for navigating the WWW

4.1 General search facilities

The problem with the WWW is not so much one of accessing information, but of knowing where to look for it. A number of web sites therefore offer search facilities, allowing the user to specify keywords or phrases, and returning a list of links to sites whose description matches these criteria. The effectiveness of such searches depends both on the choice of keywords and on the quality of the indexing and cross-referencing at the search site. Sites providing general search facilities include:

- Europe: `http://www.ebi.ac.uk/searches/netsearch.html`
- North America: `http://home.netscape.com/Internet-directory.html`
  `http://home.netscape.com/home/internet-directory.html`
  `http://www.yahoo.com/`
  `http://www.lycos.com/`

Other excellent general references for computing, particularly in the UK, can be accessed via NISS (`http://www.niss.ac.uk/`).

4.2 Bioinformatics search facilities

The 'general' search sites in Section 4.1 do not provide the most direct route to relevant bioinformatics sites. Those seeking biological information will do better to start from one of the bioinformatics 'starting points' listed below. These sites all provide well-maintained links to a wide range of biological servers, often genome-oriented. The most successful approach is usually to choose one or two such sites and to use them regularly. This can help in avoiding the common problem of reaching the same destination from several different starting points! Recommended starting points for bioinformatic searches are:

- Europe: HGMP-RC (`http://www.hgmp.mrc.ac.uk/`)
  Expasy (`http://expasy.hcuge.ch/`)
- North America: GDB (`http://gdbwww.gdb.org/`)
  Harvard (`http://golgi.harvard.edu/biopages/all.html`)
  Baylor College (`http://gc.bcm.tmc.edu:8088/home.html`)
- Japan: GenomeNet (`http://www.genome.ad.jp/`)
- Australia: ANGIS (`http://morgan.angis.su.oz.au/`)

4.3 Online bioinformatics centres

The Internet means that access should no longer be restricted by geographical locality. Indeed, most resources can be accessed via the WWW by any user in any country. However, some transnational and international connections can be painfully slow at times.

Bioinformatics centres attempt to address this problem, by providing preferential online access for biologists to resources such as programs and

Table 1. The national nodes of EMBnet

Country	Site	WWW address
Austria	Vienna Biocenter, EMBnet Austria, Dr. Bohrgasse 9, A-1030 Vienna	
		`http://www.at.embnet.org/`
Belgium	BEN, Universite Libre de Bruxelles CP300, Paardenstraat 67, B-1640 Sint Genesius Rode	
		`http://www.be.embnet.org/`
Denmark	Biobase—The Danish Biotechnological Database, The Danish Human Genome Centre, Institute for Medical BioChemistry, Ole Worms Alle, Building 170-171, University of Aarhus, DK-8000, Aarhus C	
		`http://biobase.dk/`
Finland	CSC—Center for Scientific Computing, PL 405 (Tietotie 6), 02101 Espoo	
		`http://www.csc.fi/molbio/`
France	INFOBIOGEN, 7 rue Guy Moquet, BP8, 94801Villejuif CEDEX	
		`http://www.infobiogen.fr/`
Germany	GENUISnet, German Cancer Research Centre (DKFZ), Dept. of Molecular Biophysics (0810), Im Neuenheimer Feld 280, D-69120 Heidelberg	
		`http://genome.dkfz-heidelberg.de/biounit/`
Greece	Institute of Molecular Biology and Biotechnology (IMBB), Foundation for Research and Technology Hellas, PO Box 1527, GR-711 10 Heraklion, Crete	
		`http://www.imbb.forth.gr/`
Hungary	Agricultural Biotechnology Center, Szent-Gyorgi u. 4, PO Box 410, H-2101 Godollo	
		`http://www.abc.hu/`
Ireland	INCBI—Irish National Centre for BioInformatics, Department of Genetics, Trinity College, Dublin 2	
		`http://acer.gen.tcd.ie/`
Israel	INN—The Israeli EMBnet Node, Weizmann Institute of Science, Department of Biological Services, Rehovot	
		`http://dapsas1.weizmann.ac.il/inn.html`
Italy	CNR Area di Ricerca, Via Amendola, 166/5, 70126, Bari	
		`http://area.ba.cnr.it/`
The Netherlands	CAOS/CAMM Center, Faculty of Science, University of Nijmegen, Toernooiveld, 6525 ED Nijmegen	
		`http://www.caos.kun.nl/`
Norway	The Norwegian EMBnet Node, Biotechnology Centre of Oslo, University of Oslo, Guastadallen 21, 0317 Oslo	
		`http://www.no.embnet.org/`
Poland	Institute of Biochemistry and Biophysics, Polish Academy of Sciences, Pawinskiego 5a, 02-106 Warsawa	
		`http://www.ibb.waw.pl/`
Portugal	PEN—The Portuguese EMBnet Node, Instituto Gulbenkian de Ciencia, No Portugues da Rede EMBnet, Apartado 14, 2781 Oeiras CODEX	
		`http://www.pen.gulbenkian.pt/`
Spain	Centro Nacional de Biotecnologia—CSIC, University Autonoma, Campus de CantoBlanco, 28049 Madrid	
		`http://www.cnb.uam.es/`
Sweden	Computing Department, Biomedical Centre, Box 570, S-751 23 Uppsala	
		`http://www.bmc.uu.se/`
Switzerland	BioComputing Basel, Biozentrum der Universitaet, Klingelbergstrasse 70, 4056 Basel	
		`http://www.ch.embnet.org/`
United Kingdom	SEQNET, Daresbury Laboratory, Keckwick Lane, Daresbury, WA4 4AD	
		`http://www.seqnet.dl.ac.uk`

databases, including some which are not available to general users. In addition, they provide online assistance and advice, and offer training courses in the use of bioinformatics. The principal bioinformatics centres are detailed geographically below.

4.3.1 Bioinformatics centres in Europe

i. National EMBnet nodes

The national EMBnet nodes are biocomputing centres which have been appointed by their governmental authorities to provide databases and software tools for the molecular biology and biotechnology community. National nodes offer software and online services covering a diverse range of research fields, including sequence analysis, protein modelling, genetic mapping, and phylogenetic analysis. They also offer user support and training in the local language. These nodes are listed in *Table 1*.

ii. Specialist EMBnet nodes

Specialist EMBnet nodes, detailed in *Table 2*, are oriented toward specific areas of computational molecular biology. The nodes of particular relevance to genome mapping are the Sanger Centre (oriented toward *C. elegans*, human pathogens, and yeast genome analysis); EBI (which collates and archives many genome-related databases, and provides many other useful services); and HGMP-RC (which provides access to many useful genome-related programs and databases).

4.3.2 Bioinformatics centres in North America

Bioinformatics resources in North America are provided mainly by the local

Table 2. Specialist EMBnet nodes

Site	Location/WWW address
The Sanger Centre	Hinxton Hall, Hinxton, Cambridge, CB10 1SA, UK
	`http://www.sanger.ac.uk/`
HGMP-RC	Hinxton Hall, Hinxton, Cambridge, CB10 1SB, UK
	`http://www.hgmp.mrc.ac.uk/`
EBI (outstation of EMBL)	Hinxton Hall, Hinxton, Cambridge, CB10 1SD, UK
	`http://www.ebi.ac.uk/`
SWISS-PROT	Dept. of Medical Biochemistry of the University of Geneva, CMU, 1, Rue Michek Servet, 1211 Geneva 4, Switzerland
	`http://expasy.hcuge.ch/`
MIPS	Max-Planck Institut fuer Biochimie, Am Klopferspitz 18a, 82152 Martinsried, Germany
	`http://www.mips.biochem.mpg.de/`
ICGEB	Trieste, Italy
	`http://www.icgeb.trieste.it/`
F. Hoffman-La Roche, Ltd.	Pharma Preclinical Research, CH-4002 Basel, Switzerland
	`http://www.uk.embnet.org/brochure/html/hoffman.html`

university or institute. There are also many excellent biomedical sites which provide online access to specific services, such as NCBI (`http://www.ncbi.nlm.nih.gov`, providing *Entrez* (2), *BLAST* (3), *OMIM*, and *Gen-Bank*), John Hopkins (`http://gdbwww.gdb.org/`, providing *GDB*), Jackson Laboratories (`http://www.informatics.jax.org/`, providing *MGD*), CHLC (`http://www.chlc.org/`, with a variety of linkage mapping information) and Baylor College's Biologists' Control Panel (`http://gc.bm.tmc.edu:8088/bio/bio_home.html`).

Most training courses are again provided by local universities or institutes. There are training courses available for specific bioinformatics areas nationally. The Wisconsin GCG (4) group (`http://www.gcg.com/`) provide courses on their comprehensive sequence analysis package, and Jurg Ott (`http://linkage.rockefeller.edu`) runs various linkage analysis courses.

4.3.3 Bioinformatics centres in Australia

The Australian National Genomic Information Service (ANGIS) (`http://morgan.angis.su.oz.au/`) offers a national service to genome researchers.

4.3.4 Bioinformatics centres in Japan

GenomeNet (`http://www.genome.ad.jp/`) provides a wide range of databases and data analysis systems, distributed over three main sites. It also provides regular training courses in bioinformatics.

5. Data resources

In the following sections, we have outlined the principal data resources available via the Internet for genome analysis and mapping. Although new resources are added almost daily, most will be located at (or accessible via links from) sites related to those detailed below. It pays to check relevant sites regularly for details of the latest additions.

In many cases, 'wet' resources such as clones, PCR primers, or cell lines are made available to researchers; catalogues of such materials often accompany, or are integrated with, genome data. (More extensive information on biological resources will be found in Appendix 1.)

5.1 Human genome-wide mapping databases

Under this heading, we include major databases which provide mapping information throughout the human genome. Some (such as GDB) are repositories for information from a wide variety of sources, whilst others (for example CHLC), provide access to data from a particular institute or group of researchers.

Table 3. Human genome databases

Database	Address

GDB (Genome DataBase)
http://gdbwww.gdb.org/
OMIM (Online Mendelian Inheritance in Man)
http://www3.ncbi.nlm.nih.gov/Omim/searchomim.html
Human Genome Map Search (Pennsylvania University)
http://agave.humgen.upenn.edu/cpl/mapsearch.html
The Genetic Location Database (LDB)
http://cedar.genetics.soton.ac.uk/public_html
The dysmorphic human and mouse homology database
http://www.hgmp.mrc.ac.uk/DHMHD/dysmorph.html
BodyMap—Anatomical Expression Database of Human Genes
http://www.imcb.osaka-u.ac.jp/bodymap/welcome.html
CEPH-Genethon integrated map
http://www.cephb.fr/ceph-genethon-map.html
CEPH Genotype database
http://www.cephb.fr/cephdb
YAC Data searches
http://gc.bcm.tmc.edu:8088/bio/yac_search.html
Cooperative Human Linkage Center (CHLC)
http://www.chlc.org/
Radiation Hybrid Mapping data (RHdb)
http://www.ebi.ac.uk/RHdb
dbEST Expressed Sequence Tag Database
http://www.ncbi.nlm.nih.gov/dbEST/index.html
UniGene—Unique Human Gene Sequence Collection
http://www.ncbi.nlm.nih.gov/Schuler/UniGene/
dbSTS Sequence Tagged Site Database
http://www.ncbi.nlm.nih.gov/dbSTS/index.html
Whitehead Institute/MIT Genome Center
http://www-genome.wi.mit.edu/
Histo Home Page
http://histo.cryst.bbk.ac.uk/
V BASE: A Directory of Human Immunoglobulin V Genes
http://www.mrc-cpe.cam.ac.uk/imt-doc/vbase-home-page.html
Human CpG Island database
http://biomaster.uio.no/cpgisle.html
Human population genetics database (Genography)
http://lotka.stanford.edu/genography.html
Organellar Genome Megasequencing Program
http://megasun.bch.umontreal.ca/
MITOMAP A Mitochondrial DNA database
http://infinity.gen.emory.edu/mitomap.html

The databases are listed in *Table 3*; two of the most widely used, GDB and OMIM, are described below. A further major database, IGD, has been under development for some time, but is not yet generally available. However, it is a promising attempt to integrate many genomic databases in a readily accessible form, and may become of major importance in the next few years.

5.1.1 Genome Data Base (GDB)

The Genome Data Base (GDB) supports the human genome project by serving as the central public repository for human genome mapping data. It is an international collaboration sponsored by biomedical funding agencies. The database stores information on the locations of important landmarks in the human genome, the reagents used to identify and locate them, and supporting data such as literature citations, and contacts in the genomic research community.

GDB is accessible both via the WWW and by direct online access from GDB's home site at John Hopkins and many mirror sites. The web addresses for GDB are:

- Australia: ANGIS http://morgan.angis.su.oz.au/gdb/gdbtop.html
               WEHI http://wehih.wehi.edu.au/gdb/gdbtop.html
- France: INFOBIOGEN http://gdb.infobiogen.fr/
- Germany: DKFZ http://gdbwww.dkfz-heidelberg.de/
- Israel: Weizmann Institute http://gdb.weizmann.ac.il/
- Japan: JICST http://www.gdb.gdbnet.ad.jp/
- Netherlands: CAOS/CAMM Center http://www-gdb.caos.kun.nl/gdb/gdbtop.html
- Sweden: Biomedical Center http://gdb.embnet.se/gdb
- UK: HGMP Resource Centre http://www.hgmp.mrc.ac.uk/gdb/gdbtop.html
- US: John Hopkins http://gdbwww.gdb.org/

GDB version 6.0 represents a considerable reorganization of database architecture and data structures, as compared to earlier versions. This has been necessary both to faithfully represent the increasingly diverse and detailed mapping information being generated, and to facilitate its integration with other genome databases. GDB now allows submitters to curate data which they have submitted, and also maintains a consensus genome map edited by respected researchers in the field.

The WWW interface to GDB provides query forms allowing complex searches. Entries matching query specifications are returned from the database, with links which allow the user to easily locate an item of interest and trace related items. For example, searching for a gene will give links to all associated polymorphisms, citations, maps, and probes.

5.1.2 Online Mendelian inheritance in man (OMIM)

OMIM is a catalog of human genes and genetic disorders authored and edited by Dr V. A. McKusick and colleagues at Johns Hopkins and elsewhere, and developed for the World Wide Web by the National Center for

Biotechnology Information, through whom it is accessible (see *Table 3*). The database contains text, pictures, and reference information. It also contains copious links to NCBI's *Entrez* database of MEDLINE journal articles and sequence information.

Users can search OMIM articles for keywords in any of the following fields:

- OMIM Number
- Allelic variants
- Text
- References
- Clinical synopsis
- Gene map disorder
- Contributors

The OMIM Gene Map can be viewed by either entering the gene symbol or chromosomal location. In addition to the NCBI interface, some online informatics centres provide other WWW interfaces to the OMIM database, including SRS and DBGET data searching and retrieval tools. Another search tool, IRX, can also be used to perform keyword searches of the OMIM data.

5.2 Human chromosome-specific and mitochondrial databases

Databases exist for most of the human chromosomes, and for the human mitochondrial genome. These databases are often created and maintained by research institutes or groups heavily involved with a particular chromosome, or by chromosome workshops. They frequently provide more detailed information on a particular chromosome than is readily accessible from genome-wide databases. The principal databases in this category are detailed in *Table 4*.

5.3 Rodent genome databases

The genome of the mouse and, increasingly, that of the rat has been mapped extensively in recent years. Much of this data is directly relevant to the analysis of the human genome, particularly in areas where it would be impractical or impossible to obtain the corresponding human data. The major databases for mouse and other rodents are described below.

5.3.1 The Mouse Genome Database (MGD)

MGD provides a comprehensive source of information on the experimental genetics of the laboratory mouse. It includes information on mouse markers, homologies to other mammals, probes and clones, PCR primers, and experi-

Table 4. Human Chromosome-Specific Databases

Chr	Database/Address
1	Chromosome 1 Home Page
	`http://linkage.rockefeller.edu/chr1`
2	ICRF
	`http://www.icnet.uk/axp/cmg/chr2`
3	University of Texas, San Antonio STSs and YACs.
	`http://mars.uthscsa.edu/`
4	Stanford Human Genome Center RH Map
	`http://shgc.stanford.edu/cgi-bin/4srch`
8	Baylor 1994 Workshop results
	`http://gc.bcm.tmc.edu:8088/chr8/home.html`
9	Galton Laboratory
	`http://www.gene.ucl.ac.uk/chr9`
10	Genome Therapeutics Corporation
	`http://www.cric.com/htdocs/chr10-mapping/index.html`
11	Texas University
	`http://mcdermott.swmed.edu/datapage/`
11	Imperial College
	`http://chr11.bc.ic.ac.uk/`
12	Yale Genome Centre
	`http://paella.med.yale.edu/chr12/Home.html`
13	Columbia
	`http://genome1.ccc.columbia.edu/~genome/`
16	LANL
	`http://www-ls.lanl.gov/dbqueries.html`
17	Weizmann Institute
	`http://bioinformatics.weizmann.ac.il/chr17/`
18	Registry and Research Society
	`http://mars.uthscsa.edu/Society/`
19	LLNL Human Genome Center Physical maps
	`http://www-bio.llnl.gov/bbrp/genome/genome.html`
21	LBL P1 & cDNA mapping and Syndb database
	`http://www-hgc.lbl.gov/Genome/C21Proj.html`
21	USDA browse ACeDB-style data
	`http://probe.nalusda.gov:8300/cgi-bin/browse/hch21`
21	USDA query ACeDB-style data
	`http://probe.nalusda.gov:8300/cgi-bin/qselect/hch21`
21	Phenotypic Mapping Project
	`http://www.csmc.edu/genetics/korenberg/phenomap.page.html`
21	CEPH
	`http://www.cephb.fr/chromosome21.html`
22	Philadelphia Genome Center Physical mapping data
	`http://www.cbil.upenn.edu/HGC22.html`
22	Sanger Centre
	`http://www.sanger.ac.uk/HGP/chr22`
X	USDA browse ACeDB-style data
	`http://probe.nalusda.gov:8300/cgi-bin/browse/hchx`
X	USDA query ACeDB-style data
	`http://probe.nalusda.gov:8300/cgi-bin/qselect/hchx`
X	Virtual Workshop
	`http://gc.bcm.tmc.edu:8088/chrx/home.html`
X	SIGMA data
	`http://www.ibc.wustl.edu:70/0h/CGM/human_x.html`
Mitochond.	MITOMAP
	`http://infinity.gen.emory.edu/mitomap.html`

mental mapping data. MGD is available via the World Wide Web at two sites:

- US: http://www.informatics.jax.org/
- Europe: http://mgd.hgmp.mrc.ac.uk/
- Japan: http://mgd.ni.ai.affrc.go.jp

MGD provides a set of query forms, which can be used by a number of WWW browsers. Each query form is related to a particular kind of information:

- References
- Genetic markers and Mouse Locus Catalog (MLC)
- Probes
- PCR primers
- Mammalian homologs
- Maps and mapping data
- Combined mouse/human phenotypes (MLC/OMIM)

The user selects a form, enters information in the form fields to build a query, and executes the query. Any records matching the criteria are listed, along with hypertext links which give access to further details. For those users whose web browser does not support query forms, an alternative simplified query option is available.

MGD is extensively integrated with the Encyclopedia of the Mouse Genome; the integration features are optional. The Mouse Locus Catalog (MLC), formerly a separate database, has been incorporated with MGD; users can now search for MLC records using the Genetic Markers And Mouse Locus Catalog Information query form. MLC records are incorporated into MGD tables, with full-text searching capability. In addition, integrated MLC/OMIM searching capability is included in MGD, allowing simultaneous searching of both MLC and Online Mendelian Ineritance in Man.

The Gene Expression Information Resource for Mouse Development is being integrated with MGD. This will result in a tool for examining patterns of gene expression during mouse embryonic development, elucidating biologically important networks, and exploring the genetic programs that underlie normal development and disease.

In January 1995, the Genomic Database of the Mouse (GBASE) was decommissioned. The incorporation of GBASE's data and functions into MGD is, at the time of writing, almost complete.

5.3.2 Whitehead Institute/MIT genome center mouse genetic map

This database (http://www-genome.wi.mit.edu) holds the latest release of the Whitehead Institute/MIT Center for Genome Research mouse genetic map. The map currently consists of 6183 polymorphic mouse microsatellite repeats mapped on a C57BL/6J-ob/ob × CAST F2 intercross using PCR probes.

5.3.3 MBx: mouse backcross database

This holds experimental data derived from the European Collaborative Interspecific Backcross (EUCIB) project, and is accessible at http://www.hgmp.mrc.ac.uk/MBx/MBxHomepage.html. EUCIB provides an international resource for the high-resolution genetic mapping of the whole mouse genome as a prelude to physical mapping using an STS-based YAC contig approach.

5.3.4 The dysmorphic human–mouse homology database

This consists of three separate databases (at http://www.hgmp.mrc.ac.uk/DHMHD/dysmorph.html) of human and mouse malformation syndromes together with a database of mouse/human syntenic regions. The mouse and human malformation databases are linked through the synteny database. The purpose of the system is to allow retrieval of syndromes according to detailed phenotypic descriptions and to be able to carry out homology searches for the purpose of gene mapping. Thus the database can be used to search for human or mouse malformation syndromes in different ways:

(a) By specifying specific malformations or clinical features, or chromosome locations.

(b) By homology.

(c) By asking for syndromes in one species located at a chromosome region syntenic with a chromosomal region in the other.

5.3.5 Whitehead Institute/MIT genome center rat SSLP data

This data (http://www-genome.wi.mit.edu/ftp/distribution/rat_sslp_releases/) consists of polymorphic rat microsatellite repeats mapped on a SHR × BN F2 intercross using PCR probes. The average spacing between markers on these maps is approximately 3.7 cM.

5.3.6 RATMAP

This covers genes physically mapped to rat chromosomes. It provides chromosomal localization of genes and anonymous DNA segments, and brief descriptions of genes (http://ratmap.gen.gu.se/).

5.3.7 Mouse and rat research home page

This serves as a central database (http://www.cco.caltech.edu/~mercer/htmls/rodent_page.html) detailing resources of particular interest to scientific researchers using mice or rats. In addition, it includes an extensive list of especially useful Internet resources.

5.4 Genome databases for other species

An increasing number of organisms now have major genome projects
devoted to them. It is not possible to give here detailed descriptions of each
database, but *Tables 5*, *6*, and *7* list databases for vertebrates (excluding
man); for invertebrates; and for plants, fungi, and prokaryotes, respectively.

Table 5. Other vertebrate genome databases

Database	Address

Cattle
Cattle Genome Map
　　　http://sol.marc.usda.gov/genome/cattle/cattle.html
Japan Cattle Cytogenetic Map
　　　http://ws4.niai.affrc.go.jp/dbsearch2/cmap/cmap.html

Sheep
SheepMap　　　http://dirk.invermay.cri.nz/
USDA Sheep Genome Map
　　　http://sol.marc.usda.gov/genome/sheep/sheep.html

Pig
USDA Swine Genome Map
　　　http://sol.marc.usda.gov/genome/swine/swine.html
NAGRP Pig Gene Map
　　　http://www.public.iastate.edu/~pigmap/pigmap.html
Roslin PiGMaP Data
　　　http://www.ri.bbsrc.ac.uk/pigmap/pigmap.html
Pig Cytogenetic Map (Japan)
　　　http://ws4.niai.affrc.go.jp/dbsearch2/pmap/pmap.html

Chicken
Chicken Genome Map
　　　http://www.ri.bbsrc.ac.uk/chickmap/ChickMapHomePage.html
NAGRP Chicken Gene Map
　　　http://poultry.mph.msu.edu/

Dog
The Dog Genome Project
　　　http://mendel.berkeley.edu/dog.html
The FHCRC Dog Genome Project
　　　http://tiberius.fhcrc.org/home/dog.html
DogMap　　　http://ubeclu.unibe.ch/itz/dogmap.html
The Dog Genome Project at the University of Oregon
　　　http://mendel.berkeley.edu/dogs/george.html

Horse
Horse Genetics　http://www.vgl.ucdavis.edu:80/~lvmillon/
Zebrafish
The Zebra Fish Server
　　　http://zfish.uoregon.edu/
Zebra Fish Information
　　　http://golgi.harvard.edu/zebra.html

Fugu
Fugu Landmark Mapping
　　　http://fugu.hgmp.mrc.ac.uk/

General
Agricultural Genomes
　　　http://probe.nalusda.gov:8300/animal/index.html

Table 6. Invertebrate genome databases

Database	Address

Caenorhabditis elegans

The C. elegans Genome Project
```
http://www.sanger.ac.uk/Projects/C_elegans
```
ACeDB A Caenorhabditis elegans Database
```
http://probe.nalusda.gov:8300/other/index.html
```
wEST C. elegans Expressed Sequence Tag Database
```
gopher://gopher.gdb.org/77/.INDEX/west
```
C. elegans repetitive DNA
```
http://www.ibc.wustl.edu/rpt/elegans
```
CGC Caenorhabditis Genetics Center at Minnesota
```
gopher://elegans.cbs.umn.edu:70/1
```
Caenorhabditis elegans WWW Server at UTSW
```
http://eatworms.swmed.edu/
```
The Blumenthal Lab at Indiana
```
http://www.bio.indiana.edu:8080/
```
The Strome Lab at Indiana
```
http://sunflower.bio.indiana.edu:80/~sstrome/
```
David Baillie's Worm Lab
```
http://darwin.mbb.sfu.ca/imbb/dbaillie/baillielab.html
```
Rose Worm Lab `http://genekit.medgen.ubc.ca/roselab.html`

Drosophila

FlyBase (Drosophila)
```
http://www.ebi.ac.uk/flybase/
```
Nagoshi Lab Home Page (Drosophila)
```
http://fly2.biology.uiowa.edu/
```
WWW Virtual Library: Drosophila
```
http://www-leland.stanford.edu/~ger/drosophila.html
```
Berkeley Drosophila Genome Project
```
http://fly2.berkeley.edu/
```
FlyView—A Drosophila Image Database
```
http://pbio07.uni-muenster.de/
```
The Interactive Fly `http://sdb.bio.purdue.edu/fly/aimain/1aahome.htm`

Others

National Insect Genetic Resources
```
gopher://gopher.ars-grin.gov/11/nigr
```
Mosquito Genomics WWW Server
```
http://klab.agsci.colostate.edu/
```
Sericultural Science, the University of Tokyo
```
http://www.ab.a.u-tokyo.ac.jp/shimada.html
```

5.5 Comparative genome databases

Some of the species-specific databases described in the previous sections include information on homologies with other species. However, a number of databases exist which are concerned primarily with cross-species homologies, particularly between mammals (*Table 8*). Such databases allow scientists to integrate data from the different genome projects and to study syntenic

343

Table 7. Plants, fungi, and prokaryotic genome databases

Database	Address
Arabidopsis (MIPS)	`http://speedy.mips.biochem.mpg.de/mips/athaliana`
AtDB An Arabidopsis Thaliana Database	
	`http://genome-www.stanford.edu/`
Arabidopsis Information Management System (AIMS)	
	`http://genesys.cps.msu.edu:3333/`
Dendrome: A Genome Database for Forest Trees	
	`http://s27w007.pswfs.gov/index.html`
BeanRef	`http://scaffold.biologie.uni-kl.de/Beanref/`
USDA Agricultural Plant Genome	
	`http://probe.nalusda.gov:8300/plant/index.html`
Japan Rice Genome Research Program	
	`http://www.staff.or.jp/`
MaizeDB Maize Genome Database	
	`http://www.agron.missouri.edu/`
National Corn Genome Initiative	
	`http://www.inverizon.com/ncgi`
SoyBase: Metabolic Database	
	`http://cgsc.biology.yale.edu/metab.html`
Yeast Genome Project	`http://speedy.mips.biochem.mpg.de/mips/yeast`
Saccharomyces Genomic Information	
	`http://genome-www.stanford.edu/`
Yeast Protein Database (YPD)	
	`http://www.proteome.com/YPDhome.html`
NIH Fission yeast Schizosaccharomyces pombe information	
	`http://www.nih.gov/sigs/yeast/fission.html`
Mycological Resources	`http://www.keil.ukans.edu/~fungi`
Candida albicans	`http://alces.med.umn.edu/Candida.html`
Normalized Gene Designation Database (NGDD)	
	`http://www.hgmp.mrc.ac.uk/local-data/Ngdd_form.html`
MycDB Mycobacterium Database	
	`http://www.biochem.kth.se/MycDB.html`
NRSub: Bacillus subtilis Database	
	`http://acnuc.univ-lyon1.fr/nrsub/nrsub.html`
Non-Redundant Database for Bacillus subtilis	
	`http://ddbjs4h.genes.nig.ac.jp/`
EcoCyc: Encyclopedia of E. coli Genes and Metabolism	
	`http://www.ai.sri.com/ecocyc/browser.html`
The WWW Virtual Library: Microbiology	
	`http://golgi.harvard.edu/biopages/micro.html`
DOE Microbial Genome Initiative (MGI)	
	`http://www.er.doe.gov/production/oher/mig_top.html`

relationships. Many of these resources are in the early stages of development, but are expected to expand dramatically in the near future.

5.6 Mutation databases

A number of databases have recently been made available over the WWW which deal primarily with mutations (*Table 9*). Many of these are specific to

Table 8. Comparative genome databases

Database	Address
Seldin/Debry Human/Mouse Homology Map	
	`http:/www3.ncbi.nlm.nih.gov/Homology/`
The dysmorphic human and mouse homology database	
	`http://www.hgmp.mrc.ac.uk/DHMHD/dysmorph.html`
XLocus (all species)	`http://cgsc.biology.yale.edu/xlocus.html`
AGsDB A Genus Species Database	
	`http://keck.tamu.edu/cgi/agsdb/agsdbserver.html`
The Homeobox Page	
	`http://copan.bioz.unibas.ch/homeo.html`
Gene Family Database (human, mouse)	
	`http://gdbdoc.gdb.org/~avoltz/home.html`
XREFdb (mammals)	`http://www.ncbi.nlm.nih.gov/XREFdb/`
Online Mendelian Inheritance In Animals (OMIA)	
	`http://morgan.angis.su.oz.au/BIRX/omia/omia_form.html`
GeneQuiz (automated analysis)	
	`http://www.sander.ebi.ac.uk/genequiz/`
Vertebrate Comparative Biology Database	
	`http://www.hgmp.mrc.ac.uk/compdb`

Table 9. Mutation databases

Database/Address

Human Gene Mutation Database
> `http://www.cf.ac.uk/uwcm/mg/hgmd0.html`

GDB (Genome DataBase)
> `http://gdbwww.gdb.org`

PAH Genes and alleles (PAHDB)
> `http://www.mcgill.ca/pahdb/`

Breast Cancer Mutation Data Base (BIC)
> `http://www.nchgr.nih.gov/Intramural-research/lab_transfer/bic/`

Protein Mutation Database
> `http://www.genome.ad.jp/htbin/bfind_pmd`

TBASE Transgenic and Targeted Mutation Animal Database
> `http://www.gdb.org/Dan/tbase/tbase.html`

Transgenic Systems for Mutation Analysis
> `http://darwin.ceh.uvic.ca/bigblue/bigblue.htm`

ORNL Transgenic and Targeted Mutant Animal Database
> `http://www.ornl.gov/TechResources/Trans/hmepg.html`

MITOMAP: A Mitochondrial DNA database
> `http://infinity.gen.emory.edu/mitomap.html`

NIGMS Human Genetic Mutant Cell Repository
> `telnet://online@coriell.umdnj.edu`

one class of genes, one disease, or one species, but it is expected that more wide-ranging mutation databases will emerge in the future.

5.7 PCR primer and probe databases

A number of databases contain sequences of PCR primers or details of probes. In many cases, the 'wet' resources (i.e. oligonucleotides or probe

DNA) are themselves available in addition to the data (see Appendix 1 for details of such resources). Those databases containing extensive primer and probe information include:

- The EBI's PCR primers database (human, mouse):
  http://www.ebi.ac.uk/primers_home.html
- UK HGMP Primers Database (human, mouse):
  http://www.hgmp.mrc.ac.uk/local-data/Primers.html
- UK DNA Probe Database (human, mouse):
  http://www.hgmp.mrc.ac.uk/local-data/Probes_form.html
- Genome DataBase (human): GDB
- Mouse Genome Database (mouse): MGD
- Oligonucleotide Probe Database (all species): OPD

5.8 Clone catalogues

A great many clones from a wide variety of species are available, often at little or no cost. In most cases, catalogues detailing these clones are available via the WWW. As these are primarily 'wet' resources, rather than pure data, details of them will be found in Appendix 1.

5.9 Sequence databases

The sequences produced by genome projects and other investigators are put into a variety of databases, which are rapidly expanding (many are doubling in size every 14–18 months). Many universities and biocomputing centres subscribe to the larger databases, and receive regularly updated copies which may be searched using local software facilities.

Where the relevant databases or the necessary search tools are not locally available, or where the most recent information is essential, the databases can be accessed through a variety of search facilities provided via the WWW. Such facilities include *SRS* (Sequence Retrieval System), *DBGET*, and *Entrez*, all offering a range of search facilities (such as by species, by key-words in the annotation, and so forth). Many sites provide these facilities, and most sites allow searches to be implemented on all major sequence databases.

Table 10 summarizes the principal sequence databases, and some of the sites through which each may be accessed.

The principal nucleic acid databases (GenBank, EMBL, and DNADB of Japan) all collate and share data, though there is a delay of a few days in the transfer of sequences between them. The EST and STS sequences are put into their own sections within the databases, often making it possible to search just these sections.

SwissProt is by far the best annotated protein sequence database, with extensive links to the nucleic acid sequence databases, Medline, enzyme databases, and so forth. PIR contains a larger number of protein sequences—currently about twice as many as SwissProt—but the annotation is poorer and there is considerable duplication. GenPept and TREMBL are translations of GenBank and EMBL coding sequences, respectively.

Those databases which specialize in a particular type or source of sequence, though smaller, are often particularly well curated. An example is UniGene (the Unique Human Genome Sequence Collection), which holds clusters of human expressed sequence tags (ESTs) which represent the transcription products of distinct genes. These sequences are being used for transcription mapping in collaboration with several genome mapping centres, and some of the clusters have already been localized to chromosomes.

5.9.1 Sequence comparison facilities

A number of software packages exist which allow the user to search the databases for sequences with similarity to a chosen sequence. Such programs include *BLAST, FASTA*, and *blitz*, as well as facilities included in the GCG and Staden packages. Additional software exists for 'filtering' the output of these programs to eliminate spurious or unwanted hits.

A detailed description of this software is beyond the scope of this chapter, but local biocomputing centres will often provide details of, and access to, the necessary software. In the US, the BLAST WWW facility is located at http://www.ncbi.nlm.nih.gov/BLAST. Many bioinformatic centres also provide details of e-mail servers, which allow search queries to be submitted (and their results returned) via e-mail.

6. Software resources

The Internet offers a vast array of software to those involved in genome analysis and mapping, much of it available freely or at nominal cost. It is essential that biologists have an awareness of what software can and cannot be expected to do. An understanding of the working principles behind the software will help the user to select the appropriate program, to understand its capabilities and limitations, to use it to its full potential, and to interpret its results.

Much of this software is available for downloading via the Internet, for installation on the user's own computer system. This is normally done using anonymous FTP (see Section 2). It is, of course, essential that the software be compatible with the system on which it is to be installed. A number of sites exist which provide catalogues of available software for PC, Macintosh, Unix, and other systems, and have much of this software available for downloading. The principal such sites of interest to biologists are:

- UK: HENSA (Higher Education National Software Archive) http://www.hensa.ac.uk/

Table 10. Sequence and sequence-related databases

Database	Type[a]	Species	Description	Access[b]
EMBL	N	Any	Comprehensive sequence collection	*BLAST* (EMBnet), *SRS*
GENBANK	N	Any	Comprehensive sequence collection	*SRS, BLAST* (NCBI)
DNA DB of Japan	N	Any	Comprehensive sequence collection	?
DBEST	N (G)	Any	Expressed sequence tags	*SRS, BLAST* (NCBI)
UNIEST	N (G)	Human	All human EST sequences, clustered into sets	*SRS,* NCBI
DBSTS	N (G)	Any	Sequence-tagged sites	*SRS, BLAST* (EMBnet), *BLAST* (NCBI)
DBGSS	N (G)	Mammal	Genome survey sequence	*SRS, BLAST* (NCBI)
IMGT	N (E)	Human	Immune system-related	*SRS,* http://www.ebi.ac.uk/imgt
EPD	N (E)	Eukaryotes	Eukaryotic promoters	*SRS, BLAST* (NCBI)
CPGISLE	N (E)	Human	Descriptions of genes and associated CpG islands	*SRS*
YEAST	N	*S. cerevisiae*	Complete genome sequence	*BLAST* (NCBI) http://speedy.mips.biochem.mpg.de/mips/yeast
REPBASE	N (G)	Eukaryotes	Repeated sequence families	*SRS*
Entrez	N + P (G)	Any	Genetic subset of Medline, plus associated sequences	NCBI (*Entrez*)
SwissProt	P	Any	Well-annotated protein database with many cross-references	*SRS, BLAST* (EMBnet), *BLAST* (NCBI), http://expasy.hcuge.ch/sprot-top.html
PIR	P	Any	Protein Identification Resource	*SRS, BLAST* (EMBnet), *BLAST* (NCBI)
TREMBL	P (E)	Any	Translated coding regions of EMBL	*SRS, BLAST* (EMBnet)
GenPept	P (G)	Any	Translated coding regions of GenBank	*SRS*

	P + N	Human		
Kabat	P + N	Human	Sequences of proteins of immunological interest	*BLAST* (NCBI), `http://immuno.bem.nwu.edu/`
PROSITE	P (S)	Any	Dictionary of sites and patterns in proteins	*SRS*
BLOCKS	P (Pr)	Any	Highly-conserved protein segments	*SRS*
PFAM, SFAM	P (Pr + S)	Any	Protein family alignments	*SRS*
PRODOM	P (S)	Any	Protein families	*SRS*
SWISSDOM	P (S)	Any	Shows domains identified in SwissProt entries	*SRS*
PIRALN	P (Pir)	Any	Protein sequence alignments	*SRS*
PRINTS	P	Any	Protein motif fingerprints	*SRS*, `http://www.biochem.ucl.ac.uk/bsm/dbbrowser/PRINTS.html`
SBASE	P	Any	Annotated protein domains	*SRS*, `http://base.icgeb.trieste.it.sbase/`
PDB	P	Any	3D protein structures	*SRS*, *BLAST* (NCBI), `http://www.pdb.bnl.gov/`

[a] N = DNA and/or RNA sequences; P = protein sequences. (E/G/S/Pr/Pir): these databases provide extra annotation for a specialized portion from EMBL, Gen-Bank, SwissProt, Prosite or PIR respectively.

[b] *SRS* (Sequence Retrieval System) allows keyword searching and link-following between most of the above databases. SRS is available from many sites e.g. `http://iubio.bio.indiana.edu/`, `http://www.sanger.ac.uk/`, `http://www.hgmp.mrc.ac.uk/` and `http://wehih.wehi.edu.au/`. The exact subset of databases accessible varies from site to site. Local EMBnet nodes in Europe usually provide *BLAST* and/or *FASTA*. NCBI provides access to *BLAST*, *Entrez*, etc.: `http://www.ncbi.nlm.nih.gov/`

- Europe: `http://www.ebi.ac.uk/biocat/biocat.html`
- North America: `http://ftp.bio.indiana.edu/`

In addition, in the UK, there is an archive of teaching material at NISS (`http://www.niss.ac.uk/`). In many cases, however, the software is available on its 'host' computer, and data are transferred to and from it via the Internet. This has several advantages to the user, such as access to powerful computing resources not available locally. Increasingly, this remote access is provided via a WWW interface.

The following sections cover some of the principal software applications relevant to genome mapping.

6.1 Linkage analysis software

The tremendous expansion of linkage mapping (Chapters 1–3), and of the size of the data-sets to be analysed, has posed formidable computational problems. Fortunately there have also been some dramatic programming and algorithmic improvements resulting in programs such as *Fastlink* (3) and *Vitesse* (4). In addition, various simulation programs such as *Slink* are available which can be used to determine whether a proposed study will be capable of detecting linkage. *Mapmaker* (5) is a useful linkage analysis package designed to help construct primary linkage maps of markers segregating in experimental crosses.

Linkage analysis is an area in which an understanding of the problem and of the algorithms is particularly important. People should bear in mind that many of the programs were designed before the analysis of multifactorial loci was commonplace, and may not cope efficiently with such problems.

Descriptions and archives of linkage analysis and mapping programs can be found at:

- Europe: `ftp://ftp.ebi.ac.uk/pub/software/linkage_and_mapping/`

- US: `http://linkage.rockefeller.edu/`

6.2 Contig and fragment assembly software

Extensive software for the assembly of contigs based on the marker or restriction fragment content of clones is available, as discussed in Chapter 10. In addition, the GCG (6), EGCG (7), and Staden (8) packages provide a variety of facilities for the assembly of sequences. The home page for the commercial GCG software is `http://www.gcg.com/`, whilst that for the Staden package is `http://www.mrc-lmb.cam.ac.uk/pubseq/`. Access to the GCG, Staden, and (increasingly) the EGCG packages is provided by many of the major bioinformatic centres and universities.

6.3 Software for designing primers for PCR and sequencing

There are a number of good programs for designing effective primers, including: GCG's (6) *Prime*, The Whitehead Institute's *Primer* (9), and LaDeana Hillier's *OSP* (An Oligo Selection Program) (10). These should all be available from your local biocomputing centre.

It is often a good idea to search your proposed primer against sequence databases, allowing for some mismatches to take into account sequencing errors or mutations. Primer sequences are usually too small for use by the *BLAST* (11) and *FASTA* (12, 13) programs, so you will need to use a different program, e.g. *findpatterns –mismatches = 1* in the GCG package.

References

1. Bishop, M. J. and Rawling, C. J. (1996). *DNA and protein sequence analysis*. IRL Press, Oxford.
2. Schuler, G. D., Epstein, J. A., Ohkawa, H., and Kans, J. A. (1996). In *Methods in enzymology* (ed. R. F. Doolittle), Vol. 266, p 141. Academic Press, London.
3. Cottingham, Jr. R. W., Idury, R. M., and Schaffer, A. A. (1993). *Am. J. Hum. Genet.*, **53**, 252.
4. O'Connell, J. R. and Weeks, D. E. (1995). *Nature Genet.*, **11**, 402.
5. Lincoln, S., Daly, M., and Lander, E.(1992). *Constructing genetic maps with MAPMAKER/EXP*, 3rd edn. Whitehead Institute Technical Report.
6. Genetics Computer Group, Inc. (1994). *The Wisconsin sequence analysis package, version 8.0*. University Research Park, 575 Science Drive, Madison, Wisconsin, 53711, USA.
7. Rice, P., Lopez, R., Doelz, R, and Leunissen, J. (1995). Embnet.news 2:5-7. (EGCG home page: http://www.sanger.ac.uk/~pmr/egcg.html)
8. Staden, R. (1994). In Methods in molecular biology (ed. A. M. Griffin and H. G. Griffin), Vol. 25, pp. 9-170. Humana Press Inc., Totawa, NJ.
9. Primer was written by Steve Lincoln, Mark Daly, and Eric Lander at the MIT Center for Genome Research.
10. Hillier, L. and Green, P. (1991). *PCR Methods Appl.*, **1**, 124.
11. Altschul, S. F., Warren, G., Webb, M., Myers, E. W., and Lipman, D.(1990). *J. Mol. Biol.*, **215**, 403.
12. Pearson, W. R. and Lipman, D. J. (1988). *Proc. Natl. Acad. Sci. USA*, **85**, 2444.
13. Pearson, W. R. (1990). In Methods in Enzymology (ed. R. F. Doolittle), Vol. 183. Academic Press, London.

List of suppliers

Aldrich

Aldrich Chemical Co., Inc, 1001 W. St. Paul Avenue, Milwaukee, WI 53233, USA.

Sigma-Aldrich Ltd., The Old Brickyard, New Road, Gillingham, Dorset SP8 4XT, UK.

American Type Culture Collection (ATCC), 12301 Parklawn Drive, Rockville, MD 20852, USA.

Amersham

Amersham Corporation, 2636 South Clearbrook Drive, Arlington Heights, IL 60005, USA.

Amersham International plc, PO Box 1139, Slough SL1 6PL, UK.

Applied Biosystems (ABI)

Applied Biosystems, Inc. (ABI), 9108 Guildford Road, Columbia, MD 21046, USA.

Perkin Elmer (Applied Biosystems division), Kelvin Close, Birchwood Science Park North, Warrington, Cheshire WA3 7PB, UK.

Appligene Oncor, Pinetree Centre, Durham Road, Birtley, Chester-le-Street, County Durham DH3 2TD, UK.

Beckman Instruments

Beckman Instruments UK Ltd., Oakley Court, Kingsmead Business Park, London Road, High Wycombe, Bucks HP11 1J4, UK.

Beckman Instruments Inc., PO Box 3100, 2500 Harbor Boulevard, Fullerton, CA 92634, USA.

Becton Dickinson

Becton Dickinson and Co., Between Towns Road, Cowley, Oxford OX4 3LY, UK.

Becton Dickinson and Co., 2 Bridgewater Lane, Lincoln Park, NJ 07035, USA.

Bellco Glass Inc., 340 Edrudo Rd., PO Box B, Vineland, NJ 08360, USA.

Bibby Sterilin Ltd., Tilling Drive, Stone, Staffordshire ST15 0SA, UK.

Bio-Rad Laboratories

Bio-Rad Laboratories Ltd., Bio-Rad House, Maylands Avenue, Hemel Hempstead HP2 7TD, UK.

Bio-Rad Laboratories, Division Headquarters, 3300 Regatta Boulevard, Richmond, CA 94804, USA.

BIOS Laboratories Inc., 5 Science Park, New Haven, CT 06437, USA.

Biozyme Laboratories

Biozyme Laboratories International Ltd., 9939 Hibert St., Ste.101, San Diego, CA 92131, USA.

Biozyme Laboratories Ltd., Unit 6, Gilchrist-Thomas Estate, Blaenarvon, Gwent NP4 9RL, UK.

Boehringer Mannheim

Boehringer Mannheim UK (Diagnostics and Biochemicals) Ltd, Bell Lane, Lewes, East Sussex BN17 1LG, UK.

Boehringer Mannheim Corporation, Biochemical Products, 9115 Hague Road, P.O. Box 504 Indianapolis, IN 46250–0414, USA.

Boehringer Mannheim Biochemica, GmbH, Sandhofer Str. 116, Postfach 310120 D-6800 Ma 31, Germany.

British Drug Houses (BDH) Ltd.: *see* Merck.

BRL: *see* Gibco BRL.

Calbiochem-Novabiochem

Calbiochem-Novabiochem Corporation, PO Box 12087, La Jolla, CA 92039-2087, USA.

Calbiochem-Novabiochem UK Ltd., Boulevard Industrial Park, Padge Road, Beeston, Nottingham NG9 2JR, UK.

Carl Zeiss Oberkochen Ltd., PO Box 78, Woodfield Road, Welwyn Garden City, Hertfordshire AL7 1LU, UK.

Cherwell Scientific Publishing Ltd., The Magdalen Centre, Oxford Scientific Park, Oxford OX4 4GA, UK.

Corning Costar Corp., One Alewife Center, Cambridge, MA 02140, USA.
  UK distributor: Bibby Sterilin.

Coulter

Coulter Electronics Ltd., Northwell Drive, Luton, Bedfordshire, LU3 3RH, UK.

Coulter Corporation (Cytometry Div.), 440 Coulter Way, PO Box 169015, Miami, FL 33116, USA.

Cytomation Inc., 400 E. Horsetooth Road, Fort Collins, CO 80525, USA.

Decon Laboratories Ltd., Conway Street, Hove, East Sussex BN3 3LY, UK.

Denley Instruments, Inc., PO Box 13958, Research Triangle Park, NC 27709-3958, USA.
  UK distributor: Life Sciences Intl.

Difco Laboratories

Difco Laboratories Ltd., P.O. Box 14B, Central Avenue, West Molesey, Surrey KT8 2SE, UK.

Difco Laboratories, P.O. Box 331058, Detroit, MI 48232–7058, USA.

Dow Corning

Dow Corning Corp., PO Box 0994, Midland, MI 48686-0994, USA.

Dow Corning Ltd., Cardiff Road, Barry, South Glamorgan, CF63 2YL, UK.

Dow Corning GmbH, Pelkoven Strasse 152, Werk Munchen, D 80992 Munchen, Germany.

Drummond Scientific Co., 500 Parkway, Broomall, PA 19008, USA.
  UK distributor: Lazer Laboratories.
Dynal
Dynal Inc., 5 Delaware Dr., Lake Success, NY 11042, USA.
Dynal Ltd., Station House, 26 Grove St., New Ferry, Wirral L62 5AZ, UK.
Dynatech
Dynatech Laboratories, Inc., 14340 Sullyfield Cir., Chantilly, VA 22021, USA.
Dynatech Ltd., Daux Road, Billinghurst, Sussex RH14
Eurogentech Belgium, Parc Scientifique de la Cense Rouge, rue Bois Saint Jean, 14, B-4102 Seraing, Belgium.
European Collection of Animal Cell Cultures (ECACC), Centre for Applied Microbiology and Research, Salisbury, Wiltshire SP4 0JG, UK.
Fisher Scientific Co., 711 Forbest Avenue, Pittsburgh, PA 15219–4785, USA.
Flowgen Instruments Ltd., Lynn Lane, Shenstone, Staffordshire WS14 0EE, UK.
Flow Laboratories, Woodcock Hill, Harefield Road, Rickmansworth, Herts. WD3 1PQ, UK.
Fluka
Fluka-Chemie AG, CH-9470, Buchs, Switzerland.
Fluka Chemicals Ltd., The Old Brickyard, New Road, Gillingham, Dorset SP8 4JL, UK.
FMC BioProducts
FMC BioProducts, 191 Thomaston St., Rockland, ME 04841, USA.
FMC BioProducts Europe, Risingevej 1, DK-2665 Vallensbaek Strand, Denmark.
Flowgen Instruments Ltd., Lynn Lane, Shenstone, Staffordshire WS14 0EE, UK.
Genetic Research Instrumentation Ltd., Gene House, Dunmow Road, Felsted, Dunmow, Essex CM6 3LD, UK.
Gibco BRL
Gibco BRL (Life Technologies Ltd.), Trident House, Renfrew Road, Paisley PA3 4EF, UK.
Gibco BRL (Life Technologies Inc.), 3175 Staler Road, Grand Island, NY 14072–0068, USA.
Gravatom Ltd., Claylands Road, Bishops Waltham, Hampshire SO32 1BH, UK.
Greiner Labortechnik Ltd., Station Road, Cam, Dursley, Gloucestershire GL11 5NS, UK.
Hamilton
Hamilton Co., 4970 Energy Way, Reno, NV 89502, USA.
Hamilton G.B. Ltd, Kimpton Link Business Centre, Kimpton Rd., Sutton, Surrey SM3 9QP, UK.
H. T. Biotechnology Ltd., Unit 4, 61 Ditton Walk, Cambridge CB5 8QD, UK.

Hybaid
Hybaid Ltd., 111–113 Waldegrave Road, Teddington, Middlesex TW11 8LL, UK.
Hybaid, National Labnet Corporation, P.O. Box 841, Woodbridge, NJ. 07095, USA.
ICN
ICN Biomedicals Ltd., Unit 18, Thame Park Business Centre, Wenman Rd., Thame, Oxfordshire OX9 3XA, UK.
ICN Pharmaceuticals, Inc., 3300 Hyland Avenue, Costa Mesa, CA 92626, USA.
Invitrogen Corporation
Invitrogen Corporation 3985 B Sorrenton Valley Building, San Diego, CA. 92121, USA.
Invitrogen Corporation c/o British Biotechnology Products Ltd., 4–10 The Quadrant, Barton Lane, Abingdon, OX14 3YS, UK.
Lazer Laboratories Ltd., PO Box 166, Sarisbury Green, Southampton SO3 6YZ, UK.
Life Sciences Intl., Unit 5, The Ringway Centre, Edison Road, Basingstoke, Hampshire RG21 2YH, UK.
Life Technologies Inc., 8451 Helgerman Court, Gaithersburg, MN 20877, USA.
Merck
Merck Industries Inc., 5 Skyline Drive, Nawthorne, NY 10532, USA.
Merck, Frankfurter Strasse, 250, Postfach 4119, D-64293, Germany.
Merck Ltd., Merck House, Poole, Dorset BH15 1TD, UK.
Millipore
Millipore (UK) Ltd., The Boulevard, Blackmoor Lane, Watford, Herts WD1 8YW, UK.
Millipore Corp./Biosearch, P.O. Box 255, 80 Ashby Road, Bedford, MA 01730, USA.
Nalge Nunc Intl., 2000 N. Aurora Rd., Naperville, IL 60563-1796, USA.
Narishige USA Inc., 404 Glen Cove Ave., Sea Cliff, NY 11579, USA.
NBL Gene Sciences *see* Northumbria Biologicals Ltd.
New Brunswick Scientific
New Brunswick Scientific Co. Ltd., 44 Talmadge Rd., PO Box 4005, Edison, NJ 08818-4005, USA.
New Brunswick Scientific Ltd., 163 Dixons Hill Road, North Mymms, Hatfield, Herts AL9 7JE, UK.
New England Biolabs
New England Biolabs Inc., 32 Tozer Rd., Beverly, MA 01915-5599, USA.
New England Biolabs (UK) Ltd., 67 Knowl Piece, Wilbury Way, Hitchin, Hertfordshire SG4 0TY, UK.
Northumbria Biologicals Ltd., South Nelson Industrial Estate, Cramlington, Northumberland NE23 9HL, UK.

Nunc: *see* Nalge Nunc Intl.
   UK distributors: Life Technologies Ltd.
Oncor Inc., 209 Perry Parkway, Gaithersberg, MD 20877, USA.
   See also Appligene Oncor
Orbis Labortechnik, Altenkesseler Strasse 17, D-66115, Sarrbrücken, Germany.
Perkin Elmer (Applied Biosystems division), Kelvin Close, Birchwood Science Park North, Warrington, Cheshire WA3 7PB, UK.
Pharmacia Biotech Europe Procordia EuroCentre, Rue de la Fuse-e 62, B-1130 Brussels, Belgium.
Pharmacia Biosystems
Pharmacia Biosystems Ltd. (Biotechnology Division), Davy Avenue, Knowlhill, Milton Keynes MK5 8PH, UK.
Pharmacia LKB Biotechnology AB, Björngatan 30, S-75182 Uppsala, Sweden.
Photometrics
Photometrics Ltd., 2440 E. Britannia Drive, Tucson, AZ 85706, USA.
Photometrics Gmbh, Sollner Strasse 61, D-81479 Munich, Germany.
Promega
Promega Ltd., Delta House, Enterprise Road, Chilworth Research Centre, Southampton, UK.
Promega Corporation, 2800 Woods Hollow Road, Madison, WI 53711–5399, USA.
Qiagen
Qiagen Inc., c/o Hybaid, 111–113 Waldegrave Road, Teddington, Middlesex, TW11 8LL, UK.
Qiagen Inc., 9259 Eton Avenue, Chatsworth, CA 91311, USA.
Research Genetics Inc., 2130 Memorial Parkway, SW, Huntsville, AL 35801, USA.
Riedal de Haen
Riedel de Haen, Wunstorfer Strasse 40, D-30926, Seeize, Germany.
Riedel de Haen (UK) Ltd., Lynn Lane, Shenstone, Lichfield, Staffs WS14 0EE, UK.
Robbins Scientific
Robbins Scientific Corp., 814 San Alese Ave., Sunnyvale, CA 94086-1411, USA.
Robbins Scientific Ltd., Suite B, Greville Court, 1665 High Street, Knowle, Solihull, West Midlands B93 0LL, UK.
Schleicher and Schuell
Schleicher and Schuell Inc., Keene, NH 03431A, USA.
Schleicher and Schuell Inc., D-3354 Dassel, Germany.
Scientific Laboratory Supplies, Unit 27, Nottingham South and Wilford Industrial Estate, Ruddington Lane, Wilford, Nottingham NG11 7EP, UK.
Scientific Imaging Systems, 36 Clifton Road, Cambridge CB1 4ZR, UK.

Scotlab

Scotlab Inc., 30 Controls Dr., Shelton, CT 06484, USA.

Scotlab Ltd., Kirkshaws Road, Coatbridge, Strathclyde ML5 8AD, UK.

J. L. Shepherd & Associates, Inc., 1010 Arroyo Ave., San Fernando, CA 91340, USA.

Sigma Chemical Company, PO Box 14508, St. Louis, MO 63178 USA.

Sorvall DuPont Company, Biotechnology Division, P.O. Box 80022, Wilmington, DE 19880–0022, USA.

Stratagene

Stratagene Ltd., Unit 140, Cambridge Innovation Centre, Milton Road, Cambridge CB4 4FG, UK.

Strategene Inc., 11011 North Torrey Pines Road, La Jolla, CA 92037, USA.

Tropix, Inc., 47 Wiggins Ave., Bedford, MA 01730, USA.

U.K. HGMP Resource Centre, Hinxton, Cambridge CB10 1SB, UK.

United States Biochemical, P.O. Box 22400, Cleveland, OH 44122, USA.

Vector Laboratories

Vector Laboratories, 16 Wulfric Square, Bretton, Peterborough PE3 8RF, UK.

Vector Laboratories, 30 Ingold Road, Burlingame, CA 94010, USA.

Vysis Inc., 3100 Woodcreek Drive, Downers Grove, IL 60515, USA.

Whatman

Whatman Inc., 9 Bridewell Place, Clifton, NJ 07014, USA.

Whatman Intl., St Leonards Road, Maidstone, Kent ME16 0LS, UK.

Zeiss: *see* Carl Zeiss.

Index

AAT selection 131
ACeDB 247, 248, 249, 251
Achilles' heel cleavage of DNA 309
adenine methylase 309
adenine phosphoribosyl transferase 131
African green monkey
  cell lines 320
  cosmid libraries 320
agarose-embedded DNA
  see also DNA isolation
  end-labelling of 307–308
  restriction digestion of 291–3
agarose microbeads 282, 287–8, 296
agarose strings 97–9
agricultural genomes
  see also entries by species name
  databases 342, 344
alkaline lysis 203, 233–4, 242
Alu banding 207
Alu elements
  association with microsatellites 6
  in 5′ and 3′ SINE-PCR 33, 34
Alu-PCR
  see also
    IRS-PCR
    repeat element-mediated PCR
    SINE-PCR
  of cloned DNA 112–13, 243–6
  fingerprinting 246–7
  of flow-sorted chromosomes 112–13, 180–1
  of human genomic DNA 112–13
  of human/rodent hybrids 112–13, 142,
    153–4
  for isolation of microsatellite motifs 7
  for pre-amplification of HAPPY mapping
    panels 111–12, 114–16
  as source of chromosome paints 142–3
  as source of inter-*Alu* sequence-tagged sites
    112–14
  walking 243–6
  of YAC clones 203–4, 246–7, 273–7
Alu-PCR walking 243–6
American Type Culture Collection 7, 128,
  157, 216, 316, 318, 320, 325, 326, 328
amplified fragment length polymorphisms
  36–7, 58
anchor loci 41
animal genomes
  see also non-human genomes *and entries by*
    species name
  comparative mapping 27, 28, 39, 42–43

synteny between 42–3, 44, 341, 343–4
anti-avidin, biotinylated 209, 210
anti-avidin, FITC-conjugated 209, 210
anti-avidin, fluorescein-conjugated 210
APRT *see* adenine phosphoribosyl
  transferase
Arabidopsis thaliana
  BAC library 327
  cDNA libraries 327
  cell cultures 326
  chromosome number 29
  databases 344
  expressed sequence tag library 327
  gene number 29
  genome size 29
  genomic libraries 327
  microsatellite markers 30
  mutant lines 327
  P1 library 327
  physical mapping 28, 39
  recombinant inbred lines 326
  stocks 326–7
  YAC libraries 327
ATCC *see* American Type Culture Collection
Australian National Genomic Information
  Servis (ANGIS) 335
autoradiographs, scanning of 247–8
auxotrophic selection 132
avidin-FITC 209–10
avidin Texas Red 211
5-azacytidine 310
5-aza-2′deoxycytidine 310

backups, of computer data 331
Bacillus subtilis
  databases 344
  sequence 28
bacterial artificial chromosomes (BACs)
  contig assembly using 232
  coverage of genome 228
  DNA isolation 233–4
  end-sequence isolation
    by hybridization 267–70
    by vectorette PCR 256–60
    by vector religation 270–3
  filters for hybridization 239–40
  as FISH probes 203, 206
  libraries
    Arabidopsis 327

bacterial artificial chromosomes (*cont.*)
   human 317
   pBAC108L 258
   pBELOBAC 258
   screening by hybridization 239–40
bacterial genomes
   see also entries by species name
   restriction cleavage 289–90
bacteriophage libraries
   from flow-sorted chromosomes 177–9
   human chromosome-specific 318
   Saccharomyces cerevisiae 327
barley 39
beans, genome database 344
bee
   chromosome number 29
   genome size 29
   RAPD mapping 31
bioinformatics
   see also
     World Wide Web
     Internet
     databases
   centres 332–5
   definition 329
   hardware requirements 330–1
   search facilities 332
   software requirements 330–1
biotinylated PCR primers 13
biotinylation of DOP-PCR products 194–5
bivariate flow karyotype 165–166
BLAST 347, 348, 349
Bluescript-mediated isolation of cosmid ends
   265–7
Bluescript SK⁺ 265, 266
bromodeoxyuridine (BrdU) 130, 200, 212

'C' programming language 331
Caenorhabditis elegans
   chromosome number 29
   cosmid transgenics 325
   databases 343
   G/C content 229
   gene number 29
   genetic stocks 325
   genome size 29
   physical map 227
   sequence analysis 227
   stocks 325
Candida albicans database 344
canonical clones 249
cattle
   backcross 40
   chromosome number 29
   databases 342
   genetic map 40

genome size 29
   mapping QTLs 63
   synteny with human 43
cDNA clones
   as FISH probes 203
   libraries
     Arabidopsis 327
     human 318–19
     IMAGE consortium 319
     mouse 322–3
     rice 325
     Saccharomyces cerevisiae 328
     zebrafish 323
cell lines
   Arabidopsis 326
   human 316 *see also* somatic cell hybrids
   mouse 321
   rice 325
centiMorgan 1
centiRay 90
CEPH 16, 17, 316, 317, 336, 339
chicken
   chromosome number 29
   databases 342
   genome size 29
chimerism, in YACs 74, 199, 213, 255, 316
   checking, with FISH 199
Chinese hamster cells 74, 77, 129, 132, 140,
   146, 321
CHLC *see* Co-operative Human Linkage
   Center
chromomycin A3: 165, 166, 167, 168, 169
chromosome banding 139, 141, 189–90, 207,
   212
chromosome, human
   1: 1, 157, 207, 212, 339
   2: 157, 339
   3: 157, 339
   4: 157, 339
   5: 34, 157, 158
   6: 157, 215, 221, 230
   7: 157, 194
   8: 157, 339
   9: 157, 166, 179, 212, 221, 339
   10: 43, 157, 166, 339
   11: 153, 157, 166, 192, 339
   12: 157, 166, 339
   13: 157, 230, 339
   14: 105, 157
   15: 19, 157
   16: 157, 179, 212, 339
   17: 157, 339
   18: 157, 339
   19: 24, 157, 207, 339
   20: 157, 339
   21: 157, 174, 179, 339
   22: 157, 172, 214, 228, 231, 232, 247, 250,
    253, 339

X 130, 157, 172, 214, 215, 221, 228, 247, 290, 339
Y 157, 212, 229
chromosome packing 221
chromosome-specific databases 339
chromosome-specific libraries
  bacteriophage 318
  cosmid 169–77, 318
chromosome walking *see also* contig assembly
  end-sequence isolation
    Bluescript-mediated 267
    by hybridization 267–70
    by vectorette PCR 256–64
    by vector religation 270–3
  for gap closure in contigs 239, 241
  hybridization of probes to clone filters 277–9
  principle 255
  with YAC clones 274–7
clone fingerprinting
  see
    restriction fingerprinting
    Alu-PCR fingerprinting
    contig assembly
clone libraries
  see also :
    bacterial artificial chromosomes
    bacteriophage
    cDNA libraries
    chromosome-specific libraries
    cosmid clones
    CpG islands
    P1 artificial chromosomes
    yeast artificial chromosomes,
    and entries by name of species cloned
  chromosome-specific 7
  from flow-sorted chromosomes 169–79, 182
  from microdissected chromosomes 196–7
  screening for microsatellites 8–10
colcemid 126, 136, 137, 140, 148, 167, 168, 200
comparative mapping 42
Contig9 248–52
contig assembly
  see also
    Alu-PCR fingerprinting
    chromosome walking
    restriction fingerprinting
  of bacterial clones overlying YAC contig 240–1, 252
  complications 252
  coverage 230
  gaps
    causes of 230,241
    closure of 241–6, 255
  integration of different clone types 227, 229–30
  on 'scaffold' of mapped markers 232, 238–40, 250–2

software 247–52
strategy, choice of 231–2
Co-operative Human Linkage Center 320, 335, 336
cosmid clones
  alkaline lyis 233–4
  coverage of genome 228
  DNA isolation 233–4
  end-sequence isolation
    Bluescript-mediated 265–7
    by hybridzaition 267–70
    by vectorette PCR 256–60
    by vector religation 270–3
  filters for hybridization 239–40
  as FISH probes 203, 206, 207, 213, 222
  from flow-sorted chromosomes 169–77
  host bacterial strain 177
  as hybridization probes 242–43
  labelling, for use as probe 242–43
  Lawrist 170, 267
  libraries
    African green monkey 320
    Drosophila 325
    Formosa monkey 320
    Fugu 323
    gibbon 320
    gorilla 320
    human chromosome-specific 169–77, 318
    orangutan 320
    Saccharomyces cerevisiae 327
    Schizosaccharomyces pombe 328
    zebrafish 323
  ligation and packaging 176–7
  Optical Mapping 311
  pWE 265, 267
  restriction fingerprinting of 232–8
  screening by hybridization 239–40, 241–3
  stability of inserts 170
  'Supercos' vector 265, 267
  vectors 170, 265, 267
    preparation of 169–72
cosmid transgenics 325
CpG dinucleotides 290
CpG islands
  clone libraries, human 320
  distribution in mammalian genomes 290
  restriction cleavage within 291
Cyrillic 3
cytidine methylation 290–1 *see also* DNA methyltransferases
cytochalasin B 126, 136, 137

dam methylase 173, 174, 175, 176
databases, genomic
  see also bioinformatics
  agricultural genomes 342, 344

databases, genomic (*cont.*)
  Arabidopsis 344
  Bacillus subtilis 344
  bean 344
  Caenorhabditis elegans 343
  Candida albicans 344
  cattle 342
  chicken 342
  comparative 343–4, 345
  DNA sequence 346–9
  dog 342
  Drosophila 343
  Escherichia coli 344
  expressed sequence tag (EST) 346, 347
  Fugu 342
  fungi 344
  horse 342
  human
    genome-wide 335–8
    chromosome-specific 338, 339
    mitochondrial 338, 339
  insects 343
  invertebrates 343
  microbial 344
  mitochondrial 3 38, 339
  mosquito 343
  mouse 338, 340–2
  mutation 344, 345
  mycobacteria 344
  PCR primer 345–6
  pig 342
  plants 344
  probe 346
  protein 346–9
  rat 341
  rice 344
  rodent 338–42
  Saccharomyces cerevisiae 344
  Schizosaccharomyces pombe 344
  sequence 346–9
  sheep 342
  trees 344
  yeast 344
  zebrafish 342
degenerate oligonucleotide primed PCR
    (DOP-PCR)
  see also whole-genome PCR
  biotin labelling of products 194–5
  cloning of products 182
  of flow-sorted chromosomes 181–2
  of microdissected chromosomes 191–3
DGMap 121, 122
DIG labelling 298–300
di-haploids, in QTL mapping 64
dinucleotide repeats
  see microsatellites
distance geometry 121–122
DNA/GUI 86

DNA haloes 222
DNA isolation
  bacterial clone 233–4
  genomic
    in agarose blocks 282–7
    in agarose microbeads 282, 287–8
    in agarose strings 97–9
    bacterial 282, 287
    mammalian 97–9, 282–4
    plant 282, 285–6
    yeast 287, 288
DNA methyltransferases 173, 308–9
DNaseI, titration of activity 204
dog
  chromosome number 29
  databases 342
  flow sorting of chromosomes 166
  genome size 29
Drosophila
  chromosome number 29
  cosmid clones, mapped 325
  cosmid library 325
  databases 343
  genome size 28
  P-element lethal lines 324
  P1 library 324
  sequencing 227
  stocks 323–4
  YAC library 324

email 331
EMBnet
  national nodes 333, 334
  specialist nodes 334
Escherichia coli
  chromosome number 29
  databases 344
  gene number 29
  genome size 29
  sequence 228
EST *see* expressed sequence tag
EUCIB mouse backcross 40, 322, 341
expressed sequence tags
  see also cDNA
  Arabidopsis 327
  database entries 346
  as markers in radiation-hybrid maps 84, 92
  UniGene 347

FASTA 347, 348, 349
Fastlink 350
fibre-FISH *see* fluorescence *in situ*
    hybridization
file transfer protocol 330

fingerprinting
  see Alu-PCR fingerprinting
    contig assembly
    LINE amplification fingerprinting
    restriction fingerprinting
flow karyotype 165–6
flow-sorted chromosomes *see also* flow sorting
  amplification by DOP-PCR 181–2
  amplification by IRS-PCR 180–1
  amplification by linker adaptor PCR 181
  cloning in bacteriophage 177–9
  cloning in cosmid vectors 169–78
  cloning of DOP-PCR amplification products
    182
  cloning in yeast artificial chromosomes 179
  microsatellites markers derived from 7
  partial digestion of 173–6
  purity of 166, 172
  purity testing 172–3
  restriction digestion of 173–6
  screening of, for STSs 179–80
flow sorting
  of aberrant chromosomes 179
  chromosome preparation 167–8
  chromosome sources 166–7
  of human chromosomes 166
  instrumentation 169
  of mammalian chromosomes 166
  principle 165–6
  purity attainable 166, 172
  of somatic cell hybrid chromosomes 166–7
fluorescence *in situ* hybridization (FISH)
  to aberrant chromosomes 215
  ageing of metaphase spreads 201
  applications 199
  background signal 213
    in interphase FISH 217
    from YAC probes 203
  band assignment of signal 213
  chromosome banding 212
  chromosomes, sources of 200
  on DNA fibres ("fibre FISH") 222–3
    preparation of fibres 222–3
    variability 223
  using DOP-PCR products 195–6
  image registration 223
  instrumentation 223
  interphase FISH
    distance measurements 219–20
    distance range 218
    effect of background signal 217
    effect of chromosome packaging 221
    hybridization 217
    interference between probes 221
    interpretation of results 218, 221
    molecular vs. measured distances 215,
      220–1
    order analysis 218

  preparation of nuclei 215–16
  preparation of slides 216
    probes for 216–17
  interpretation 213
  interspecific hybridization ('zoo-FISH') 41
  metaphase spreads, preparation 201–2
  using microdissected, amplified material
    193–6
  microscopy 223
  multi-colour 211, 216–18, 223
  ordering of multiple probes 213–15
  probes
    choice of 203
    hybridization 207–9, 217
    for interphase FISH 216
    labelling 204–6
    from microdissected chromosomes 193–6
    preparation of 203
    size, effect on hybridization 203
    small 206–7
    sources of 203–4
    denaturation 207–8
  resolution 199, 222
  signal detection
    multiple colour 211
    single colour 209–11
  signal displacement from origin 213
  signal, secondary 213
  of somatic cell hybrids 142
  spot-counting 213
  'zoo-FISH' 41
Formosa monkey *see Macaca cyclopisis*
fosmids 170, 178, 227, 229, 241
  purification of clone DNA 233–4
FPC 248–52
FTP 330
Fugu
  chromosome number 29
  cosmid libraries 323
  database 342
  genome size 29
fungi
  see also
    yeast
    Saccharomyces cerevisiae
    Schizosaccharomyces pombe
    Candida albicans
  databases 344

G3 panel (radiation hybrids) 83
G–11 (Giemsa, pH11) staining 139, 142
gap closure *see* contig assembly, chromosome
    walking
G-banding 139, 142, 189–90
GDB *see* Genome Database
Genebridge4 panel (radiation hybrids) 83

genetecin 130, 131
genetic map *see* linkage map
Genome Database (GDB) 335, 336, 337
GenomeNet 335
genome sizes 29
genome size, variation in 28–29
genomic DNA *see* DNA isolation
genomic libraries *see entries by vector type and
by species*
genotyping
see
HAPPY mapping
microsatellites
radiation hybrids
randomly amplified polymorphic DNAs
5′ and 3′ SINE-PCR
Genstat 68
gibbon, genomic clone libraries 320
Giemsa staining 141, 189–90
gorilla, genomic clone libraries 320
guanine phosphoribosyl transferase 130

Haemophilus influenze
chromosome number 29
gene number 29
genome size 29
sequence 28
HAPPY mapping
data analysis 119–22
data entry 119
DNA fragment size 97
mapping function 121–2
markers 109–10, 112
panels
DNA preparation 97–9
of irradiated DNA 104–8
pre-amplification
using *Alu*-PCR 111, 114–15
using nested PCR 109–10
using whole genome PCR 118
quantifying DNA content 100–3, 108
screening
with inter-*Alu* sequences 116–7
using nested PCR 111
of sheared DNA 99–100
principle 95–6
range 97
resolution 97
HAT selection 130–1
heritability 49
heterochromatin 212
heterogeneous nuclear RNA, isolation from
somatic cell hybrids 154
heterokaryon 125
heterozygosity
definition 1–2
evaluating 15–16

useful range 2, 16
HGMP *see* UK Human Genome Mapping
Project
hnRNA *see* heterogeneous nuclear RNA
Hoescht 33258: 166
horse database 342
HPRT *see* hypoxanthine phosphoribosyl
transferase
HTML *see* Hypertext Markup Language
Human Genome Mapping Project *see* UK
Human Genome Mapping Project
hybridization *see also*
Southern blotting
fluorescence *in situ* hybridization
to bacterial clone arrays 239–40, 241–3,
277–9
of clone end-sequences to clone arrays
277–9
cosmid-to-cosmid 241–3
cosmid-to-PAC 241–3
of PAC-derived probes 243–6
probes
databases 345–6
human 320
maize 326
microsatellite 9–10
vector-specific 267–70
wheat 326
of vector-specific oligonucleotides 267–70
hygromycin 130, 131
Hypertext Markup Language 330
hypoxanthine phosphoribosyl transferase 130

IMAGE consortium 319
Image 247–8
insect genome databases 343
Internet 329–31*see also* bioinformatics, World
Wide Web
interphase FISH *see* fluorescence *in situ*
hybridization
intron-encoded nucleases 289, 290
irradiation
of cells 78, 81–2
of DNA 104, 106
IRS-PCR
see also
repeat element-mediated PCR
Alu-PCR
LINE amplification
of flow-sorted chromosomes 180–1
of HAPPY mapping panels 111–2
isozyme analysis 147

karyotyping
of human-rodent hybrids 139–42
by PCR 143

Lawrist 170, 267, 268
libraries *see* clone libraries
LINE amplification fingerprinting 143–5
LINKAGE 3, 24
linkage maps
   see also
      microsatellites
      quantitative trait loci
   alignment with physical maps 24–5
   analysis of results 23–4
   comprehensive 7
   framework 7
   integration of new markers 23–4
   lod scores 23–4
   planning studies 2–3
   pedigrees, drawing 3
   of rice 30
   software 3, 24, 350
linker adaptor PCR 181
lipopolysaccharide 167
lod score 23–4, 87–90, 120
lymphoblastoid cells 167, 200
lysolethicin 133

Macaca cyclopisis, genomic clone libraries 320
maize
   chromosome number 29
   databases 344
   genome size 29
   hybridization probes 326
mammalian anchor loci 41
Mapmaker/QTL 68, 350
Mapmaker/SIBS 68–9, 350
Mapper 90
Mapping function 121–2
MapQTL 69
marker-assisted selection 50
metaphase chromosome spreads
   ageing of 201
   preparation
      for fluorescence *in situ* hybridization
         201–2
      from hybrid cells 140–1
      for microdissection 188–90
microbeads 282, 287–8, 296
microbial genomes, databases 344
microcell hybrids *see also* somatic cell hybrids
   principle of formation 125–6
   production of 136–8
   retention of donor chromosomes 125, 148
   stability in culture 148
microcell-mediated chromosome transfer
   125–6, 136–8
microdissected chromosomes
   amplification using DOP-PCR 191–3
   cloning of fragments

      directly 196
      following DOP-PCR 196–7
   DNA content per fragment 185, 191
   FISH probes from 193–6
   microsatellite markers from 7
microdissection
   equipment 186–8
   metaphase spreads for 188–90
   of plant chromosomes 185
   procedure 190–1
   resolution 185
microneedles 187
micronuclei 136, 137
micropipettes 187
'microprep' of bacterial clone DNA 233–4
microsatellites
   allele frequencies, population bias 15–16
   in *Arabidopsis* 30
   assocoation with *Alu* elements 6
   classes of 4
   compound 4
   conservation between species 41
   description 4
   dinucleotide repeats 4
   evaluation of polymorphism 15–16
   frequency and distribution 4–6, 30
   genotyping of
      analysis of results 23
      continuity of allele assignment 16–17,
         22–3
      using end-labelled primers 20–1
      errors 23
      using fluorescently-labelled primers 21–3
      multiplex 14, 20, 22
      optimization of PCR conditions 14–15,
         20
      using radioactive labelling 17–21
      scoring results 23
      'shadow' bands 5, 6, 20
      software for 22
   imperfect 4
   linkage disequilibrium 5
   as markers for mapping quantitative trait
      loci 57
   origins of 5
   PCR primers for
      design of 13–14
      end-labelled 20–1
      fluorescently labelled 14
      human 320
      pig 321
   in plant genomes 30
   polymorphism of 4, 6, 15–16
   population bias of allele frequencies 15–16
   in rat 341
   region-specific sources of 7–8
   repeat number, effect on polymorphism
      4, 6

microsatellites (*cont.*)
  screening clone libraries for 8–10
  sequencing of flanking regions 13
  'shadow' bands when genotyping 5, 6, 20
  in sheep 41, 321
  tetranucleotide repeats 5–6
  transfer between species 41
  trinucleotide repeats 5–6
millet 43
mitochondrial genome databases 338, 339
Morgan *see* centiMorgan
Mosaic 330, 331
mosquito database 343
mouse
  backcrosses 40, 322, 341
  cDNA libraries 322–3
  cell lines 321
  chromosome number 29
  databases 338, 340–1
  DNA stocks 322
  embryos 321
  EUCIB backcross 40
  flow sorting of chromosomes 166
  genetic map 40
  genome databases 338, 340–1
  genome size 29
  homology with human 341
  live animal stocks 321
  Mouse Genome Database 338, 340
  Mouse Locus Catalogue 340
  P1 library 322
  somatic cell hybrids 321–2
  stocks, live animals and embryos 321
  YAC libraries 322
MQTL 69
Multimap 24
mutation databases 344, 345
mycobacteria database 344
mycophenolic acid 130
Mycoplasma genitalium
  genome size 29
  chromosome number 29
  gene number 29
  sequence 28

nematode *see Caenorhabditis elegans*
neomycin resistance 130
Netscape 330, 331
newsgroups 331
nick translation 204–6
non-human genomes
  see also comparative mapping *and entries by*
    species name
  resources for 39
  directed breeding programs 40
  transfer of markers between species 40–2
nucleases, intron-encoded 289, 290

oats
  genomic libraries 326
  recombinant inbred lines 326
oligonucleotides *see also* polymerase chain
    reaction
  as hybridization probes 268–70
  labelling, radioactive 268–70
  T_m of 10, 14, 147
Online Mendialian Inheritance in Man
    (OMIM) 336, 337–8
optical disc drives 331
Optical Mapping 311
orangutan, genomic clone libraries 320
ouabain selection 132
'Oxford' grid 43

P1 artificial chromosomes
  Alu-PCR of 243–6
  contig assembly using 232
  coverage of genome 228
  DNA, purification of 233–4
  end-sequence isolation
    by hybridization 267–70
    by vectorette PCR 256–60
    by vector religation 270–3
  filters, for hybridization 239–40
  as FISH probes 203, 206
  libraries
    Arabidopsis 327
    Drosophila 324
    human 317–18
    mouse 322
    Schizosaccharomyces pombe 328
  pCYPAC 258, 271, 317
  probes derived from 243–6
  purification of clone DNA 233–4
  screening by hybridization 239–40
palindromic sequences 290
pedigree analysis *see* linkage maps
PEG *see* polyethylene glycol
permutation analysis 59
pBAC108L 258
pBELOBAC 258
pCDM8 319
PCR *see* polymerase chain reaction
pCRII 273
pCYPAC 258, 271, 317
Pedraw 3
pFOS1 170
pGEM-T 273
phytohaemagglutinin 134, 167
pig
  backcross 40
  chromosome number 29
  databases 342
  flow sorting of chromosomes 166

genome size 28
PCR primers 321
PiGMaP project 40
reference family DNAs 320–1
synteny with other species 43
YAC library 321
plant DNA *see* DNA isolation
plant genomes
 see also non-human genomes *and entries by species name*
 microdissection of chromosomes 185
 synteny between 43
pokeweed mitogen 167
polyethylene glycol 125, 133
polymerase chain reaction
 see also
  Alu-PCR
  amplified fragment length polymorphisms
  degenerate oligonucleotide primed PCR
  IRS-PCR
  linker adaptor PCR
  microsatellites
  randomly amplified polymorphic DNAs
  5′ and 3′ SINE PCR
  whole-genome PCR
 'clean' conditions for 97, 188
 of clone inserts 11–13
 contamination by foreign DNA 97, 188, 192, 196–7
 fingerprinting, by LINE amplification 143–5
 of flow-sorted chromosomes 179–82
 multiplex 14, 20, 22, 109–10, 111
 'nested' 100–3, 109–10
 primers
  biotinylated 13
  databases 345–6
  design of 13, 30, 34, 114, 152, 351
  end-labelled 20–1
  fluorescently labelled 14, 21
  human 320
  in inter-*Alu* sequences 114
  for microsatellite amplification 13–14
  for nested PCR 101
  pig 321
  for RAPD analysis 30
  rice 326
  for 5′ and 3′ SINE PCR 33, 34
  software for design of 13, 351
  T_m 10, 14, 147
 products
  as probes for fluorescence *in situ* hybridization 203
  sequencing of 13
 of radiation hybrids 84–6
 of somatic cell hybrids 145–7
 vectorette 256–60
 whole genome 118
polymorphism information content

calculation of 16
definition 1
useful range 2, 16
positional cloning of QTLs 50, 67
potato
 chromosome number 29
 genetic map 43
 genome size 29
primates (non-human)
 cell lines 320
 cosmid libraries 320
primers *see* polymerase chain reaction
probes *see* hybridization, fluorescence *in situ* hybridization
progeny testing 63
propidium iodide 212
pufferfish *see Fugu*
pulsed-field gels
 blotting and probing of 296–300
 one-dimensional 293–6
 for size-fractionation of irradiated DNA 104–8
 size standards for 294–5
 Southern blotting of 296–300
 two-dimensional 300, 305–7

QTL Cartographer 69
quantitative trait loci
 clonal lines in mapping 64
 definition 49
 dihaploid individuals in mapping 64
 environmental factors, interaction with 63, 66
 environmental noise 61, 62
  'pre-correction' of data to allow for 63
 genotyping
  of pooled DNA samples 65
  selective 64–5
  verification of results 62
 'granddaughter' design 63
 interpretation of results 65–6
 interval mapping 53–5
 linkage maps for QTL studies 57–8
 linked 66
 mapping in outbred populations 55–6, 58
 marker-by-marker analysis 50–3
 multiple 66
 noise, environmental 61, 62
  'pre-correction' of data to allow for 63
 outbred populations 55–6, 58
 overestimation of effects 65–6
 positional cloning of 50, 67
 power of mapping experiment
  definition 59
  effect of marker spacing 54, 58
  effect of study population 57
  effect of population size 58, 59–61

quantitative trait loci (*cont.*)
  increasing, by use of multiple markers 56
  progeny testing 63
  use of recombinant inbred lines 57, 64
  permutation analysis 59
  population for mapping 56–7, 64
  selective genotyping 64–5
  significance threshold 58–9
  software 68–9
  trait measurement 61–2
radiation hybrids
  construction of 77–80
  data analysis 86–91
    diploid situation 89–90
    haploid situation 87–9
  database 336
  DNA isolation from
    preparative 81
    small-scale 80–1
  donor cells 76
  genotyping 84–6
  map construction 86–91
  mapping, principle of 74
  markers for mapping 83–4
  panels 40, 83
  physical distance, relation to centiRays 90
  preliminary assessment 81–2
  principle 74
  radiation dose 81–2
  recipient cells 76
  retention frequency
    dependence on radiation dose 76
    evaluation of 82
    useful range 76
  selectable markers 75–6
  single chromosome 74–5
  software 90–2
  stability during culture 77
  STS markers for
    chromosomal assignment 84
    from coding regions 84
    design of 83–4
    standardization of 84
  whole genome 75–6
randomly amplified polymorphic DNAs
  (RAPDs)
  dominance of 31, 57
  genotyping 31–2
  haploid individuals for genotyping 31
  in plant genome mapping 30, 57
  polymorphism of 31
  primers, design of 30
  principle 30
  reproducibility 31, 57
rat
  chromosome number 29
  databases 341
  flow sorting of chromosomes 166

  genome size 29
Ray *see* centiRay
R-banding 212
recombinant inbred lines *see* quantitative trait
  loci
repeat element-mediated PCR
  see Alu-PCR, IRS-PCR, LINE amplification
restriction enzymes
  see also
    restriction fingerprinting
    restriction mapping
  digestion in agarose 291–3
  and DNA methylation 290–1, 308–10
  modification of cleavage pattern
    by 5-azacytidine 310
    by DNA methyltransferases 308–9
  rare-cutting 289
restriction fingerprinting of clones
  see also contig assembly
  complications 252
  of cosmids
    coverage 232
    digestion 235
    electrophoresis 237–8
    purification of clone DNA 233–4
  digestion procedure 235
  electrophoresis system 230, 237–8
  enzymes, choice of 229–30
  fragments per clone 229–30
  gap closure 231, 241–6
  integration of different clone types 227,
    229–30
  use of multiple fingerprints 229
  principle 228–9
  size standards 236
  software 247–52
  strategy 231–2
  'whole chromosome' 232
restriction fragment length polymorphism 1,
  30, 39, 41, 43, 57
restriction mapping
  CpG islands, effect on cleavage 290–1
  cytidine methylation, effect of 290–1,
    308–10
  digestion of DNA in agarose 291–3
  DNA isolation 202–8
  electrophoresis
    pulsed-field, one-dimensional 293–6
    pulsed-field, two-dimensional 300,
    305–7
  enzymes, choice of 288–90
  Optical Mapping 311–12
  partial-complete mapping 302–3
  partial digestion 293
  principle 281–2
  reciprocal digest mapping 303–5
RFLP *see* restriction fragment length
  polymorphism

RHMap 90
rice
    cDNA clones 325
    cell cultures 325
    chromosome number 29
    databases 344
    genome size 28
    linkage map 30
    as model for other cereals 28, 43
    PCR primers 326
    synteny with other spp. 28, 39, 43
    YAC library 325
rodent genome databases 338, 340–1

Saccharomyces cerevisiae
    cDNA library 328
    chromosome number 29
    databases 344
    DNA isolation 287
    genome size 29
    genomic libraries 327
    sequence 28, 227
S-adenosyl-L-homocysteine 310
SAMapper 90–1
SAS 68
scanning of autoradiographs 247–8
Schizosaccharomyces pombe
    cosmid library 328
    databases 344
    DNA isolation 287
    P1 library 328
    sequencing 227
    YAC library 328
segregation analysis 49
Sendai virus 132
sequence comparison 347
sequence databases 346–9
Sequence Retrieval System 346
sequence-tagged sites (STSs)
    see also
        expressed sequence tags
        microsatellites
    database entries 346
    for HAPPY mapping 101, 112, 114
    for radiation hybrid mapping 83–84
    as 'scaffold' for contig assembly 232,
        238–240
sequencing of PCR products 13
'shadow' bands *see* microsatellites
shearing of DNA 99, 100
sheep
    chromosome number 29
    databases 342
    genome size 29
    microsatellites 321
    PCR primers 321

short tandem repeats (STRs) *see*
    microsatellites
5′ and 3′ SINE-PCR
    design of PCR primers for 33, 34
    genotyping 34–6
    polymorphism of products 33
    principle 33
single strand conformation polymorphisms
    (SSCPs) 29
Slink 350
slippage mispairing 5
software
    *see also entries under program name or
        function*
    available via Internet 347
    contig assembly 247–52, 350
    linkage analysis 24–5, 350
    PCR primer design 351
    pedigree drawing 3
    QTL analysis 68–9
    radiation hybrid analysis 90–2
somatic cell hybrids
    for chromosomal assignment of STSs 151
    chromosome content, assessing
        by chromosome painting 142–3
        by FISH 142
        by isozyme analysis 147
        by karyotyping 139–42
        by PCR analysis of STSs 145–7
        by PCR fingerprinting 142–5
    chromosome loss during proliferation 125,
        127, 146, 148
    chromosome-specific microsatellites from 7
    clone isolation 138–9
    DNA isolation
        high purity 150–1
        by mitotic shake-off 143–4
    donor cells 128
    as donors for radiation hybrids 74
    flow-sorting of donor chromosomes
        166–167
    heterogeneity of 145–146
    isolation of donor coding sequences 154
    isolation of donor genomic sequences
        112–3, 153
    microcell-mediated chromosome transfer
        125–6, 136–8
    monochromosomal 155–6
    mouse 321–2
    panels
        monochromosomal 155–6
        mouse 321–2
        regional 156–9
        whole-chromosome 154–6
    PCR typing of 151–3
    pooling of monochromosomal hybrids 152
    principle of construction 125–6
    recipient cells 128–9

somatic cell hybrids (*cont.*)
  regional panels 156–9
  selection systems 129–32
  Southern analysis of 148–51
  stability in culture 125, 127, 146, 148
  for sub-chromosomal localization of STSs
    156–9
  whole-cell fusion 133
    of monolayer cells 133–4
    of suspension with monolayer cells 134–5
    of suspension cells 135
sorghum 43
Southern blotting
  of cosmid restriction fragments 267–70
  of pulsed-field gels 296–300
  of somatic cell hybrid DNA 148–51
soybean, genome database 344
spruce
  chromosome number 29
  genome size 29
  RAPD analysis of 31
SRS 346
Stoffel fragment 191, 193
sugar cane 43
synteny 28, 41, 42–43, 341

T7 polymerase 191, 193
tape streamers 331
Telnet 331
temperature-sensitive selection 132
tetranucleotide repeats *see* microsatellites
6-thioguanine 130
6-thioxanthine 130
thymidine kinase 76, 129, 130–1
T_m of PCR primers 10, 14, 147
TNG panel (radiation hybrids) 83
tomato
  chromosome number 29
  genetic map 43
  genome size 29
trees, genome database 344
trinucleotide repeats *see* microsatellites

UK Human Genome Mapping Project 7, 24,
  317, 318, 319, 320, 322, 323, 330, 332,
  334, 337, 341, 346
UniGene 347
universal mapping probes (UMPs) 41–2

vectorette PCR
  of bacterial clones 256–60
  principle 256–7
  of YAC clones 261–4
Vitesse 24, 350

wheat
  chromosome number 29
  genome size 29
  genomic libraries 326
  hybridization probes 326
  recombinant inbred lines 326
whole genome PCR 118
  see also degenerate oligonucleotide primed
    PCR
World Wide Web *see also* bioinformatics
  access to 330–1
  description 330
  search facilities
    bioinformatics 332
    general 331

X chromosome *see* chromosome, human
X-windows 331
YAC *see* yeast artificial chromosome

Y chromosome *see* chromosome, human
yeast
  see also
    Saccharomyces cerevisiae
    Schizosaccharomyces pombe
  base composition 290
  databases 344
  DNA isolation 287, 288
  restriction cleavage 290
yeast artificial chromosomes (YACs)
  Alu-PCR of 203–4, 246–7, 273–7
  bacterial contig assembly on 240–1
  chimerism in 74, 199, 213, 255, 316
    checking, with FISH 199
  end-sequence isolation 261–4
  fingerprinting, by *Alu*-PCR 246–7
  as FISH probes 206, 207, 213, 216
  internal sequences, isolation of 273–7
  libraries
    Arabidopsis 327
    Caenorhabditis elegans 227
    Drosophila 324
    from flow-sorted chromosomes 179
    human 316–17
    pig 321
    rice 325
    Schizosaccharomyces pombe 328
  Optical Mapping of 312
  as probes for fluorescence *in situ*
    hybridization 203
  sub-cloning into bacterial vectors 241
  as substrates for bacterial contig assembly
    240–1
  vectorette PCR 261–4

zebrafish
  cDNA libraries 323
  chromosome number 29
  cosmid library 323
  databases 342
  genome size 29
  genomic libraries 323
  mutant strains 323
  RAPD analysis 31
  'zoo-FISH' 41